By the Rasping in My Lungs Something Evil This Way Comes

The Chronicle of an Independent Air Quality Field Researcher and Activist Presenting a Personal Case History of the Clash Between Energy Science and Energy Politics in the State of Wyoming

by

R. Perry Walker

The contents of this work, including, but not limited to, the accuracy of events, people, and places depicted; opinions expressed; permission to use previously published materials included; and any advice given or actions advocated are solely the responsibility of the author, who assumes all liability for said work and indemnifies the publisher against any claims stemming from publication of the work.

All Rights Reserved
Copyright © 2019 by R. Perry Walker

No part of this book may be reproduced or transmitted, downloaded, distributed, reverse engineered, or stored in or introduced into any information storage and retrieval system, in any form or by any means, including photocopying and recording, whether electronic or mechanical, now known or hereinafter invented without permission in writing from the publisher.

Dorrance Publishing Co
585 Alpha Drive
Pittsburgh, PA 15238
Visit our website at *www.dorrancebookstore.com*

ISBN: 978-1-4809-8265-9
eISBN: 978-1-4809-8241-3

By the Rasping in My Lungs Something Evil This Way Comes

The Chronicle of an Independent Air Quality
Field Researcher and Activist Presenting a Personal
Case History of the Clash Between Energy Science
and Energy Politics in the State of Wyoming

R. Perry Walker

Contents

Dedication xiii
Introduction xvii

Chapter One

The Problems 1
    Citizen and Media Attention 8
    Ozone 12
    Larger Forces at Work 15

The Feds

Chapter Two

The Environmental Protection Agency (EPA) 19
    First Contact 19
    Lobbying Region 8 23
    Region 8 Weighs In 25

Citizens' Petition to Nowhere 30
Up Close and Personal 33
Region 8 Doubles Down 35
Casualties 39
Doubtful Emissions Accounting 41
Serendipity 44
View From the Top (Sort Of) 48
Closer to the Top 52
Rear View 55

Chapter Three

The Bureau of Land Management (BLM) 57
    Prologue 57
    Into the Present 59
    The State Office of the BLM 62
    A Dance Macabre 65
    Intrigues of the Inner Sanctum 70
    The Pinedale BLM Regional Office 76
    Kabuki Theater 77
    The Pinedale Anticline Working Group (PAWG) 83
    Demise of the PAWG 87
    Looking Back 90

Chapter Four

The U.S. Forest Service (USFS) 91
    Beginnings 91
    Another Agency Weighs In 93
    My First Day and Beyond 94
    A Voice in the USFS Administrative Wilderness 95
    My Own Voice in the Same Wilderness 97
    Battle Over the Anticline 103
    Battle Over the Wyoming Range "44,000" 106
    Battle Over the Wyoming Range Noble Basin 111

Victories 123

Chapter Five

The Air Quality Task Group(AQTG) 127
    The AQTG Legacy 128
    The Meetings 129
    2004 130
    2005 150
    2006 185
    2007 189
    2009 190
    Through a Glass Darkly 203

THE STATE

Chapter Six

The Wyo. Dept. of Environmental Equality (WYDEQ) 227
    The Early Days 228
    A new Guy Enters the Fray 238
    Emissions Accounting – A shell Game 250
    By the Rasping in My Lungs - Ozone 252
    Tales of the Dark Side 254
    Dance of the Regulators 257
    Repel All Invaders 271
    Trojan Horse 277

Chapter Seven

The State Apparatchiki 279
    The County Commissioners 279
    My Backchannel to the Governor 288
    Walking a Tightrope 297
    The State Legislature 304

The WOGCC 311
The State's Washington Delegation 317

THE OTHERS

Chapter Eight

The Operators 323
    Ultra 323
    Questar 328
    EnCana and Engineered Concepts, LLC 330
    Rocky Mountain Shell 335

Chapter Nine

The Enviros and the Media 339
    The Enviros 339
    The Wyoming Outdoor Council 340
    The Upper Green River Valley Coalition/Alliance (UGRVC/A) 347
    The Greater Yellowstone Coalition 350
    The Wilderness Society and Trout Unlimited 353
    The News Media 355

THE SCIENCE

Chapter Ten

Real vs. Virtual 361
    Visible Haze 364
    Winds and Models 368
    Optical Spectroscopy 379
    Infrared Imaging 383
    Ozone 385
    Emissions Statistical Legerdemain 393
    Missing Gas 401

    Burned Gas 403
    Wasted Years? 405

Epilogue 407

Terms and Abbreviations 413

APPENDICES

Appendix 1 – Agency Roles and Authorities 417
    Related to Air Quality
Appendix 2 – Final Draft – Air Quality Monitoring Plan 421
Appendix 3 – Recommendations from the AQTG to the PAWG 461
Appendix 4 – Ozone History 465
Appendix 5 – Example API Well Emissions Tables 473
Appendix 6 – Fracking Additives Exempt 477
    by WOGCC from Disclosure

Sources and Endnotes 481

*"Wyoming should require a passport
of anyone seeking entry…"*

Anecdotal quote attributed by Edna Tichac to James Michener as told to my mother-in-law Peggy Butler who was her neighbor in Laramie. Ms. Tichac was a member of a circle of friends that included Gov. Mike Sullivan, Michener, and herself.

Dedication

I was but one person in Sublette County who tried to challenge the feds, the State, and the operators to do a better job of protecting our environment in general and our air quality in particular. Over the course of my battles, I came into contact with and on several occasions, partnered with dedicated individuals having diverse backgrounds. These wonderful folks have my thanks and respect that I try to convey as I tell my tale in this book. They are:

Linda Baker: An independent environmental activist who set up shop not long after I started my own optical spectroscopy fieldwork. I contacted her to open a dialogue and that led to a close working relationship that has endured. Like me, she too gradually became worn down by the relentless issues, not the least of which was professional environmental group meddling in our local initiatives.

William Belveal: An unforgettable character. He was also a charter member of the Air Quality Task Group. As such, he had a way about asking embarrassing questions of the BLM that laid bare our sense of futility. Toward the end of the Group's existence, he fell increasingly ill but kept up a humorous dialogue with me via email. When he passed away, I was greatly saddened and resolved that his contributions would figure strongly in this book.

Susan Caplan: My first BLM air quality scientist at the state level with whom I opened a dialogue. She was conscientious and professional. She tried hard to perform her job description despite interference from her superiors but finally their mistreatment forced her to transfer to a more research oriented position in the Denver division of BLM.

Evie Stura-Carr: My volunteer assistant who invested at least 1000 labor hours per year for five years keeping my data base of well statistics up to date. At least weekly, she tapped into the WOGCC website to download well production figures and insert them into my spreadsheets. Her work created a database of literally tens of thousands of monthly production numbers.

Glenn and Linda Cooper: The professional interaction and friendship I developed with these two folks is hard to describe in a few words. We teamed up in a successful joint effort to challenge a proposed drilling project in the Wyoming Range. When I worried aloud to them that I feared my ten years of combat with the BLM and the State would vanish in the mist of time, Glenn proposed that I write a paper about it. That went nowhere but the paper did become the core of this book.

Lloyd Dorsey: A field worker for the Greater Yellowstone Coalition tasked with monitoring wildlife issues. He soon took up my cause with the GYC Board of Directors, and consequently was assigned the added duty of keeping the Board aware of my activities. He also managed to pry some modest funding from GYC to help me establish my ozone monitoring station.

Mary Flanderka: My first contact on Governor Freudenthal's staff. In her position of Planning Coordinator, she quickly became my strongest advocate in Cheyenne. I was welcome in her office whenever I visited the halls of WYDEQ and more than a few times found myself in extemporaneous conversations with other highly placed governor staffers who would drop by to consult with her.

Paul Hagenstein: A long time rancher south of Pinedale who participated in the Pinedale Anticline Working Group during most, if not all of its existence. In so doing he exhibited the patience of a saint and the resilience of granite. His practical based wit showed in a public meeting when WYDEQ was reciting the dysfunction between themselves and BLM. Paul spoke up by asking "when are you going to get your horses to pull together?" The WYDEQ guy had no realization of what he had just been asked. Sadly, Paul passed away during the publication of this book.

Carmel Kail: One of the most tenacious citizen environmentalists I had the pleasure of knowing. She kept me aware of County maneuvering designed to deflect local air quality concerns and she accompanied me on a few field trips to monitor ozone levels in our area. Her enthusiasm was graduate-level in its quality.

Jeffry Allan Lockwood: A Professor of Natural Sciences and Humanities and Director of Creative Writing at the University of Wyoming Department of Philosophy. In 2014 he interviewed me for his own book *Behind the Carbon Curtain* and became an unintended mentor who suggested I write my own book. He was courteously tolerant of my amateurish requests for guidance and steady in his encouragement as I experienced for myself the frustrating process of finding a publisher willing to even review my manuscript.

Ted Porwoll: A US Forest Service field operative working in the Pinedale field office. He has an excellent background that goes back to his early days in Colorado. He has a knack for stating what should be obvious but was often missed in our Air Quality Task Group Meetings. He invited me to accompany him on a field inspection to one of his IMPROVE sites and I learned much about the analysis protocol for samples we retrieved from that site.

Terry Svalberg: Now-retired US Forest Service air quality scientist who worked in the Pinedale field office. He had far more oil and gas projects to monitor than any one person should have been tasked to follow. Despite the resistance I felt he received from his superiors, he continued to produce quality and sternly worded air quality assessments right up to his retirement.

R. Perry Walker

The Wagon Wheel Information Committee: This core group of Pinedale citizens started the resistance movement in Sublette County in the early 1970s that successfully blocked an Atomic Energy Commission plan to explode a series of small nuclear explosives below what became the Jonah Field. The idea was to frack the deep strata but too many serious details were left unaddressed. Committee members won support from Wyoming's Washington Delegation and eventually a critical mass of opinion, added to technical problems ended the project. Remaining members asked me to sit down with them in 2005 when they elected to disband and contribute their remaining committee bank balance to me as a gesture of support. They were Doris Burzlander, Sally Mackey, Mary Anne Steele, Ken Perry, Daphne Platts, Phyllis Birr, John Perry Barlow, and A.B. Cooper. Sadly, several of these folks have also passed away.

Barbara Ann Butler-Walker: My wife of 50 years who raised our son during my many absences that were part of my Air Force career. She followed my fieldwork and kept pushing me to get this book project completed in spite of many setbacks.

Introduction

Writing this book promised to be a strenuous exercise in self-discipline for many reasons. First, I didn't know how. Second, I doubted any publisher would have a moment's worth of interest in accepting it for publication. Third, I doubted anyone would be interested in reading such narrow sounding subject matter.

However, I decided that even if all this were true, just maybe, the manuscript and all the research data I have accumulated would find their way into some university library archive. In fact, Craig Thomas a faculty member with the University of Western Wyoming first suggested that notion to me.

Furthermore, my wife and a close friend and occasional collaborator, Lloyd Dorsey, who had been in the employ of the Greater Yellowstone Coalition both insisted that my work and experience needed to be set down in a permanent record. This would hopefully benefit those who might undertake their own effort of coping with the many environmental problems that inevitably come with natural gas development.

I had spent almost ten years of my life trying to make a difference as an individual. I naively got into the game believing that honest, credible, scientific based field monitoring of the growing impacts to air quality in my home county of Sublette in western Wyoming would be inevitably successful.

I thought the gas developing companies, the regulators, and the environmental groups would all come together on the basis of sound scientific data and craft environmentally smart ways to get at the gas hidden under the prairie and still protect the air, water, dirt, and wildlife. In retrospect, I should have known better.

By the end of my decade of activism, I had decided to fold my tent and call it quits. The Jonah and Pinedale Anticline gas fields were pretty much drilled to completion and the operators were bailing out to attack the newest gas El Dorados in the Marcellus Formation gas fields in the eastern U.S. and the Bakken Formation in North Dakota.

As I was boxing up my files, I discovered old news clippings I had long forgotten. Reading through them I was discouraged by the prophetic nature of their content. Article after article quoted the operator's plans and determination to drill at specific rates, in particular locations, and achieve ultimate well counts. With little variation those all had come true despite the huge effort expended by citizens and environmental groups trying to bring some sanity to the headlong rush to drill.

I sat back and concluded that on a personal level I had not influenced the environmental outcome in any useful manner. Instead, I concluded that my experiences had contributed to the body of proof that one person cannot fight city hall nor can one person often make a difference. My friend Terry Svalberg who is the regional U.S. Forest Service air quality scientist challenges these conclusions. Also, a number of citizens of Pinedale insist I changed some outcomes in a material way but I am hard pressed to see any evidence of their being correct.

Other dark discoveries were many. I learned that the professional environmental groups can be their own worst enemies. They vie for funding from contributors in such an intense manner that they will take credit for work done by individuals with whom they associate, no matter how loosely, in order to appear as effective environmental activists in the eyes of their contributors.

Furthermore, these enviro groups are ill equipped to act on the basis of physical science. In my experience with them they avoided talking about that.

By the Rasping in My Lungs

They seem most comfortable when dealing with natural science issues and wildlife biology. However, when I tried that approach in attempting to make groups like the directors of the Greater Yellowstone Coalition and the Wyoming Outdoor Council understand the threat to wildlife and forests from air pollution generated by gas development, it was a bridge too far.

The government entities charged with regulating development and protecting the environment presented their own class of frustrations. I challenged the Wyoming Department of Environmental Quality almost constantly but received little more than bureaucratic stonewalling during the entire ten years of our interaction. In the early years, I became such a thorn in their side that I was asked by a staff director for the governor to "cool it" with my public criticism because whenever my name came up in staff meetings, the Department directors would simply lock up.

The Federal Environmental Protection Agency was equally unresponsive in the early years. In its defense, this was when the Bush-Cheney Administration was in power. At that time the EPA and the Department of Interior were both being run by administration appointees who were unsympathetic toward anything they perceived to be road blocks to energy development in the West. Direct participation by EPA Region 8 headquarters in Denver during deliberations with a group of stakeholders formed by the Bureau of Land Management was never more than perfunctory. When the Region 8 director did start to weigh in with sharply critical warnings about gas development shortcomings, he soon thereafter retired from his position under politically charged circumstances.

As for the Bureau of Land Management, its involvement in managing the development of gas fields in Sublette County was a classic study in non-responsive, bureaucratic doublespeak and stonewalling. From the beginning, BLM-Pinedale was outclassed and outgunned by the drilling industry and was in a constant state of catch-up. During the decade I dealt with the Pinedale office I went through four directors. After the first director, it appeared the Department of Interior in Washington D.C. had elected to make Pinedale the dumping ground for flaccid managers nearing retirement.

Perhaps the ultimate irony is that gas development in Sublette County ground to a halt primarily because the collapse in oil prices had rippled into the gas industry and particularly shale gas where costs are higher. I had to feel a little satisfaction when I learned from Terry that Shell Rocky Mountain Production had sold all of its Pinedale gas field holdings during the 2011-2014 oil price plateau because the company had concluded there was more money to be made in oil. The price of oil dived not long after that. However, that too reversed five years later.

Along the way, familiar news stories from the Bakken gas field became common on the nightly news. At the same time, Terry received calls from his friends in the Marcellus Formation region asking him for guidance about how to deal with the invasion of gas guys and their environmental impacts. This convinced me that I had to take on the book project.

My work generated thousands of pages of data in the form of tables, graphs, charts and imagery. Also, I authored scores of public comments in reply to environmental impact statements and I wrote a number of opinion editorials for local and state news media. Interested readers may view those materials in the archive of the American Heritage Center at the University of Wyoming. They are referenced as the Ronald Perry Walker Papers at https://rmoa.unm.edu/docviewer.php?docId=wyu-ah12688.xml#idp3752592

Chapter 1

The Problems

The issue of environmental impacts from natural gas development in Sublette County, Wyoming first loomed into my consciousness in 1998 as drilling activity began to literally get in the way. I was on the local county airport board when an issue arose involving a driller who wanted to set up on a site close to the centerline of the approach path to our runway.

Safety concerns became the topic of discussion rooted in worry that flaring of drill rig waste gas could prove so bright as to obscure a pilot's vision in the critical moments before he was about to touch down. We on the board were surprised this had not occurred to the drilling operator and as I recall, conversations with them resolved the issue. However, this presaged a decade of conflict with drillers that would follow.

This conflict was multifaceted and not confined to citizens versus the drilling industry. It rapidly evolved into a contest between citizens and state government and citizens versus federal government. Adding to the mix were citizens versus elected politicians who listened more to the industry than to those who elected them. There were citizens versus citizens who wanted the economic boom more than quality of life. There were citizens versus County Commissioners who stubbornly sought to conjure badly flawed explanations for the eventual decline in air quality. Finally, there were citizens versus the glacial, labyrinthine regulatory process.

It is fair to say that most people don't realize the collateral damage that comes with drilling for natural gas and subsequent development of the gas field, nor does the natural gas industry talk about the scope of the impacts. The average citizen likely envisions a gas field to be a large stretch of territory from which little innocuous pipe stubs protrude from the ground to deliver the gas beneath.

The truth is much uglier. Sublette County has seen the sage grouse population decline to such an extent that it has been a recurring candidate for the Endangered Species list, the native mule deer herd has been reduced to half its original size, and the last largest big game migratory route for pronghorn antelope has been seriously disrupted.[1,2,3]

This is largely due to drilling and the subsequent enormous support infrastructure that covers the 30,000-acre Jonah gas- field and the 198,000-acre Pinedale Anticline gas field. This infrastructure includes steel buildings of all sizes, fenced pipe valve yards, tank farms, and waste gas flaring stacks by the thousands. In addition, there is a network of roads between all of these sites that challenges every living thing crossing them. In one incident in 2007, a tank truck drove through a heard of pronghorn antelope, killing 21 of them.[4]

The impact from roads was first revealed in a news release in the summer of 2005.[5] Drawing upon travel route data from the BLM and wildlife data from the state, a 2.9 million acre stretch in the Upper Green River region was found to harbor an extensive transportation network. The Jonah and Anticline gas fields were cited in particular where over 8000 wells had already been drilled and 10,000 more were thought possible in the coming decade. These roads were cutting through migration routes causing changes to wildlife behavior and interfering with preferred habitat.

In 2005 the Salt Lake City Tribune took notice of the down side that accompanied the gas boom around Pinedale. A former city manager for Pinedale was quoted as lamenting that a better balance was needed between the headlong rush to drill and the strain on local infrastructure and services that was taking place.

Rural roads were being overwhelmed by the massive growth of gas development trucking activity. New subdivisions were straining the town's ability to provide enough sewer and water lines and to construct curbs and gutters for new streets. Classrooms were strained by the rise and fall in the student population that matched the spring-through-fall drilling season.[6,7]

Crime also boomed with the arrival of those I called "the gas guys." DUIs, batteries and drug-related arrests spiked. Methamphetamine use became rampant because it was the narcotic of choice by rig workers putting in 12-hour, 14-consecutive-day shifts. Ultimately, the energy companies had to step up to the issue and crack down with more frequent drug testing.[8]

What the Tribune missed was the prowling for women that rig crews engaged in. Pinedale had a limited supply of unattached members of the fair sex so the gas guys improvised. I was told by a friend who was in a bar with them one night that they wanted to know where the high school was so they could take a shot at picking up a date there.

Another problematic issue had to do with the gas companies paying so well. Field wages of $25-$40 per hour was the norm and that drained the labor pool from other Sublette businesses. I heard this personally from several businessmen. Not discussed in the Tribune story was the economic punishment the gas guys were willing to impose on unfriendly businesses and politicians.

I was told by more than a few individuals that they wanted to back my campaigns to make the energy industry clean up its act by addressing air pollution but they felt they dare not. The industry offered veiled threats that its business would be withheld from unfriendly proprietors. On the political side, a local auto dealer told me that an operator had "suggested" that his state legislator father who had made recent critical remarks about the industry be urged to tone down his comments. The gas guy was informed this would be a grave mistake and only inflame his father so the suggestion was dropped.

The Tribune story acknowledged a fact I knew from personal experience. Many of us faulted the Bureau of Land Management for failing to foresee the consequences of approving so many well permits over such a short time period. The drilling boom actually began in 1999 under the Clinton administration, but really took off under relaxed leasing and permitting rules of the Bush administration. I explore this situation in fine detail in the chapter on the BLM.

It is sufficient here to simply acknowledge the Tribune's quote from Prill Mecham, the Pinedale BLM field office manager then: "When I got here [in 1998], we issued 75 drilling permits. This past year it was 400." Readers are urged to remember this quote when you read about her role in the BLM chapter of this book.[9]

Finally, the Tribune offered a cogent observation and quote about the non-visible aspect of drilling on the Jonah and Anticline fields. *"[They] lie southeast of Pinedale, well out of sight from the area's major highways. But seen from the air, it looks like the subdivision from hell."*[10]

In July 2005, the Casper Star-Tribune published a story that described a situation that remains basically unchanged today. It observed that BLM was charged with identifying impacts to the sage grouse that were underway due to natural gas development in the Upper Green River region. These impacts were described as a big question mark and a more extensive picture of populations of the bird were needed to craft good protection recommendations.[11]

A wildlife biologist with the BLM Pinedale office commented that attempts to limit harm to the bird may have been "spinning their wheels." He further observed that there was no coordinated effort in the Upper Green or across Wyoming. Therefore, more active protection measures would likely be needed in the future to stay ahead of the declining bird population problem developing at that time in 2005.[12]

These revelations followed release of findings by a University of Wyoming Ph.D. candidate. He presented research incorporating some of the most exhaustive studies of the grouse in our area. He stated flatly that development on the Pinedale Anticline was causing a decrease in their productivity there.

It was unclear if the birds were not breeding or just abandoning the area but regardless, the numbers of birds were in decline. At a minimum, he suspected the heavy industrial clamor was responsible in part for their departure. The importance of this aspect lies in the fact that, in the words of a Wyoming Fish and Game Commission wildlife biologist, *"sage grouse are loyal to historic breeding areas, so if there is displacement of grouse on the Mesa, there's a chance we won't get grouse back on the Mesa."*[13]

In light of the 2014 refusal of the Federal Fish and Game Commission to list the sage grouse as endangered, it was relevant that a Wyoming Fish and

Game Commission wildlife biologist made the following statement: "*The question is what kind of displacement and impacts people are willing to endure before mitigation measures should be enacted.*"[14]

Apparently those limits remain undiscovered because yet again the bird was considered for endangered protection and yet again in September 2015, that protection was deemed unneeded. Despite a recorded 90 percent decline in population due to mining and drilling, Secretary of the Department of Interior Sally Jewell, issued a perfunctory statement. The grouse did not meet required standards because of the unified efforts by federal and state agencies, and ranchers and industry and environmentalists to restore its breeding grounds.[15]

The Audubon Society expressed confidence in that decision by citing the success of the Wyoming sage grouse management plan since its inception in 2010.[16] The director of the Cornell Lab of Ornithology praised the collaboration but warned that intensive monitoring would be "absolutely necessary" to prove effectiveness. On this I agree because my history with the feds and the state regarding broken promises on air quality protection justify skepticism.

One of the more bizarre conflicts associated with gas development on the Pinedale Anticline involved anthrax. One day following a County Commissioners meeting, I was chatting with Pinedale's family physician Doctor Tom "Doc" Johnston. He soon told me a story I almost could not believe. A lady who came from of a local multi-generational ranch family with a long local history had contacted him. She was very worried that drilling and road construction on the flank of the Anticline Mesa were going to disturb an anthrax infected cattle burial site.

Taken aback, Doc Johnston asked for details whereupon she explained that in 1956 there had been an incident when several cattle had been infected and were disposed of by burning and burial alongside the Mesa. He immediately contacted state authorities and within 24 hours the Department of Homeland Security swooped in and permanently placed it off limits for any drilling. He said he had never seen the state react to anything so fast.

A local friend and news reporter remembered going on an oil and gas bus tour of the Pinedale Anticline a number of years back and on the return trip to town someone pointed out that "over there was where that anthrax incident happened and they (the oil and gas guys) aren't supposed to even go over there."[17]

I was able to find circumstances that match this event in a local news story from 1956 but it told a story that was more limited in extent than another local ranch family daughter recalled.[18] The paper cited a loss of three head of cattle but the historical memory of these witnesses cited many more. Regardless, in my mind, this was a classic example of how one human activity could potentially go very wrong because of a loss of local historical memory about another activity.

Pinedale management's sleight of hand and obfuscation continued through 2011. In a Wildlife Annual Planning Meeting that was convened in February of that year in a full-to-capacity conference room, the public was treated to an "update" on the process in place to mitigate surprisingly significant declines in mule deer herds due to gas field operations. An additional stated purpose was to solicit public input on the issue. I had my doubts and suspected that public input was only being solicited in an effort to merely appear compliant with NEPA requirements.

The BLM wildlife staff revealed that the mule deer population in their wintering range, which was right in the middle of the Anticline drilling project area, had declined 60 percent in the period of 2001 and 2009 when drilling was being pushed at a break neck rate. A lot of public comment in a previous meeting had been very critical of this trend and demands for corrective action were numerous. Here is where the newest local BLM manager, Shane DeForest showed he was in lock step with his superiors' wishes.

He proceeded to explain the "confusion" about what mitigation was and was not, by clarifying that mitigation was a tool to lessen impacts (scope, intensity, frequency and/or duration), not to reverse them. He said mitigation meant finding ways to avoid, minimize, rectify, reduce or compensate for undesirable results on local game herds.

Not surprisingly, this did not go down well with the audience and one person in particular, Rollie Sparrowe. I came to know Rollie as we fought the BLM together on our respective issues. He had over 35 years' experience in state and federal wildlife management and received the Meritorious Service Award from the Department of the Interior in 1991. His reply to Mr. Deforest was blunt and to the point.

He called winter drilling the "elephant in the room." He reminded BLM that he had been following this for a long time and was running a wildlife task force for the PAWG in 2005, making critical recommendations. His task force said to the PAWG and the PAWG relayed to BLM, *"what's happening with the deer herd is already unacceptable in 2005, so get the right people together to do something about habitat and population"* but BLM refused. The warning continued: *"During the years we've been jousting with this, we have a deer decline and we've had an unwillingness of the management agencies involved to address the real issue (winter drilling)."* [19]

DeForest claimed the Anticline ROD drilling arrangements had been designed to offset the impact associated with wintertime drilling exceptions but those in the meeting were unimpressed. They cited the decision as being poor at best and potentially catastrophic at worst due to the impacts on the wintering wildlife there. Rollie opined that anyone hoping those responsible for allowing winter drilling will admit the decision was a mistake shouldn't hold their breaths.

DeForest then dismissed the public challenges with a classic piece of bureau-speak. He claimed that although the ROD specifically provided for cessation of winter drilling if negative impacts were demonstrated, a process would have to take place whereby various attempts to remedy the problem would be implemented. However, he proceeded to offer the caveat that this was a question of timing and implementation of the ROD.

This was a BLM long practiced tactic here and also in my air quality challenges; it was in fact an endless stall. He then really seemed to misrepresent the public's concern by accusing them of *"… aiming to basically redefine the provisions of the ROD and say well, we don't need to do any on- or off-site mitigation; just change the ROD (to prevent winter drilling), and that's off the table."*

Rollie shot back with his characteristically blunt assessment: *"What we continue to hear from BLM and G&F and industry is 'Hey, let's not jump to conclusions (about why mule deer population is down). We're not going to address operations until the bottom falls out. Well, you can kiss this deer herd goodbye if this continues even just a little while longer."* [20]

R. Perry Walker

Citizen and Media Attention

Activities in the Jonah and Anticline gas fields were heavily reliant upon internal combustion engine power that was not subject to emission regulation. In addition, vehicular traffic was very heavy and simple observation suggested that big-rig diesel trucks traveling throughout the area were notoriously ill maintained as evidenced by the carbon soot exhaust plumes they emitted. I wondered if this could be relevant on a local scale to climate change research at Lawrence Livermore National Laboratory revealing potential significant climate heating contributions from carbon particulates of this type.[21]

Furthermore, the Jonah field had become interlaced by miles and miles of dirt roads. Heavy traffic operated on these roads 24 hours a day every day. This lofted large amounts of dust into the air, particularly where trucks were operating. The dust was very fine and stayed airborne for long periods of time particularly in strong winds that were prevalent there. I believed that observational evidence and photographic evidence I developed strongly suggested this dust represented a major component of a noticeable growing haze layer. Any increase in development activity could only exacerbate this already badly degrading situation.

Sadly, the complexities associated with air, water, and land protections, seemed to overwhelm the BLM in its role as the primary public land steward. To find relief, its administrators appeared to adopt a lenient posture toward strict monitoring and enforcement of all three. This was evidenced by recurring waivers of compliance with environmental rules that had been successfully requested by petitioning operators.

In 2004 the issues growing out of gas development in our county had drawn the attention of citizens in nearby Teton County. During the summer of that year a public forum was held In Jackson Hole which utilized the format of a panel of notables. It was moderated by one of our more respected former governors. The panel was made up of the newly elected governor, one of our senior senators to Washington, and a few lesser state officials.

The forum format included a requirement that questions from the audience be written on slips of paper and passed forward to the moderator for consideration. This displeased me because it served to tamp down the spontaneity of feedback from us in the audience. As a result, I initially decided to simply sit there and listen. However, after about a half-hour of questions that I didn't consider to be homing in on the problems, as well as answers from the panel that were mostly non-answers, I could no longer restrain myself.

I signaled one of the floor assistants for a question form on which I scribbled a pointed and assertive comment and passed it back. I stated that after waiting fruitlessly for state and federal regulators to take some kind of action, I would no longer sit still and wait. Accordingly, I would apply my scientific background to the task of initiating my own field research to determine the impacts of drilling on air quality.

I had no expectation this would be judged safe enough to be read aloud but I was stunned to hear my words booming out of the speakers around us. The moderator had first scanned the note, chuckled, and then stated that he liked Wyoming folks because we spoke our mind. He then read it aloud and looked over at the panel, asking "Any comments?"

There followed a very awkward silence that seemed endless but probably only lasted 10 seconds. The moderator who was known for his sharp wit commented "Very eloquent," and moved on to the next question.

After the forum broke up, I was working my way through the crowd to leave. Suddenly an activist affiliated with an environmental group pulled me aside and expressed astonishment that I had made such a direct declaration. His reaction likewise surprised me and in the coming years, I would occasionally receive verbal acclamation from such groups.

The passage of time would see Sublette County become a major focus of the news media. On July 22, 2004, ABC news ran a feature about oil and gas development versus wildlife issues called "Range Wars" on its Nightline program.[22] It drew upon the views of opponents and proponents and as would be expected, many of us who worried about the impacts felt the program badly explored the whole story. Over the next few years, news outlets from all over the nation and the world would cover our story and I would personally be interviewed and written about or videoed by print and TV news outlets.

Relentless and headlong gas development impacts upon the area's land, air, water, and wildlife continued to increase my worries and instilled in me a sense of increasing urgency. Eventually, two factors caused me to focus upon the issue of air quality. First, it had become clear that issues of air, water, and land protection were individually so far reaching, complex, and poorly researched that only one could be adequately addressed by a single person.

Second, the Pinedale BLM field office hosted a Jonah Infill Development Project (JIDP) Public Meeting on November 13, 2003, that included unwelcome comments by the meeting moderator. The agency had made the determination that it only had jurisdiction over surface disturbance. It would, therefore, confine its attention to that topic, and by implication, abdicate the air and water quality issues to state or other entities.

That didn't sit well with me and I challenged the assertion on the grounds that BLM did indeed have jurisdiction and also a chartered responsibility to execute steward-ship over air quality. The implication of the BLM disclaimer of jurisdiction over air quality seemed to signal that BLM had no interest in pursuing air quality stewardship.

I, therefore, made a tactical decision to find some point of challenge that was well founded in existing federal law. I noted that gas drill rigs were disturbing a lot of ground surface area that also had the collateral effect of displacing native wildlife. Furthermore, the clarity of the local atmosphere was exhibiting a definite decline.

I was aware that I lacked any expertise to guide me in challenging impacts from ground disturbance and the same was true about wildlife impacts. However, I had a growing familiarity with the fact that the federal government had attempted to protect air quality from industrial pollution. Specifically, I knew in non-detailed ways that there was a law called the Federal Clean Air Act which was supposed to address the subject so I elected to become active in that category of concern.

By the time BLM was calling for public comments on the pending Jonah Infill Project I had printed my own copy of the Federal Clean Air Act (CAA) and carefully read through it. I meticulously identified passages I suspected were not being enforced and crafted them into what I thought was a persuasive argument. However, this was naive and I would learn many times over the coming years that this weakness would be one of my prevailing faults.

I was concerned that a serious conflict of interest by BLM was in the making due to its advocacy of drilling in the face of statutory obligations to safeguard CAA air quality values. I proceeded to list the specific portions of the CAA I felt were not being adhered to in the draft JIDP environmental impact statement:

> *(d)(1) Each State shall transmit to the Administrator a copy of each permit application relating to a major emitting facility received by such State and provide notice to the Administrator of every action related to the consideration of such permit.*
>
> *(2)(A) The Administrator shall provide notice of the permit application to the Federal Land Manager and the Federal official charged with direct responsibility for management of any lands within a class I area which may be affected by emissions from the proposed facility.*
>
> *(B) The Federal Land Manager and the Federal officials charged with direct responsibility for management of such lands shall have an affirmative responsibility to protect the air quality related values (including visibility) of any such lands within a class I area and to consider, in consultation with the Administrator, whether a proposed major emitting facility will have an adverse impact on such values.*
>
> *(C)(i) In any case where the Federal official charged with direct responsibility for management of any lands within a class I area or the Federal Land Manager of such lands, or the Administrator, or the Governor of an adjacent State containing such a class I area files a notice alleging that emissions from a proposed major emitting facility may cause or contribute to a change in the air quality in such area and identifying the potential adverse impact of such change, a permit shall not be issued unless the owner or operator of such facility demonstrates that emissions of particulate matter and sulfur dioxide will not cause or contribute to concentrations which exceed the maximum allowable increases for a class I area.*
>
> *(C)(ii) In any case where the Federal Land Manager demonstrates to the satisfaction of the State that emissions from such facility will*

> *have an adverse impact on the air quality-related values (including visibility) of such lands, notwithstanding the fact that the change in air quality resulting from emissions from such facility will not cause or contribute to concentrations which exceed the maximum allowable increases for a class I area, a permit shall not be issued.*

I argued that a close reading of this portion of the Clean Air Act appeared to signal that the stated objectives in the Jonah Infill EIS may have been insufficient and perhaps even in violation of the CAA by both BLM and Forest Service. The provision in (C)(i) placing upon the *operator* the responsibility of demonstrating *"that emissions... will not cause or contribute to concentrations which exceed the maximum allowable increases for a class I area"* was being circumvented by the BLM. The BLM was actually the entity engaged in such demonstrations by reason of "open house" slide show presentations conveying arguments that were industry friendly.

I felt this constituted a conflict of interest and a compromise of BLM's land stewardship responsibilities in that such activity placed it more firmly in the role of energy industry advocate. BLM compounded this behavior by contracting private firms, albeit paid with industry funding, to conduct air quality modeling and presented the results in an advocacy role on behalf of industry. Finally, on the basis of those model results BLM admitted that impacts "may" indeed occur but stated in the environmental impact statement that it judged these impacts to be at an "...*acceptable level*." As usual, "*acceptable*" became a BLM non-empirical value judgment that outweighed the value judgment of a concerned public.

Ozone

The winter of 2005 began an especially difficult time for the Wyoming Department of Environmental Quality (WYDEQ), the operators, and concerned citizens in the region. That was when the first signs of a new experience was being brought to the region by gas development, that of ozone pollution. Elevated levels of ozone began to be detected in the winters of 2005 and 2006. The phenomenon didn't occur in 2007 but in 2008, winter levels became surprising.

The issue became impossible to dismiss when in February of 2008, ozone levels were monitored at a level of 122 parts per billion (ppb), well above the EPA recommended 8-hour exposure level of 75 ppb.[23] WYDEQ was compelled on five separate days to issue warnings of potential harmful ozone levels; this was apparently the first time such action had been taken in the winter anywhere in the US.

From my perspective, the history I present in the coming pages chronicles a failure of the federal NEPA process as it was carried out in western Wyoming. That process is described by the EPA: *"The National Environmental Policy Act (NEPA) of 1969 requires federal agencies to take into account the environmental impacts of federal decisions which could significantly affect the environment. In implementing NEPA, federal agencies have to assess the environmental impacts of decisions and inform the public of any impacts. The public has the right to comment on any proposed EIS before action is taken".*[24]

Unfortunately, NEPA does not mandate that the environmentally preferable choice always be chosen, only that environmental impacts be *considered*. This frees managers of public lands from having to *act* upon input from the general public, so on occasion, as in western Wyoming, this has become the Achilles heel of air quality protection while at the same time providing an advantage to the operators.

Adding to industry's advantage is its easy access to government agencies and elected politicians. In the case of the Wyoming state BLM office, I had been told by concerned BLM employees of routine access to that office by one operator in particular and of other operators' incessant threats to precipitate political consequences from the state's Washington D.C. delegation if action was not immediately forthcoming on behalf of their particular agendas.

Not to go unnoticed should be WYDEQ's behavior. Aside from its determination to invoke "voluntary" actions from the operators in its effort to avoid regulating them, its own internal processes could qualify as obstructionist. It accepted an operator proposal that traded VOC emissions for NOx emissions and over the early years it changed the format of its emissions permit application several times. That confounded my own efforts to track several gas production hydrocarbon emissions.

Overall, Sublette County has seen an endless string of BLM preliminary, draft, supplemental, and final environmental impact statements, subsequent perfunctory Records of Decision (ROD), and equally perfunctory Findings of No Serious Impact (FONSI). By 2012 these documents literally stacked up thigh high when placed in a pile. This entire process is badly weighted against opponents.

The EIS documents are numbingly long and so loaded with technical multidisciplinary tedium that no single individual can hope to comprehend the entire document. Furthermore, the BLM habitually releases these EIS documents for public comment close to holidays such as just before the Christmas-New Year's time. I saw this truncate the usual reply period of 60 days down to 45 days thereby complicating citizens' ability to respond.

It is within this framework that the BLM continued to place gas development at any cost above a well-balanced approach. In Sublette County, an extension of the Jonah Field called the Normally Pressurized Lance Field is pending which will encompass 141,000 acres and 3500 wells drilled at a proposed rate of 350 per year. Also pending is a new La Barge Platform Infill Project west of the Jonah that would have 838 new wells. Additionally, despite local citizen objections citing the area's ozone non-attainment status, WYDEQ was set to approve EnCana's petition to build a large gas processing plant because of purported favorable VOC/NOx offsets by that operator.[25,26]

On a state-wide scale, most sobering is the Continental Divide-Creston Natural Gas Drilling project east of Sublette County which will span the historic Red Desert in south-central Wyoming; it will encompass 1.1 million acres and 8950 wells on 6100 five to ten acre size pads. The destruction of land surface alone will likely be even more extreme than that of the Jonah field. It also threatens the existence of the famous Overland Trail which still exhibits visible signs of the settlers' wagon wheels that cut lasting ruts into parts of the prairie.

It appears the struggle over effective regulation of gas development methods and outcomes in Wyoming is destined to continue.

Larger Forces at Work

Until now a decade later, I never fully realized the headwinds that were blowing against me the entire time I was challenging the way drilling business was being done. The natural gas industry began to have such success in bringing to market so much gas that it seemed to realize that new markets were needed to use it. Otherwise, there would be no point in pushing for more development and production.

One of industry's first efforts early in this second decade of the 21st century focused on vehicle fuels. Questar began advertising natural gas as a plentiful fuel for cars and pickups and began circulating pamphlets to that effect. Also, news stories began to surface about natural gas advocacy groups and the Obama Administration starting to approach the big-rig truck community with the proposal that it consider conversion from diesel fuel.[27,28,29] At about the same time, the natural gas industry lobby in Washington D.C. started to press for permission to export the gas because of its abundance (or over abundance).[30]

As these initiatives made it into public view, I realized all of us who had argued for regulation and a slow-down in development until the impacts could be better controlled never had a chance. This was particularly illustrated by a revealing piece of research done by the New York Times.

In December of 2014, a story was published that exposed the role of the energy industry in co-opting states attorneys general where gas and oil development was dominant or pending, to send protest letters to leaders in Washington D.C.[31] The Times revealed that energy companies had literally drafted comment letters that were scathingly critical of environmental concerns, and passed them to attorneys general who, in some cases, then simply copied them onto their own letterhead for mailing to the Administrator of EPA and other of the highest officials in Washington D.C.

Some of these attorneys general received record sums of money for political campaigning in return. Also, some attorneys general had terminated legal investigations, altered policy, or agreed to corporate-preferred settlement provisions following intervention by energy company lobbyists and legal counsel. These practices specifically mentioned the EPA Administrator as having been targeted. The dates cited were after the period when I personally had brief interaction with Administrator Lisa Jackson but I can't help wondering if this went on during my active period as well. If so, I was surely outclassed and butting heads with the "big guys."

As I mentioned in this book's introduction, in late 2015, the oil market collapsed worldwide and took the gas industry with it. Natural gas drilling then dropped off with a collateral effect of lessening environmental impacts to air quality around Sublette County. Thus, in a mere handful of months, market forces achieved better results than the decade of my efforts and those from other concerned citizen groups opposing the way the gas industry has been doing business. Unfortunately, the ozone problem persists as a lasting legacy, and likely will into the foreseeable future.[32]

The Feds

Chapter 2

The Environmental Protection Agency (EPA)

One of the biggest players in the Sublette County gas development saga was the Federal Environmental Protection Agency. My first exposure to the agency came about as a result of the presence of one of its representatives on the Air Quality Task Group (AQTG) attached to the Pinedale Anticline Working Group (PAWG). These latter two entities are extensively discussed in separate chapters of this book so I will not go into great detail here about them. Suffice it to say as an introduction, the issue of air quality was rapidly ascending in its level of importance in the eyes and ears of the BLM and the Wyoming Department of Environmental Quality due to citizen uproar, so EPA presence became unavoidable.

First Contact

In general, I found this EPA representative, Joe Delwiche, to be somewhat bureaucratic. He seemed to have a set pattern of participation in the AQTG that often involved recitation of statutory guidance when issues came up. In his defense, I must point out that he was compelled to drive all the way from his Denver-based Region 8 Headquarters of the EPA each time we held a meeting. I suspect that at least initially, he was underwhelmed by the notion of making a 10-hour drive for the purpose of lending agency presence to a room full of citizen, operator, and government stakeholders. As time went on though he seemed to warm to his task and at one point in the second year of our group's butting heads with the BLM, he initiated a spirited debate with a state level BLM manager over the concept of implied versus explicitly stated statutory guidance. His performance in that debate won my grudging admiration.

Outside of that event, he and I tended to be on a fairly constant level of disagreement. I was often frustrated by what I viewed to be his habit of retreating into a statutory bunker when we tried to even conceptualize some kind of concrete action designed to shield air quality from gas field development impacts. It would not be unusual for him to caution us about our limits of authority under the PAWG umbrella or to explain what he perceived to be the limited scope of BLM and WYDEQ.

Ultimately a few years into the AQTG process, I became so annoyed at what I felt was EPA Region 8 foot-dragging that I initiated a citizens' petition to its director. It urgently requested more direct agency involvement with WYDEQ and BLM and the implementation of better efforts to control the growing problem of deteriorating air quality in the area. The petition apparently filtered down through the layers of Region 8 management to Joe because in a subsequent AQTG meeting he asked in a somewhat irritated tone "What is it you want from me and the agency?"

I tried to explain my perception that his presence was not proving informative as to solutions. Instead he was adding to the barriers being thrown at us by WYDEQ and BLM. These barriers took the form of endless interpretive objections challenging our view of our scope of authority. As I recall, he rejected this and tried to assure me that he was in fact doing his best to steer the group along a path that would produce a positive outcome. I had to take his word for that and be satisfied that I had at least gotten his attention. In retrospect, perhaps this was the catalyst that drove him to challenge the BLM official in the exchange I described earlier.

A few years into the air quality debate, I made the indirect acquaintance of another EPA employee at a deeper level of trench work. That employee was Cindy Beeler who worked under the staff title of "Energy Advisor" at Region 8 Headquarters. Our connection came about as a result of my having become aware of a gas well dehydration technology developed by a small company called Engineered Concepts, LLC. I discuss this technology known as the Quantum Leap Dehydrator, in some detail in Chapter 8. Basically, it removed moisture from gas coming out of the borehole. This was achieved by means of a proprietary technique that resulted in air pollution emissions being reduced from tens of tons per year (this is consistent with World War II-era technology still the norm) to mere *tens of pounds* per year.

The EPA had tested this dehydration technique a few years previous in Colorado, so Cindy had been involved in monitoring the test process. Over the course of several telephone conversations, we developed a mutual trust because we were both searching for best ways to minimize gas development impacts to the regional atmosphere without obstructing operators' production efforts.

Over the next many months I maintained contact with her for purposes of insuring that I understood the test process that had been applied to the Quantum Leap dehydrator as well as to keep her informed of my own activities in Sublette County. My hope was that perhaps I might cultivate her as an ally and thereby develop a direct pathway of information feeds to her superiors.

As an example, there arose in the public media a series of stories talking about EPA concerns regarding the effects of fracking on ground water and whether the agency was minimizing the issue for political reasons. According to media scuttlebutt this was leading to involvement by the EPA office of its Inspector General (IG). I wrote to that IG about my field findings which I felt were being overlooked. I was concerned that the IG was taking too narrow a view by considering only effects on water. I specifically warned of the lofting of fracking fluid chemicals into the air as part of the practice of well completion flaring.

I never received a response and assumed that the IG had dismissed my submission as just another citizen gripe. A year later, however, the subject hit the news outlets again and I informed Cindy of what I had done regarding the submissions to the EPA IG. To my surprise, she requested that I send her a copy of those materials which I took to mean that she might forward them to the same IG. I did send them to her but I never learned what subsequently transpired.

In May, 2005, the director of EPA Region 8, Robert (Robbie) Roberts and some of his staff traveled to Sublette County to conduct a field review of activities associated with gas development there. A husband-and-wife team of local activists invited the group to their home in the Upper Green River Valley to meet and converse with some concerned citizens from around the area. I was invited because of my growing local notoriety and leaped at the opportunity.

I came to the meeting armed with a folder containing several copies of photos and optical spectrometer graphs I had collected over the previous year in the course of my field research. As the evening progressed, I circulated among the staff members introducing myself and describing the work I was doing with my spectroscopy. I began with a brief tutorial explaining what my spectrometer was, how I was using it and why. I explained how I would locate myself a quarter mile or more from a gas well flare at night and collect optical readings with the spectrometer, which I later turned into graphs on paper. Spikes on these graphs revealed the presence of chemicals from the fracking process.

The group seemed taken aback, perhaps because they rarely - if ever - heard a citizen scientist presenting them with such advanced information. They asked a lot of questions and viewed my materials, with interest.

When we finally broke up for the evening, the snowy weather typical of springtime in the Rockies had moved in. Indeed, we had scarcely driven a mile, when we realized a full blizzard had settled down around us. The EPA team was riding in a large van but I followed to be sure of their safety until we had completed the riskier part of the route home. Even though we had only about thirty miles to go to Pinedale, it was a treacherous, painstaking drive.

Now came the waiting.

Lobbying Region 8

On October 12 and 13, 2005, the news media reported that the EPA had weighed in on BLM's environmental impact assessment for the proposed Jonah Infill Drilling Project.[1,2] The crux of the project was that EnCana was proposing to drill 3,100 to 3,200 wells on 8,316 acres of the 23,500-acre Jonah Field where it had thus far drilled 1,006 wells.[3]

Roberts was quoted as supporting the supplemental report alternative which proposed a slower pace of development unless the operator, in this case EnCana Corp., could shift to newer, cleaner diesel engine technology powering their drill rigs and thereby cut emissions by 80 percent. This was the most restrictive of five alternatives that proposed 80, 60, 40, 20 and zero percent reduction in emissions. He went on to offer that if these drill rig emissions were reduced, EnCana would be permitted to drill at a faster rate.

The reason for this response was a perceived need to protect nearby wilderness areas as well as Yellowstone and Grand Teton Parks lest they become subject to visible haze. However, it was well known locally that the target rate of development being pushed by operators was 350 wells per year and the EPA seemed to be reacting to the local citizenry who were worried that the emissions load on the local air mass would overwhelm their prized visual scenery with heavy obscuring haze.

Under the 80 percent plan, EnCana would have to reduce drilling emissions of nitrogen oxides which are precursors to visible haze from 3,206 tons per year from its 250 new wells to 642 tons per year. Randy Teeuwen, EnCana's spokesman claimed that the company had identified mitigation measures that would significantly reduce emissions. Be that as it may, BLM admitted in the same EIS that there still *might* be *some* hazy days in the Bridger Wilderness Area. It appeared that BLM was being vague about what it surely knew was going to happen.

Teeuwen sought to assuage any concerns by pointing out that the company was investing in cleaner diesel engines and was testing a natural gas fueled drill rig. He further asserted that EnCana would exceed emission goals once those technologies were in place field-wide. The coming months would reveal to those of us on the AQTG that such efforts would be minimal in both extent and effectiveness.

Crucially, there was a word in his statement that went unrecognized in import but in coming years would become a hated term by those of us seeking to force emissions controls. That word was "goals." It was not legally binding and we would learn through bitter experience that BLM would come to use it as an escape route for failing to meet environmental impact mitigation promises in its impact assessments and in its records of decision.

At the time, I was guardedly satisfied that EPA had issued these cautionary findings but I was pretty certain that the recommendations would not rest well with the Bush Administration's EPA director in Washington D.C. and I suspected what Region 8 had done would draw considerable flack.

I had been speaking with the operators regarding their environmental impacts in an attempt to steer them toward a more environmentally aware posture. These discussions had thus far been futile, but now I hoped that this newest action would have major influence on these operators. Indeed, one had recently confided in me that the company was growing concerned over WYDEQ's increasing scrutiny. I hoped this EPA finding would cause them to resist less and cooperate more.

I wrote to Director Roberts expressing these points and reminded him of his statement, as quoted in the press, that if drill rigs were cleaned up more of them could go into operation.[4] I suggested that this could run smack up against a recently released study, funded by Questar and Ultra Petroleum, showing that current activity levels had driven approximately 50% of the deer population off the Anticline.[5] However, I admitted this was not in EPA's charter to address.

I then asked that he keep his staff focused upon post-drilling activities, namely gas production emissions. In Jonah alone, several thousand wells multiplied by tens of tons per year per well of VOC and NOx emissions would be devastating to visibility in the adjacent Class I regions. This part of the equation had to receive equal attention.

I further explained that for my part I had been successful in brokering a meeting between EnCana and the inventors of a device called the Quantum Leap Dehydrator (QLD) that had been tested a year earlier and found to be a near zero-emissions device (pounds per year rather than tons as noted a few pages earlier). EnCana was planning a test through the winter of two units and was open to the possibility of retrofitting its current fleet of five hundred old dehydrators with the new technology. I was hoping that if one operator embraced the device, the others would be forced to follow suit.

I also pointed out that I had developed a strong professional relationship with the inventor and had learned that he had no doubt the units would come through the winter with flying colors. Accordingly, he was gearing up for mass production and saw a great market in Sublette County. In fact, if all went well, I would hopefully be able to say that I had personally brokered a situation that reduced well production emissions by over one thousand percent. I closed by offering the opinion that it would not hurt at all if EPA Region 8 personnel kept pressure on the operators as an incentive to embrace such newer technologies.

Region 8 Weighs In

In early 2007, Region 8 management issued its opinion of the proposed BLM development plan for the Anticline gas field. Under Director Roberts' signature, a formal letter assessing the BLM plan dated April 6 was sent to the director of the state BLM office director Robert Bennett.[6] The letter specifically explained that it was addressing the Draft Supplemental Environmental Impact Statement and also the BLM's subsequent Supplemental Ozone Analysis for the project area.

The EPA noted that a prior Draft SEIS supplement to a NEPA analysis and Record of Decision had authorized up to seven hundred producing wells in the Anticline project area. It was noted that this Draft SEIS was assessing cumulative and site-specific environmental impacts of year-round drilling, well completions, and production of up to 4,399 additional gas wells on a possible 12,278 acres of new disturbance. Also stated explicitly was the fact that the adjacent Bridger Wilderness Area is a federal Class I area under the Federal Clean Air Act requiring protection of air quality and visibility.

The letter then summarized the Draft SEIS trio of alternatives consisting of the Proposed Action, BLM's Preferred Alternative, and No Action. The proposed Action involved 4,399 additional wells on 12,278 acres of new disturbance by the year 2023 with drilling and completions happening year-round within concentrated development areas that included "…big game crucial winter habitats"; the Proposed Action alternative included waste water gathering systems in the central and lower parts of the Anticline project area; the Preferred Alternative was similar to the Proposed Action Alternative but differed with regard to the surface extent of disturbance and also included what EPA described as "important" mitigation actions to address air quality impacts.

The letter then went on to express "concern" over the range of alternatives considered and noted that while the Draft SEIS did identify a "Conservative Alternative" and a "Reduced Pace of Development Alternative," these options did not receive detailed analysis. EPA did not consider the rationale cited for eliminating these option to be sufficient and urged further consideration. EPA also wanted further refinement of the description and analysis of the "No Action" Alternative.

Citing NEPA requirements for analysis of a range of reasonable options, and the magnitude of potential impacts to the Sublette County environment, BLM was urged to fully analyze at least one more option that embraced the same number of wells but over a longer period of time. Such an analysis was deemed particularly important in light of the magnitude of the project and its close proximity to the Class I Bridger Wilderness Area.

The letter then more directly tackled the issue of air quality impacts but here the agency wavered and did a soft shoe dance. Opening comments noted that EPA had participated in the Air Quality Stakeholders group which was a reference to their representative, Joe Delwiche on the BLM Air Quality Task Group. Later in the book, I devote an entire chapter to this group.

It was next stated that their work with us had resulted in development of a mitigation strategy that would significantly reduce visibility impacts on the Class I Bridger Wilderness Area over the next five years. A two-phase strategy would be implemented that would produce zero days of visibility impairment over one deciview of *modeled* impact. This emphasis on modeled impact was the questionable element that drove everything. Many times I would rail at BLM for singing the praises of modeled results that were almost daily refuted by simply looking out the window toward that wilderness area.

The letter went on to assure that if the goal could not be met, Operators, BLM, EPA, and WYDEQ would jointly agree to a plan that met the goal by any and all means. If that failed, the rate of drilling and development would have to be revisited. It was noted that this situation had been included in the Draft SEIS and EPA was supportive of it; that said, the letter did little to assuage my concerns. The paper promises it offered would remain just that so long as the agents inside those government bureaucracies sat and stared at each other, waiting for the other guy to make the first move. Basically, the government stakeholders were guardians of process and the citizens were prisoners of it.

As for the continual standoff within agencies, it was rooted in job and/or turf protection. Sadly the worry over job protection during the Bush Administration was all too real as I will soon reveal in the case of EPA.

There was another reason the assurances were unlikely to ever gain traction. It would become necessary to identify who was responsible for the pollutants causing the visibility decrease and to what extent. This was because the operators had long demonstrated their skill at invoking procedural roadblocks. This tactic protected their ability to conduct business as usual and both WYDEQ and BLM were terrified by the prospect of legal challenges of any kind. From a practical point of view, because the pollutant emissions could not be tagged as coming from EnCana wells, or Shell wells, or Ultra wells, there would be no unassailable method of assigning blame and crafting corrective actions.

Complicating this reality was the fact that proponents for gas development in Sublette County including certain powerful members of the Board of County Commissioners went out of their way to "prove" that visibility impairment was due to emissions coming from Salt Lake City even though meteorology rarely supported that assertion.

The letter from EPA Region 8 then went on to address the subject of ozone. It acknowledged findings of the Supplemental Ozone Analysis that included observation of elevated ozone levels in winter months at several WYDEQ ambient air monitoring stations near the Anticline project area. This analysis had become necessary because ozone behavior here had run so contrary to understanding by the scientific community that it became an issue of elevated concern. In fact, this presence of ozone in Sublette County during deep winter was unnerving for reasons of science and business.

Up to this point, ozone creation had been believed to be a process limited to summer periods when sunlight was at maximum intensity to react with the gaseous precursors, breaking them down to reform into ozone molecules. In wintertime, several atmospheric variables come into play but generally speaking, cold air/warm air inversions can trap the precursor gases and concentrate them. Then if snow covers the ground, it reflects sunlight, thereby providing a double pass through those concentrated gases enabling additional ozone creation.[7]

Regulation of pollutant gases as stipulated in the Federal Clean Air Act while seemingly clear-cut to a casual reader is in fact a labyrinth of definition minutia and complex circumstance differentiation. Ozone, however, has a well-established record of detrimental human health impacts leading to specific exposure standards that are an enforceable metric overseen by the EPA. The Operators, BLM, and WYDEQ were immediately aware that the current headlong press to develop the Anticline as well as other gas fields could be brought to an unforeseen halt if the issue wasn't better understood.

This specter had begun to lift its head months earlier when locals started learning of it; some even claimed to have experienced health impacts. This drew the governor into the issue and also precipitated a request from many of these citizens that I personally develop a project to measure its presence in our local region. I will describe the nature and findings of this project in Chapter 10.

WYDEQ saw that it had to have a more reliable and comprehensive record of how ozone was being created. It embarked upon a field-sampling program that was somewhat uncharacteristic of that agency in terms of its level of emphasis and time acceleration. A large reason for this was that a major tool upon which it and BLM relied was modeling and that modeling needed actual data to feed into the process.

Seeming to take some up-front credit, EPA in its letter cited its "close work" with BLM to provide "early input" into the ozone modeling analysis. It noted that elevated ozone levels in the summer months had been predicted and the implication of this, juxtaposed against the earlier comment about actual elevated ozone levels in the winter, was that real problems could be encountered if successful mitigation measures were not developed and implemented.

EPA went on to suggest to BLM that it needed to undertake more refined modeling analyses utilizing a finer pattern of grid cells to better characterize the geography of the local region. Also, it needed to better inform the public and air quality stakeholders about what to expect regarding potential summertime ozone levels.

The letter also praised WYDEQ for its work addressing the wintertime ozone problem it was monitoring in the field. However, EPA cautioned that the work was so important as to require a level of intense field measurement that exceeded the scope of the Draft SEIS. A veiled warning concluded that future ozone exceedances attributable to Anticline gas field development would have consequences. This would include coordination between EPA, BLM, and WYDEQ in defining strict adaptive management measures that would mitigate ozone precursor emissions. EPA was offering vinegar and honey to its government partners and this was the vinegar.

Over the next several months, I would come to see this as a pattern of behavior within the EPA. I spent much of that time lobbying my contacts at Region 8 to exert more pressure on BLM and WYDEQ, and when I compared their private comments to me with those of the state office of BLM I realized that EPA was walking a fine line. It was attempting to execute its legal obligations vis-à-vis oversight of air quality protection while not antagonizing powerful voices in Washington D.C. and Wyoming.

I was told the governor and local officials had quietly requested that Region 8 management rely upon voluntary efforts being urged upon the operators by WYDEQ. This was seen as preferable to implementation of demanding mitigation steps that could limit the operators' enthusiasm for continuing to do business in Wyoming and thereby damaging mineral revenues.

The EPA letter closed with comments awarding its rating on the adequacy of the BLM Draft SEIS under guidance of Section 309, Federal Clean Air Act. That rating was *Environmental Objections – Inadequate Information (EO-2)*. The reason given was that BLM had failed to include other "…reasonable alternatives that would mitigate significant impacts to the environment." It went on to advise that if BLM chose not to more fully analyze an option for a reduced pace of development, then it had to give a more detailed explanation supporting the BLM preferred decision. Now came the EPA honey; it warmly stated that it was "…looking forward to continued work with BLM and WYDEQ…" to reduce visibility and ozone impacts.

In my mind, the light scolding in this formal notice to BLM regarding its Draft SEIS, was woefully insufficient. It would take many years, however, before I fully realized the kabuki-like theater that goes on between government agencies as they compete for dominance against each other.

I wanted to see blunt declarations that BLM was in violation of its stewardship obligations under the Federal Clean Air Act and there would be real consequences unless those violations were rectified. When the letter turned out to be a disappointment, I attempted to organize a petition drive that would force the issue.

Citizens' Petition to Nowhere

Thinking there was power in numbers, in early May 2007, I placed an advertisement in the local newspaper asking everyone concerned about visibility and ozone levels to come to a meeting and sign a petition. However, when I added up the number of respondents, I was disappointed to find that only 110 citizens out of the few thousand in and around Pinedale had joined the effort. But it was better than nothing, so I went ahead with distribution of the petition.

The recipients to whom I mailed the petition were a who's who of politicians and agency directors. They were the Wyoming representative and two senators to Washington D.C., our governor, and the directors of EPA-Washington D.C., the Department of the Interior, BLM-Washington D.C., EPA-Region 8-Denver, and the state office of the BLM in Cheyenne.[8]

I attached to the petition a cover letter that declared the EPA reply to BLM's Draft SEIS to be too little too late. I cited the EPA statement commending BLM for "leadership" with the air quality stakeholder group and the assertion that "… air quality modeling and proposed mitigation measures reflect the efforts of the stakeholder group." I explained that from our perspective, BLM had studiously avoided leadership in its attention to air quality issues and the modeling and mitigation measures invoked were only paper exercises that failed to recognize the degradation already underway since at least 2001.

I further argued that BLM's abrogation of leadership could be found in a fact that evidently had not reached beyond the Pinedale BLM Regional Office: the spring 2005 report from the Air Quality Task Group to the Pinedale Anticline Working Group, in which Region 8 invoked in its detailed comments, was not read by BLM or the PAWG until early summer of 2006, and only after the Task Group literally begged BLM to do so.

After waiting for four months for a response from any of the addressees, I wrote a formal letter dated September 17, 2007 to the director of EPA-Region 8. I opened by pointedly noting that 16 weeks had passed with no response to our petition and this seemed to be signaling a determination by the Agency to ignore us.

I listed the concerns we wanted EPA to address: (1) current insufficiently stringent regulatory efforts, (2) objection to use of 2005 impairment levels as the baseline for the future, (3) objection to reliance upon computer modeling as the predictor of the future rather than upon actual empirical measurements of gases coming out of exhaust stacks and combustors, (4) need for tighter oversight of well completion flaring, (5) need to undertake an instrument based exhaust/ combustor stack emissions measurement and reporting program, (6) dissatisfaction with the assertion that we must accept the presence of the "brown cloud"[9] because it did not breach statutorily prescribed limits, and (7) dissatisfaction with the pace of development being allowed because it was outrunning the minimal efforts being imposed upon industry to address atmospheric impacts.

In September 2007, I received a reply to our petition from Region 8 headquarters which it carbon copied to the Wyoming governor, the state senators and representative to Washington D.C., the director of the state office of the BLM and his director of the Pinedale office of the BLM, and the director of WYDEQ.[10] The Agency "appreciated" the concerns we expressed and the willingness of the petitioners to come forward. It then proceeded to immediately disagree with the notion cited in the petition follow-up cover letter questioning the usefulness of dispersion modeling.

It further asserted that there were no better methods than this technique to predict the future effects of emissions. Rather, the problem was due more to underestimated emissions resulting from a BLM failure to properly estimate the rate of gas field development which, it went on to assure, the BLM had since "quantified" and would mitigate per its promises in the current EIS. Nevertheless, EPA "shared our disappointment" that estimated emissions in the Anticline area had increased "several-fold between 2000 and 2006."

The reply then took issue with the point I raised about actual versus modeled visibility impacts. It didn't directly address this conflict but instead parsed its previous formal comments to BLM regarding the subject as well as the EIS itself. EPA pointed out that it supported comparison of modeled results to *"natural background conditions"* (emphasis by EPA). It specifically cited the BLM model for 2005 as showing 45 days per year of visibility impairment greater than the 1.0 deciview limit for the Bridger Wilderness Area relative to background levels.

It implied this was unacceptable by stating that the Agency did not support using 2005 as a reference or background level but then stated that the current EIS did not use 2005 as such a reference. While this may have been *technically* true, 2005 levels were in fact adopted as the reference in a closed door meeting between federal and state regulators (I explain this further in the chapter on the U.S. Forest Service). Consequently, that date did in fact become the baseline of reference in subsequent Sublette County gas field development EIS documents issued by BLM.

Of the seven issues cited in the petition follow-up letter to EPA, the only other point the Agency saw fit to address was that of slowing the rate of development. Here again though, Region 8 dodged. It quoted the EIS statement, "… reducing the pace of development may be used as a mitigation tool if other means fail to mitigate the modeled impacts." This was a subtle statement that modeling was being relied upon in lieu of actual field measurement. EPA closed by offering assurances of its recognition of the importance of this option and its intent to continue working with the BLM in its role as the lead agency and with WYDEQ as the project proceeded. This statement adroitly passed the buck to BLM and WYDEQ.

This formal reply from EPA was an incident of stark education for me. I had been woefully naive and idealistic. I could not miss the fact that EPA-Region 8 was the only addressee to whom the petition was sent that ever replied, formally or informally. My takeaway lesson from this adventure was that politicians had to be pressured by a much larger and more vocal critical mass of citizens than we represented. Furthermore, we had to pose a clear and imminent threat to the economic business model being pursued by the operators and state leaders whose first priority was preserving the mineral tax cash flow coming from natural gas development.

Up Close and Personal

My next opportunity to interact directly with the group from Region 8 Headquarters would come in January 2008. An environmental lawyer working for the Wyoming Outdoor Council requested that Director Roberts and his staff return to Sublette County for the purpose of conducting an in-depth ground tour of the Anticline gas field. To my surprise, I received notification from Linda Baker, the director of a local environmental organization called the Upper Green River Valley Coalition that someone on the Region 8 staff had expressed a desire for me to participate in their tour. This was music to my ears!

Once again, Roberts and several of his department leaders traveled to Pinedale for a detailed examination of the current situation on the Anticline. I met them in Pinedale early in the morning and we piled into the two cars they had rented for their tour. I was deliberately directed to the car carrying Roberts and for a period of over an hour I gave him a personal tutorial on all the observations I had documented regarding the sources of air pollution coming from the hundreds of well pads and thousands of active gas wells. At each of the well pads where we stopped, I pointed out the specific apparatus I had documented with spectroscopy and infrared imaging to be sources of methane and dehydrator combustion products. As I spoke, I noticed that he was focused on my every word for the entire period.

On the return trip home, I was shifted to the car carrying Roberts' director of his Regional NEPA Compliance and Review Division, Larry Svoboda.

Again, I launched into a tutorial of the type I had given his boss but this time I added my experiences with the director and staff of WYDEQ. I did this because I hoped that being responsible for overseeing rules of enforcement, he would want to know about my frustrating experience due to WYDEQ's resistance. When I recounted one of my meetings with the director of WYDEQ in which he coldly stated he would never accept my data and observations, Svoboda shook his head in dismay and told me point blank to "Keep doing what you are doing." That was one of my rare high moments.

When we arrived back in Pinedale, we settled down to a nice lunch and I made sure to place myself next to Roberts. I started our conversation by noting that he and I had been career Air Force officers at the same time. He surprised me by stating that he had been operating in Thailand during the Viet Nam War as an official of the Joint U.S. Military Advisory Group or "JUSMAG" as we knew it. I happened to be indirectly on the receiving end of this organization while I was stationed in Udorn, Thailand as an ordnance officer. Immediately our small world became smaller and we proceeded to swap military career stories. Linda Baker later commented that this was one of the most amazing interactions she had ever seen.

Eventually, I moved the conversation to the issue of the Pinedale Anticline and what I considered to be the BLM's dismal record of performance in executing its legal obligations to protect the regional airshed. He listened to my recitation of my fieldwork as well as my experiences associated with my attempts to challenge the too cozy relationship between BLM and the operators. I illustrated my opinion by citing BLM's use of an oil and gas industry engineer to "help" the Pinedale office clear away pending permit applications.

I also described the pressure my contact at the state BLM office was experiencing. She was under constant criticism by her superiors for not approving drilling permit applications fast enough. I discuss both of these examples in detail in the chapter about the BLM. When we concluded the day Roberts told me that I had given him a great deal to consider and that he and his staff would go back to Denver and digest it all.

Region 8 Doubles Down

To my great surprise and satisfaction, the reply from Region 8 to the BLM was released only a month later. In a letter to the Director Robert Bennett of the state office of the BLM, dated February 14, 2008, the agency presented its opinion about the Revised Draft Supplemental Environmental Impact Statement for the BLM's proposed Pinedale Anticline Oil and Gas Exploration and Project Development (Revised Draft SEIS).[11]

As in the previous comments letter that it issued on April 6, 2007, the EPA noted here again that an earlier EIS supplement and 2000 Record of Decision had authorized up to 700 producing wells in the Anticline project area. It was noted that this Revised Draft SEIS was again assessing cumulative and site specific environmental impacts of year-round drilling, well completions, and production in the development area.

It was further noted that the Pinedale Anticline Project Area (PAPA) would cover 198,037 acres that coincided with some crucial big game winter habitats and was only 11 miles from the Bridger Wilderness Area in one direction and 2.3 miles from the Bridger-Teton National Forest in another direction. It further and pointedly stated that the Bridger Wilderness Area was a Federal Class I area under the Clean Air Act which must be afforded special protection of air quality values such as visibility.[12]

The letter then observed that the Revised Draft SEIS considered five development alternatives but then addressed the proposed and preferred alternatives. This section was a bit difficult to follow because the narrative slipped back and forth in its use of the titles "Proposed Action," "Proposed Alternative," "Preferred Action," and "Preferred Alternative." This made necessary a close reading, word-by-word in order to tease out the crucial import of what was being said.

EPA's summary of the relevant options started by noting that the Preferred Alternative envisioned 4,399 more wells sited on 12,885 acres of new disturbance by year 2025. Not stated was that this extended the time of full development by two years from the initially proposed 2023. The Proposed Action was basically the same and included an added liquids gathering system for water coming out of the gas called *produced water*, and condensate hydrocarbon liquids. This would complement another such system in place in other parts of the field. Also, Tier 2 equivalent emissions mitigation would be applied to 29 of the 48 rigs anticipated at the anticipated peak drilling period in 2009.[13,14]

The Proposed Alternative was deemed similar to the Preferred Action because it involved the same project elements including the same number of wells on the same amount of disturbed acreage. However, under the Preferred Alternative, the core drilling area was "spatially different" from that of the Proposed Action and involved an additional year-round "potential development area" of 70,200 acres which represented a 60 percent greater expanse than the core area under the Proposed Action. This comment section ended with the blunt observation that EPA was concerned over potential impacts involving visibility, and ozone that were explained in more detail in following sections.[15]

The section on air quality impacts began with stark recognition that the Revised Draft SEIS revealed significant and unforeseen visibility consequences since implementation of the year 2000 Pinedale Anticline ROD. This declaration was supported by a quote from the Revised Draft SEIS: *NOx emissions from the PAPA in 2005 were five time higher than the analysis* (read that as *modeled*) *threshold established in the year 2000 Pinedale Anticline ROD; 2005 emissions produced 45 modeled days of visibility degradation greater than 1.0 deciview at the Class I Bridger Wilderness Area, five days at the Class I Fitzpatrick Wilderness Area to the northwest, and additional days at other regional Class I areas.*

EPA noted that even the modeled No Action option which pursued development under the 2000 ROD stipulations, predicted 2007 visibility impacts would exceed 2005 predictions with 62 days above the 1.0 deciview at the Bridger Wilderness Area, eight days at the Fitzpatrick Wilderness Area, and additional days at other Class I and Class II areas.

As a result of the unforeseen effects resulting from the development of 642 wells under the 2000 Pinedale Anticline ROD, EPA pressed for inclusion of meaningful and enforceable corrective measures in the Revised Draft SEIS that would ensure environmental protection as the additional 4,399 wells were developed. Lastly, EPA expressed its wish that the Revised Draft SEIS include plans to mitigate the already significant air quality impacts from the existing PAPA development.[16]

EPA then went on to cite its participation in the Air Quality Stake Holders group made up of government agencies that crafted early guidance and suggestions to the BLM, all of which had been included in the December 2006 Draft SEIS. It was acknowledged that this work had been carried forward into the Revised Draft SEIS but EPA was dismayed that the mitigation part of that guidance had exhibited modifications to the original commitments. These modifications weakened the ultimate goal and introduced uncertainty about reaching the target of zero days of visibility impairment at the Bridger Wilderness Area.

EPA further commented that those modifications suggested reluctance by BLM to commit to the full mitigation plan and that in turn had reduced EPA's confidence that zero days could ever be reached. As a result, EPA felt that the absence of specifics from BLM about corrective strategies to reach zero days of decreased visibility would result in at least ten days of impairment at the Bridger Wilderness Area. EPA ruled this to be significant and implication was that it would be unacceptable.[17]

Next came the section where EPA evaluated the issue of ozone. It acknowledged that BLM had included updates in the Revised Draft SEIS to an ozone analysis (again, read *modeled*) and agreed that this was needed because of recent high levels of ozone being detected by ambient air monitoring stations associated with the PAPA. EPA cited predicted ozone levels of 78 parts per billion (ppb) near the PAPA under the Proposed Action which approached the EPA's National Ambient Air Quality Standard (NAAQS) of 80 ppb in effect at that time.

Even under the 80 percent drill rig emissions reduction scenario which was similar to the Preferred Alternative, levels were expected to reach 76 ppb around the PAPA. Furthermore, EPA seemed unhappy that the SEIS analysis failed to model ozone during the initial five years ahead of full implementation of the 80 percent rule under the Preferred Alternative.[18]

This all was a flag of concern because EPA went on to explain that while it had confidence in the ozone model's reliability in predicting ozone, there were known issues. Specifically, it was known that there existed a bias toward under-estimation. This all added up to concern by EPA that when such bias was included in the prediction results already approaching the

NAAQS, the actual levels of ozone could pose potential environmental and human health impacts.

All of this was further reinforced by the fact that in 2005 and 2006, ozone levels were measured near the PAPA that exceeded the NAAQS. Furthermore, development under the Preferred Alternative was expected to continue into the year 2065 so the implication was that these conditions could not be permitted to go unabated.[19]

The EPA letter concluded this ozone commentary by offering recommendations. It wanted BLM to develop an air quality mitigation plan that would correct the cited possible health impacts. Furthermore, it wanted BLM to "demonstrate" through *modeling* that the proposed development would not breach the NAAQS. Lastly, EPA warned that it was currently reviewing the standards for ozone that could result in revision of those standards and consequently prompt a revisit of the BLM analysis.[20, 21]

To those of us who had waged a long battle to make BLM more attentive to environmental impacts from the Jonah and the Anticline, the final section of the EPA comment letter was very surprising and gratifying. It declared EPA's rating of the adequacy and thoroughness of BLM's Revised Draft SEIS. The second sentence summed it up by awarding a rating of *Environmentally Unsatisfactory– Inadequate Information (EU–3)*.

The reasons given were that EPA felt it had identified important long-term adverse impacts to air quality. Also, Class I areas were open to possible adverse visibility impacts due to inadequate mitigation. Furthermore, EPA warned it might find it necessary to return to the issue with additional comments in the event that it revised the national ozone standards to more stringent levels.

Then came the real shocker. The letter stated bluntly that the Proposed Action should not proceed due to the magnitude of the environmental impacts. Additionally, the "EU" designation qualified the project for referral to the Council on Environmental Quality (CEQ) should the identified unsatisfactory impacts not be corrected.

The CEQ was established within the Executive Office of the President by the same act that established NEPA in 1969. The purpose was to insure coordination of federal environmental efforts and the CEQ Chair acts as the main environmental policy advisor to the President.[22] The warning seemed to be placing BLM on notice that the matter could be elevated to the office of the President.

Next, air quality modeling to reveal predicted ozone levels under the different development scenarios was deemed inadequate. The "3" rating signified that the Revised Draft SEIS was inadequate to meet NEPA requirements and should be formally revised and resubmitted for public comment.[23]

Casualties

It is probably safe to say that BLM and WYDEQ were displeased by this polite but scathing indictment of their hard work to push the PAPA development to its final approval. In fact, it was likely unacceptable to the office of the Vice President of the United States, then occupied by Dick Cheney.

His willingness to influence environmental conflicts had been illustrated by his efforts to alter a situation on the Klamath River in 2001. He personally called a 19th ranking official within the Department of the Interior to successfully exert pressure challenging the science that protected two species of threatened fish. His intervention, for better or worse, prevented shutting off irrigation water to agricultural businesses that would have been hurt as a result of compliance with environmental law known as the Endangered Species Act. The result was the largest fish loss the West had experienced.[24]

Perhaps more relevant to Sublette County, was when in August 2001 Cheney signaled disdain for the Clean Air Act to the newly appointed EPA Director Christine Todd Whitman. In an interview years later, she told of receiving a call directly from Cheney, angry over what he considered foot dragging over the easing of pollution rules regarding power and oil refinery facilities. The contest was over a Clean Air Act stipulation that required aged facilities in these categories to undergo modernized emission control upgrades during scheduled refurbishment.[25] This was to address increased emissions presumably due to anticipated higher production output.

Whitman cited Cheney task force meetings where she heard strong opinions that EPA regulations were the main impediment to construction of new power plants. She concluded over the course of many battles with Cheney and his staff that President Bush had given Cheney full authority to deal with energy-related issues and at the end of the day, Bush and Cheney would dictate final policy. Finally, the White House version of the EPA pollution rule was crafted and placed before her for signature in 2003. She asserted in her interview that this version was so awful that she resigned because of it and not because she wanted to spend more time with family as her press release claimed.[26]

It should be no surprise then that his home state would not escape his oversight. In fact, I received a revelation in 2007 from a retired biologist formerly with the Pinedale BLM office who still had contacts on the inside. They were informing him of weekly calls from Cheney's office demanding removal of environmental roadblocks to approval of the PAPA and demanding that the project be approved.

My main concern now was how the Region 8 evaluation of the PAPA EIS would be received in Washington D.C. My worries were not misplaced because a few weeks later I learned through my contacts in USFS and BLM that Director Roberts had been summoned to D.C. I immediately called his office and found myself speaking to his secretary. I asked point blank if the rumor was true and she said it was. I then asked if he was being called on the carpet because of the PAPA ruling. She replied in a worried tone that no one in the Denver office knew the answer.

In May 2008 EPA announced that he would be undergoing knee surgery and would be taking a month off. Then on June 17, he suddenly retired. This was the same month that the EPA director of Region 5 resigned. Region 5 headquarters oversaw the Midwest where that director had been in a contest with Dow Chemical over dioxin contamination in Michigan.[27] These events were feared to have a chilling effect on other EPA regional directors and my own experience with Roberts' successor Carol Rushin would soon seem to support that.

A post-script to the Roberts retirement is in order. Given the sequence of events prior to and subsequent to Roberts' critical opinion about BLM's revised draft SEIS, I have to wonder if I am partly to blame for his departure. I will always worry that I had a role in what appeared to be his politically motivated removal from his post.

Doubtful Emissions Accounting

By 2008 I was becoming increasingly convinced that serious shortcomings likely existed in the manner and content of the methodology by which WYDEQ maintained its emissions inventory. This inventory was of crucial importance to the regulation of gas development in Sublette County because it was the fundamental basis upon which both WYDEQ and the BLM had argued that ongoing development would not prevent the attainment or maintenance of any ambient air quality standard required by the state's version of NAAQS.

I disputed that on the basis of projected emissions statistics I had amassed from hundreds of well drilling permit applications archived by WYDEQ. Also, I had built up a massive database of all wells and their production statistics from the Wyoming Oil and Gas Conservation Commission (WOGCC). These data tracked back as far as 2004 for permit applications and the late 1990s for WOGCC data.

I had constructed massive spreadsheets listing every well by various state identifiers, their latitude and longitude, gas production volumes, wastewater production, and oil production. I came to suspect I had identified a number of discrepancies that when added together, presented a body of evidence demonstrating the "actual" emissions inventory claimed by WYDEQ was very likely in error by a significant amount.

I wrote a letter dated April 8, 2009 to Larry Svoboda at the Denver Region 8 headquarters. As I mentioned earlier, he was the director of the regional NEPA Compliance and Review Division. I requested that he undertake three initiatives.

First, call for an audit of the emission inventory process practiced by WYDEQ. I asked that the audit address the methods of data collection, data reconciliation, and incorporation into a final empirical emissions inventory. Also, I urged that the audit ascertain to what extent WYDEQ applied emissions estimates contained in operators' permit applications to the construct of its emissions inventory. Furthermore, how did it reconcile changes in those emissions over the years as well sources evolved, declined, and new wells came online?

Second, send to Sublette County a team of technicians with the mission of conducting empirical instrumented measurements of a statistically representative sample-set of all emitting sources of the types cited within operator facility permit applications to WYDEQ. The purpose for this field exercise would be to validate the five most cited modeling tools by operators as the sources for their emissions estimates submitted to WYDEQ as part of their operating permit applications.

Third, require WYDEQ to reconcile its emissions inventory with the findings of task number two, adjust the inventory accordingly, and make public the outcome of this exercise.

As a result of my participation in the PAPA tour with Region 8 officials, I was now well known on a personal level and regarded as being credible and unusually well informed in the technical specifics of air quality protection. This now bore fruit in the form of a personal email letter directly to me from Joyel Dhieux, the Regional NEPA reviewer who had been one of the members on the PAPA tour.[28]

She gave many assurances that the air quality staff personnel in WYDEQ were sincere in their dedication to protecting air quality in my region and she went on to state that she was working with them to accomplish that goal. In that spirit, she wanted to show my statistics to those staffers because she agreed that my report raised questions that she and her fellow NEPA compliance staff wanted to place before WYDEQ.

She then advised that my report had been circulated to EPA Region 8's Air Program staff members who were working on ozone and emission inventory issues in Pinedale. She stated that her Air Program staff understood my concerns and that previously, EPA had reviewed emission documentation for southwestern Wyoming which initially reported annual volatile organic emissions (VOCs) as 200 tons per year. This estimate was later changed to 125,000 tons of VOCs per year and the estimate was currently undergoing further refinement! This further refinement of the emission inventory was in her estimation not unlike the first action item I described in my report.

She went on to explain that an entity called the Western Regional Air Partnership (WRAP) was at that time updating the emissions inventory for the region and was in what she called Phase III. This reflected the fact that it was the latest exercise at updating the database following two previous such exercises in which emissions were under estimated. Data collection, methodology, and reconciliation were major components of this latest Phase III effort.

She further explained that it was her understanding that WYDEQ air quality personnel were actually reviewing the permits they had accumulated from operators and were comparing the number of sources listed in the permits to the number of sources and their emission estimates that would be included in the Phase III emission inventory. She suggested that this activity had the potential to address many of my concerns regarding emissions estimates.

Joyel then addressed a probable limitation her people faced. She began by expressing the desire to see the results of the Phase III effort because the outcome of this exercise was being anticipated with keen interest, especially due to the outbreak of oil and gas development in the region. The resulting revelations would serve to inform future mitigation activities but until that time, funding for another emissions inventory review would likely not be forthcoming. That said, she assured me that her people intended to conduct a detailed critical review of the Phase III results and would keep me informed.

She closed her letter by informing me of another EPA initiative which she felt addressed my concern about the absence of empirical measurements of emission sources in the field. EPA's Office of Research and Development was currently examining emissions from oil and gas development in the form of an emissions study examining emissions from oil and gas sources, such as water treatment ponds. Testing had recently taken place in Colorado but the report had not yet been issued so again, she would keep me informed.

I was generally quite satisfied with this unusually personal feedback and I wrote to her to express my appreciation. However, I also expressed my feeling that the discussion of empirical field measurements of emissions sources had been a little weak.

I explained to her that in the aerospace business from which I had come, the practice was to model a response, measure the actual stresses and strains, and compare them to the model. If divergence was observed, the model was adjusted until an acceptable level of agreement was achieved. That seemed to be missing here.

I restated my belief that reliance upon a modeled inventory both regional and local would likely be non-credible until such validation was demonstrated. My concern was that, for political reasons within EPA, this would not be undertaken because discovery of serious inaccuracy with the computer codes in our local region would be too embarrassing and/or because the cost of fielding a team to conduct such measurements would not be funded due to current federal fiscal belt tightening.

Sadly, the promises of keeping me in the loop and informed never actually happened. I suspect that future events and maneuvering by BLM, WYDEQ, and the timidity of the new Region 8 director all played a part in bringing about a weakening of resolve and actions aimed at reigning in the operators' air quality impacts.

Also, the rate of movement by EPA, rendered glacial by the operators' lobby efforts in Washington D.C., figured into the picture. This was evidenced by its release for public comment in 2015 a proposed update of oil and gas development air protection rules that had been under development as far back as the early years I have been chronicling in these pages.[29]

Serendipity

My involvement with EPA was soon to develop in a direction I never could have foreseen. In the period of 2006 and on, I befriended a husband and wife team that was spinning up a citizens group to challenge proposed drilling on 44,720 acres of prime national forest in the Wyoming Range on the west side of Sublette County. They hosted meetings at their home located off the Hoback Canyon and I decided to attend despite my hesitation over involving myself in another big gas-versus-environment fight. Happily, I did because I became close friends with my hosts Glenn Paulson and his wife Linda.

Glenn, Linda, and I soon developed a working relationship that led to our authoring jointly crafted science-based comment documents for submission to the U.S. Forest Service administrator in Jackson, Wyoming, who was overseeing the project. Over the course of this collaboration, we learned of many common interests. He and I had been involved with the subject of nuclear power plant science and engineering as well as the health science aspects from different perspectives. This was a source of hours of riveting conversations and I gradually learned that his background was the stuff of legends.

Glenn is a giant but for a very long time I had no idea because as the old idiom goes, "He hides his light under a bushel." He started out with a bachelor's degree in chemistry from Northwestern University In 1963 and a PhD in environmental sciences and ecology from Rockefeller University in 1971. He began his career as a staff scientist at the Natural Resources Defense Council in New York.

He next served as the assistant commissioner for science at the New Jersey Department of Environmental Protection in Trenton, during which he was the primary author of New Jersey's superfund law that preceded the federal law by several years. Also, he had become a central figure and right hand man to the governor of that state when the infamous Three Mile Island nuclear power station meltdown occurred.

He later became a professor of occupational and environmental health at the former University of Medicine and Dentistry at the New Jersey-School of Public Health (now known as the Rutgers School of Public Health), then served as the school's first associate dean for research. In May 2012, he was appointed as the Science Advisor to the Administrator of the EPA where he advised Administrator Lisa Jackson and her deputy administrator in areas of environmental assessment and exploration of unconventional energy resources for the nation.[30] This EPA involvement would become my greatest source of support and encouragement.

As the months passed he and Linda learned of my many years of fieldwork and written comment documents about the Jonah and Anticline gas field projects. One day while enjoying dinner at their home, I lamented that all my work was going nowhere and having little effect in moving WYDEQ, BLM and the Denver EPA office. Furthermore, Lisa Jackson had recently visited Sublette County under the close escort of industry and pro-drilling political leaders whom I opined would blow a lot of smoke at her.

He reassured me that Ms. Jackson was a good "smoke detector" then suggested that I write my thoughts in a letter directly to her. When I scoffed that it would never pass through her palace guard, he said I should give it to him and he would deliver it to her personally. This was a development I had always fantasized about and immediately chose to exploit!

I went home and started work on the idea. The letter had to be succinct, credible, and directly to the point. Above all, it had to be interesting; nothing would be accomplished if it put readers to sleep. Over the next several weeks, I pulled together my materials and selected those writings I considered to address the most pressing of the many issues that had developed in connection with management of the Sublette County gas boom. I delivered the letter, dated June 9, 2009, to Glenn a few weeks later.

I began the narrative by citing my growing concern that Wyoming regulators, aka, those in WYDEQ, were failing to execute responsibilities delegated to them by the USEPA. I further worried that her recent visit to the Sublette County gas fields had her too closely shadowed by representatives from the drilling industry and its political allies without presence of citizens having a more critical perspective.

I proceeded to explain my personal efforts to expose shortcomings in regulatory actions by EPA, BLM, USFS, and WYDEQ. These shortcomings were facilitating air quality impacts in seven surrounding Class I federal air protection zones from aggressive natural gas development in the Jonah and Pinedale Anticline gas fields. I argued that my research indicated the above named agencies had been crafting their entire approach to air quality degradation solely upon the basis of what I believed to be flawed computer modeling.

I argued that through my field observations, these models seemed to be seriously failing in part because of the use of wind field data not indicative of our complex topographical situation. I also pointed out that dependence upon modeling was argued by DEQ and BLM to be validated by a rather sparse network of WYDEQ air quality monitoring stations scattered around the gas fields. However, I insisted that these stations could only measure the generalized regional atmospheric loading of criteria pollutants with no capacity to link the loading to specific source gas fields.

I further maintained that attempts to claim validation of the computer modeled emissions sources (i.e., gas dehydrators, storage tanks, fugitive leaks) using data from these general monitoring stations were suspect. This was so because there had been no efforts to empirically measure those source emissions. I felt that use of computer models to report what the operators and WYDEQ called "actual" criteria pollutant emissions to generate an "actual emissions inventory" was in fact creating a virtual reality.

Furthermore, I commented that I had petitioned EPA Region 8 management to ascertain the data quality and accounting specifics that justified WYDEQ confidence in its emissions inventory. It seemed essential that EPA step into this issue and settle once and for all the credibility of that emissions inventory. Furthermore, it seemed necessary to determine in what manner if any, the computer models generating this inventory needed to be modified to agree with the actual field measurements I wanted the Region 8 administrator to undertake.

I also described what I considered to be a more threatening development. WYDEQ had, at the behest of EnCana, the largest operator in the Jonah Field, embarked upon crafting a plan by which the operators could address a documented violation of EPA ozone standards in the region. This plan would create "credits" for reduction of VOC emissions and apply those credits in a manner that would allow increased emissions of NOx, a different ozone precursor.[31,32] It seemed to me this concept violated two BLM records of decision mandating reduction of NOx in the gas fields and would, I believed, exacerbate an already obvious increase in visible haze in the nearby Bridger-Teton Class I airshed.

I next described my own effort to solve the VOC problem. I told my story of facilitating the Quantum Leap Dehydrator field test program and the subsequent barrier thrown up by WYDEQ. I pointedly explained how that WYDEQ behavior worried EnCana sufficiently to halt the test program. The QLD inventor ultimately proved WYDEQ resistance to be unjustified but to no avail. WYDEQ just dug in and stood firm.

I ended by expressing my fear based upon conversations with various persons within EPA, USFS, and BLM that WYDEQ, supported by our governor, was holding Region 8 management at bay under the notion of state's rights. That notion maintained that Wyoming should be free of federal interference and be allowed to pursue its own internal approach to regulation. Additionally, Region 8 management seemed limited in its enthusiasm to exert oversight due to still present conditioning by the Bush Administration that championed minimal federal interference in a state's right of self-determination.

Finally, I warned that the Wyoming legislature and our Washington delegation were dedicated to seeing, hearing, and speaking no evil about the energy industry. All of this had shut out the rest of us and rendered the NEPA process a paper exercise. Now all I could do was wait for a reply.

View From the Top (Sort Of)

To my great surprise and discouragement, the EPA response reached me after only two months. The discouragement was due to the fact that my letter had been bucked down the chain of command to Region 8 and Director Roberts' successor Carol Rushin.

The letter dated July 14, 2009, first thanked me for my letter to Administrator Jackson and explained that the Administrator's office had tasked Ms. Rushin with responding.[33] What followed was what I often received from all the government bureaucracies and that was a recital of statutory scripture. However, this time the Agency provided a lot of reassuring information as well as cautious and indirect support.

The section addressing visibility impairment started with such a recital. It stated the obvious fact that Federal Land Managers (FLMs) have an obligation to protect air quality including visibility in Class I areas. It was next stated that the EPA had been involved with the FLMs to deal with impacts from gas and oil development and cited unspecified "strong mitigation measures" that had been invoked to reduce NOx emissions.

Also, ambient air monitoring was ongoing to assess impacts. The Pinedale Anticline Record of Decision was invoked because of its provisions targeting an 80 percent reduction in NOx emissions over the next five years. Left unstated, however, was that this reduction was to be performed through modeled oversight and would be more virtual than real.

This visibility impairment section closed with an assurance: the U.S. Forest Service was installing an additional ambient air monitor (additional to those operated by WYDEQ) to ensure that visibility would be protected and that the mitigation strategy was effective.

Unbeknownst to Ms. Rushin, thanks to my contacts in the USFS, I had known about this monitor since it was in the planning stage. Those same sources also told me that WYDEQ was most displeased that a monitoring station would be brought online over which it had no control. This was but one indicator of the ongoing turf battles that hindered interdepartmental cooperation between the state, the feds, and even amongst the various federal agencies.

The EPA reply also addressed my issues with air quality modeling. It began by acknowledging the uniqueness of the problems associated with our cold climate and its creation of ozone "… on a scale never previously observed in the United States." Next cited were meteorological variables due to complex terrain that are hard to model accurately, limited understanding of emission sources from gas development in tight rock strata, and fast changing emissions control technologies for drill rigs, compressor stations, dehydrators, and flares. This list seemed clearly to go directly to the specifics I cited to Administrator Lisa Jackson.

It went on to acknowledge my concern that computer modeling was "… less than perfect" and seemed to agree without overtly stating as much. Non-specific concerns were admitted about many analyses in the Upper Green River region regarding wind fields, emission inventories, and atmospheric chemistry. However, these flaws notwithstanding, modeling was deemed necessary to predict future changes in emissions levels and provide estimates in places where air monitoring stations were not present.

EPA was at that time waiting for improved data sets to use in validating the models but in the interim it was recommending the use of conservative modeling tools. An example cited was the CALPUFF ozone model which had been a staple of many NEPA analyses in our area and was believed to be a conservative estimation tool that was "unlikely" to under-predict visibility degradation.

Nevertheless, EPA was concerned about the ability of grid modeling on a regional scale to accurately depict the situation in our region citing winter ozone prediction uncertainty in particular. To address this weakness, WYDEQ was engaged in further field measurement studies and the resulting data was being applied to performance testing of the models.

Two years into the future, during a Pinedale Anticline Working Group (PAWG) meeting on air quality, I would see this again. WYDEQ's Cara Keslar,[34] in response to a question from a group member about emissions mitigation progress, rambled on about the need for ever more measurements in order to "understand" what was taking place. After so many years of hearing this person obfuscate about their understanding of the impacts from the gas fields, I lost it.

When the meeting facilitator asked for comment I jumped in and asked point blank to all in the room: "Did any of you hear what I just heard? We have been presented yet another delaying tactic hidden inside the call for ever more study. WYDEQ very well understands the problem. ACT!" The facilitator asked her for a response but there was only many long seconds of silence. Then she spoke. "Not at this time." I had won the debate.

But back to the EPA letter responding to my letter to Lisa Jackson. The letter next tackled the issue of emission inventory. The obvious was stated, which was that while many gas field emissions sources were small, in aggregate they became numerous and, therefore, were having serious impact. Recent high ozone levels were an example. All the players in the Pinedale production area were aware of deficiencies in the emission inventories. EPA was working with these players to improve matters using data from the WRAP Phase III gas inventory but admittedly VOC emissions accounting needed improvement. This reference to WRAP and the Phase III study nicely supported what Joyel Dhieux had written to me back in April; so here there seemed to be agreement between upper and middle management.

Number four on the discussions list of the reply was the topic of NOx and VOC reductions. This section contained the assertion that the 2006 and 2008 RODS for the Anticline development included mitigation strategies to reduce NOx but did not enumerate. It merely stated that it was EPA's "understanding" that operators had met or exceeded the requirement. Specifically mentioned was Tier II drill rig NOx reduction requirements through the use of natural gas fueled drill rig engines which exceeded Tier IV engine requirements. The problem with this statement was always evident in our Air Quality Task Group meetings. The use of such rigs was never implemented on a large scale. EnCana was the chief user of that technology but only to a limited extent.

Due to admitted absence of sound understanding of winter ozone events, other NOx and VOC reduction tactics were being implemented. Beginning in 2009, WYDEQ was exercising a program of voluntary participation in the curtailment of on-pad activities that created ozone precursors. Starting in 2008, WYDEQ started a voluntary permitting effort for drill rig engines the details of which were not spelled out but presumably addressed exhaust emissions cleanup.

Furthermore, WYDEQ was requiring Best Available Control Technologies (BACT) for pumping station compressor engines. However, the timeline for operator compliance also had a strong voluntary component. Starting in 2007, VOC "presumptive" BACT was required on production equipment. Again, voluntarism was implied here.

In 2008, WYDEQ initiated its Interim Permitting Policy which introduced the NOx/VOC credit trade program mentioned earlier allowing operators to increase NOx emissions in return for VOC reductions. I have more to say about this program in the chapter on WYDEQ.

In compliance with the Anticline ROD, the operators Ultra, Shell, and Questar had committed to installation of a waste liquids gathering system that would eliminate some 165,000 tank-truck trips per year. Lastly, Operators would submit annual emissions *estimates* from their operations to include NOx and VOCs in compliance with the above named strategies. Here modeling estimation was again the operative approach and voluntarism heavily invoked throughout.

Moving on to the section on emissions from dehydrators, the letter cited EPA BACTs it was advocating on a voluntary basis under its Gas STAR program. It specifically acknowledged the Quantum Leap Dehydrator technology as being included in its Gas STAR program and while adoption of such cost-effective technologies was encouraged, their use remained voluntary.

The EPA reply closed with a recital about current regulatory authority. I was reminded that EPA had delegated to WYDEQ the authority to manage air quality issues in the state and that EPA retained regulatory responsibility in such matters as attainment, nonattainment, and unclassifiable designation actions. The governor had authority to recommend ozone nonattainment designation after three consecutive years of violation of the NAAQS which he did in March of 2009. This by the way was a crucial issue that WYDEQ tried to end run in 2009 and is discussed more in the chapter on that agency.

Lastly, EPA advised that it was currently in the process of revising the 8-hour ozone NAAQS for Wyoming and was executing the appropriate public notification protocols required by law. It further advised that additional information from the state and the public would be received until January 2010 and then by March of that year issue new 8-hour ozone regulations.

Overall, this reply from Administrator Rushin signaled a few interesting points. First, my having been able to go straight to Administrator Jackson in Washington D.C. seemed to get her attention. All my points to Ms. Jackson had been addressed and for the most part in a manner that seemed forthright. On the other hand, the agency had limited latitude within which it could maneuver thanks to the toxic political environment in Washington.

Adding to that environment by the way were our own two senators and congresswoman. In fact, during one of Senator Barrasso's town hall telephone sessions with voters, I was almost stunned by the vitriol heaped on EPA by those callers and the fanning of that attitude by Barrasso. Laughably, he even made the silly declaration that Administrator Jackson was too afraid of him to testify before his committee.

Finally, I felt the response from Ms. Rushin still perpetuated the reliance upon modeling as the preferred method over that of actual measurement of visibility impairment. Her agency had handed WYDEQ a fairly sweeping blank check in that regard and would continue to place its reliance upon that agency for information to guide its future actions. Unfortunately, WYDEQ was in turn relying upon the operators to supply emissions data while lacking any real in-house expertise to critically evaluate that data. Thus, the operators were in an enviable position to influence the regulatory actions of both WYDEQ and EPA.

Closer to the Top

Three years later in May 2012, Glenn Paulson was appointed to be EPA Administrator Jackson's science advisor. A few months later he requested that I provide him with literally everything I had written to EPA-Region 8, BLM, USFS, and WYDEQ about the subject of air quality impacts resulting from natural gas development in Sublette County.

I immediately set about the task of sifting through all I had written to select the best and most technical of my comment documents. It didn't take long to settle on 19 documents I had authored over the previous six years. These included seven of my most extensive public comments. These documents were my responses to proposed projects for expansion of the Jonah gas field and development of the fields on the Anticline and in the Wyoming Range on the west side of Sublette County.

I also provided several studies I had penned. One study assembled predicted compressor stations emissions contained in EIS documents. Another looked at gas produced from the hundreds of well pads versus the gas from those pads that was actually counted as having gone to market. I discuss this in detail in the chapter on the WOGCC because there was a big discrepancy. Yet another study pulled together the chemicals used in fracking fluids, and finally, I included my published technical paper on my spectroscopic field measurements of well completion flaring.

Many months later, I happened be at Glenn's home as part of a social gathering. He was in Washington and had called on the occasion of his birthday to mark the date by chatting with his wife Linda. To my surprise, she called me to the phone because he had something to tell me.

I picked up and congratulated him on surviving to another birthday and we agreed that a fair amount of luck and good health was to blame. He then informed me that all of my materials had been reviewed by agency personnel, and were well received.

He recommended that I be on the lookout for future rule proposal actions that would come from EPA because I would see things that looked familiar. Such actions really did not start showing up until August 2015 when the Federal Register carried the first notice of intent by EPA to promulgate new rules addressing new and modified natural gas sources.[35]

I was unaware this happened until I was alerted by Cindy Beeler via an email announcement. She notified me of the public review period that was underway on a new rule.[36] The notice advised of three public hearings to be conducted at the Denver EPA headquarters and the procedure to provide input.

This was very frustrating because I could not afford the cost of travel that would be required to participate. Also, I had pretty much closed down my air quality campaign. The PAWG and its task groups had disbanded and in the intervening three years I had moved on and began to experience a fading memory of the details attached to the long air quality fight.

Recalling Glenn's comment that I would see things that looked familiar, I skimmed through the 591-page proposed rule. It addressed the phenomenon of gas leaks from equipment in the field called fugitive emissions. It emphasized the concern over methane as a greenhouse gas and also VOCs because of their contribution to the formation of smog. The rule was advertised as an update to 2012 New Source Performance Standards applicable to the oil and gas industry and added methane as a pollutant to be covered by the rule.[37]

The rule would direct operators' attention toward leak detection and maintenance of field production equipment down to the level of dehydrator and compressor pipe fitting connections, flanges, hatch seals, gaskets, joint seals, and vents, to name but a few. Inspection surveys for such leaks were proposed and the schedules for conducting them were defined. Also, the requirements were broken out according to whether the facilities in question were new or modified. Lastly, incentives designed to lessen the frequency of monitoring were included based upon the scarcity of actual leaks detected. Put another way, if leaks were fixed and stayed that way in a verifiable manner, operators could inspect less often.

Also of immense satisfaction to me was the rules' attention to the issue of well completion. It cited well completion processes utilizing fracking as significant sources of VOC and methane emissions when the flowback from the borehole is vented to the atmosphere. A great deal of discussion was dedicated to a description of the processes and the proposed requirement to utilize green completion techniques to replace open atmosphere flaring for disposal of waste methane gas coming out of the borehole.[38]

While the presence of methane and VOCs in completion flares was a clear issue, there was no mention of the elements potassium, sodium, and lithium that my field spectroscopy revealed. However, these elements are not criteria pollutants subject to regulation in the Clean Air Act so technically, they do not qualify for CAA attention.

A highly intriguing inclusion in the proposed rule was the recommended use of optical gas imaging technology or OGI.[39,40] To my eye, this was specifically referring to infrared imaging. This was an approach I believe I can correctly claim I first introduced into the debate on controlling emissions in the Jonah and Anticline gas fields. I discuss this in great detail in the chapter on WYDEQ and the operators; so I simply offer some reactions to the rule's inclusion of it here.

In the discussion of the OGI technique, it noted that EPA's cost analysis recognized that the projected magnitude of the number of wells nationwide could create problems in the use of the technique. Worries were expressed that availability of OGI instruments and personnel trained to properly use them could result in their being difficult to apply with the rule's recommended frequency. It suggested that small businesses might be a way to achieve the goals but again cited concern that the bigger operators with their larger financial resources could dominate the supply of available screening resources.

Returning to the comment from Glenn Paulson that I would recognize my influence in subsequent rules, I have to state that I do not see much in the proposed rule that qualifies. While the content addressing fugitive emissions and well completion flaring does coincide with the fieldwork I did on both, as well as the thousands of words I wrote to EPA about both, the resemblance could be described as being purely coincidental.

Rear View

Late in December of 2015, I placed a call to Cindy Beeler for the purpose of clarifying a point to be included in this book and about which I had worked with her years ago. To my pleasure, she instantly recognized my name. We recounted the basics of some of our events together in those past years and she updated me on the status of the company that developed the Quantum Leap Dehydrator technology.

She then went on to express her personal satisfaction over the progress she had witnessed in recent years regarding the protection of air quality. She opined that although we had been in an uphill fight back in the period from 2003 through 2012, attitudes inside and outside the EPA had much improved and with that, so had the progress in implementation of the regulatory framework surrounding the subject. She ended by leaving me with the encouragement that we had made a difference that should carry forward well into the foreseeable future.

However, in a seismic shift of political fortune, all of this is at risk. This is due to efforts to roll back environmental regulations by the new Trump appointee to the directorship of the EPA. That appointee, Scott Pruitt has a long history of animosity toward EPA attempts to regulate the oil and gas industry's emissions so Cindy's cited progress may be fleeting.

Chapter 3

Bureau of Land Management (BLM)

Prologue

From the days when Wyoming was a territory in the 19th century it has had a dubious relationship with the federal government and captains of industry. A fascinating example involved the Union Pacific Railroad with which the state had a checkered relationship in its push across the West. Literally.

The government decided to subsidized two railroad companies to build simultaneously from the west and from the east. They were incentivized through the granting of loans and land in the form of ten sections of public land for each mile of track laid down. These incentives were soon deemed insufficient, so a Union Pacific attorney drew up legislation that became the Pacific Railway Act of 1864 doubling the land grant to twenty sections per mile of track laid.

These would be the odd numbered sections along the route forming a strip 40 miles wide, 20 miles on either side of the track. Another crucial addition was mineral rights attached to those sections, particularly useful at a time when coal was the fuel of choice in locomotives and coal deposits often were found in those sections.[1]

At the start of the 20th century, Wyoming next developed a dubious relationship with the energy industry and the federal government. For those of us who grew up in the state, a regular school topic (not so much anymore) in our history was the infamous Teapot Dome scandal. This was an affair involving Secretary of Interior Albert B. Fall in the Warren Harding Administration and captains of the oil industry, specifically Harry F. Sinclair in the case of Wyoming.

In 1909, President Taft set aside oil producing regions in Wyoming and California for the purpose of securing their output for use by the Navy as it transitioned from coal fired propulsion for its ships to oil-based fuel. In 1921, President Harding issued an executive order transferring control of these regions from the Department of the Navy to the Department of Interior. In 1922, Interior Secretary Fall proceeded to persuade then Secretary of the Navy Edwin C. Denby to actually implement this transfer of control.[2]

Later in 1922, Secretary Fall proceeded to issue a noncompetitive lease of the Teapot Dome oil reserve to Harry F. Sinclair of Mammoth Oil, which was then a subsidiary of the Sinclair Oil Corporation. Although the terms of the lease were deemed legal under the Mineral Leasing Act of 1920, they were suspiciously favorable to the oil companies and a collateral effect was to make Secretary Fall a millionaire by today's monetary equivalent. Two years of investigation by the Senate Committee on Public Lands ultimately turned up evidence of wrong doing that put Fall in prison for a year. Sinclair also went to prison but for only six months.[3]

Fall thus became the first cabinet official to be convicted on corruption charges. Although he went to prison for his dealings, his oil executive friend in the California part of this scandal managed to land on his feet. That man, Edward L. Doheny had provided Fall a no-interest loan of $100,000 that helped get Fall convicted of bribery. Conversely, however, Doheny not only was acquitted of bribing Fall, but his company foreclosed on Fall's home for "unpaid loans," i.e., the $100,000 sum.[4]

Almost incredibly, the Teapot Dome affair continues to echo today. In January 2016, the Department of Energy finalized the sale of the field for $45 million to a company that specializes in difficult to reach oil reserves called "stranded oil."[5]

The Mineral Leasing Act of 1920, still applies to mineral, oil, and gas development in Wyoming to this day and has occasionally been invoked by at least one gas developer in public meetings held by the BLM in Pinedale. This act was passed to replace the General Mining Act of 1872 because the latter allowed citizens to freely prospect on public lands and to stake claims even to surrounding lands where a deposit might be discovered.

By 1909, the director of the U.S. Geological Survey sounded an alarm that oil tracts would be gone within mere months because of the haste with which claims were being filed. President Taft thus created the oil reserves by executive order for the Navy and Congress subsequently ratified his authority to do so by means of the Pickett Act of 1910. This prompted the Congress to follow up with the Mineral Leasing Act of 1920.

The gist of the Act was to "establish qualifications for leases, set out maximum limits on the number of acres of a particular mineral that can be leased by one lessee, and prohibit alien ownership of leases except through corporate stock ownership."[6] The Bureau of Land Management (BLM) which resides within the Department of Interior is the primary administrator of this act.

This history lesson serves to reinforce the old adage that the more things change, the more they stay the same, as I am about to describe. Here we are in the 21st century, far removed from the 19th and 20th century events of BLM malfeasance but although science and technology has changed radically, politics and human moral corruption remains as familiarly the same as ever.

Into the Present

The experience of Sublette County as a result of this BLM stewardship can be shown to have been dismal during the rush to develop natural gas here. In fact, it can be credibly argued that the BLM, up until the late twentieth century, had little more to manage than the number of cows and sheep grazing on public lands. Then gas development became the new gold rush and this formerly sleepy agency found itself outclassed and ill-prepared to deal with the political and economic pressures that roared into the state. Essentially it remains in this condition to this day.

There is ample evidence, both direct and circumstantial, that shows the Pinedale Office of the BLM was a fundamental obstacle to environmental protection from gas field development, followed closely by the State office of BLM and various entities in Washington D.C.

In 2004, the Wilderness Society published a revealing study about the Bush Administration and the BLM. This study is presented in a document called *Abuse of Trust*. The Society found that Administration guilty of having initiated a series of internal directives to state and field offices of the BLM which were known as "instruction memoranda." Developed outside public review or comment, these memoranda were found to have fundamentally altered the way public lands were being managed.[7]

Examples illustrated the Administration's strategy of exploiting grey areas of law. In August 2001, state BLM directors were directed to issue oil and gas leases as well as drilling permits on lands that had yet to have completed development plans. In February 2004, a memorandum directed state BLM directors to essentially issue leases on demand to the operators. These actions challenged the claim by then Interior Secretary Gale Norton of decision making based upon "consultation, communication, and cooperation."[8]

Pinedale was specifically mentioned in the context of ignored environmental safeguards. Records from the Pinedale BLM field office reveal that virtually all requests from operators for exemptions from lease terms intended to safeguard wildlife populations and habitats were granted. I detected this trend independently and cited it more than once in my own comments in reply to BLM's EIS documents addressing development of the Anticline. For example, of 172 exemption requests from sage grouse protection stipulations, 169 were granted.[9]

Most likely by coincidence rather than design, a policy news release from Interior was issued the following month that seemed to fix the problems. In June 2004 the assistant secretary of the Interior for Land and Minerals Management announced new guidelines intended to protect the environment from long-term gas and oil development impacts.[10]

They included a requirement for BLM project managers to consider including best management practices (BMPs) in applications for drilling permits. Operators would be encouraged to meet with BLM staff during the project planning phase to identify possible BMPs for inclusion. A pointed caution was cited, however, that BMPs were neither minimum nor mandatory standards. Also they would be invoked on a case-by-case basis as opposed to an overarching requirement.[11] This neatly skirted the usual criticism that government agencies always try to make one size fit all.

This news release was not what it appeared. It was loaded with back door exits that only a reader having had bitter experience with BLM double-speak could spot. First, "guidelines" are just that. They are not binding and they are primarily voluntary. Next, a second layer of elasticity was inserted in the form of a requirement that project managers only "consider" incorporating BMPs. This consideration could be performed by simply stating that BMPs would be nice to have. If an operator were to respond that "yes, we agree but the cost (or some other element) would be inconvenient," then all parties at the table could declare "OK, BMPs have been duly considered, next issue." And not to be overlooked, BMPs were already understood to be non-mandatory.[12]

Yet another press story released two months later presented further details about White House intervention in the western state's affairs of oil and gas leasing. This exposé began in 2001 with Vice President Dick Cheney's Energy Task Force, characterized as a complaint desk where the operators could have their concerns passed directly to federal land managers in the field. Furthermore, administration officials dispatched internal memos directing those managers to be responsive to industry and praised those who complied while criticizing those who did not.[13]

One such memo from the White House asked Cabinet officers to find ways their agencies could expedite energy related permit applications and accelerate energy project completions. The Bush administration was even referring to wildlife protections as being direct impediments to leasing. The record showed that the BLM was eager to comply. The Wyoming state director of BLM issued an award in 2002 to one field office for pushing more permits through than all the other field offices combined.[14]

BLM field employees complained of being warned that processing oil and gas leasing and permitting was their first priority. Without doubt, they were being driven by deadlines imposed on them by the operators. On top of it all, field offices were receiving calls directly from the Energy Task Force complaining about that office taking too long to process permits. The BLM routinely protested this description of events as mischaracterization of the purpose of such calls. Instead they were recast as efforts to get information in response to operator complaints and speed up coordination as well as improve administrative efficiency.[15]

All of this exactly corroborates what I had been told by my BLM contacts in Cheyenne and the Rock Springs and Pinedale field offices.

The State Office of BLM

From the beginning of the Sublette County gas development contest, the state office of the BLM was a constant source of conflict. My own dealings with the Cheyenne office took place through the air quality scientist who worked there. She was Susan Caplan and I first met her when she became a member of the initial makeup of the Air Quality Task Group about which I dedicate an entire chapter later.

It was immediately apparent that Susan did not follow the pack. She wore her BLM uniform shirt with the tails out like a shirt jacket, unlike everyone else I knew, who wore the shirt tucked into their trousers. Nevertheless, I quickly determined that she was very dedicated and knew her business well. Our respect for each other grew continuously over the following years as we worked together.

Our relationship began when I explained to her what I was doing in regard to observing and identifying air pollution I suspected was coming from the drill rigs and well-heads in the two gas fields under development in the Pinedale area. As she learned about my approach using optical spectroscopy I think she came to quickly realize I was not the normal citizen activist and that maybe I too knew what I was talking about.

She advised early on that pressure was very high within BLM to concentrate efforts upon expediting gas/oil processes. This was due in part to the fact that her operating fund was obtained from revenues at the rate of 70% from gas and oil, and 10% from agency funding allocated to her air science specialty.

This prompted me to ask for clarification, as my understanding of BLM funding was that it derived from Congressional appropriation through the Department of Interior. She affirmed that as being true *but* offered that gas and oil were being so emphasized in Washington that it was receiving disproportionate emphasis in the form of directives as to how funding was to be applied, i.e., expediting gas and oil imperatives.

In one instance, I told her of a recent news story citing pressure that was coming from directives from on high but no one knew who was actually promulgating the directives. She cryptically suggested that one should look at communications between the Wyoming State Director and Washington D.C. This seemed to be an elliptical reference to the state BLM director, a person named Bob Bennett. She went on to explain that Bennett himself was "terrified" of environmental groups.

She further explained that oil and gas lobbyists in D.C. would visit Director Gail Norton and that same day Bennett would receive a call from her. Susan went on to state that one of the operators, EnCana, would send its officers in groups of five and most of whom were lawyers, to conduct personal meetings with Bennett on a weekly basis. Their purpose was to obtain progress reports from Bennett and intelligence as to environmental challenges they needed to anticipate.

However, Susan said that when she would recommend similar meetings with environmental group representatives he refused to agree to see any more than two per year. He worried that even this was too frequent because of his fear of being accused of partisanship by his superiors. He also fretted that any such meetings might require presence of government lawyers to protect him. To my mind a key revelation here was that Bennet was firewalling himself from opposition voices that were challenging the operators.

We discussed the Sublette County situation to considerable degree during the time she worked in the state office. I cited BLM-Pinedale's frequent insistence that it had no authority to regulate air quality matters. She challenged this idea by arguing that although BLM had no *regulatory* authority it did have the authority to set qualitative criteria for visibility. Loosely stated, BLM directives contained guidance to the effect that, "air quality shall not degrade beyond thus and so." She also advised me of rumors circulating that

B.P. America may have been contemplating a lawsuit against BLM for exceeding its air quality authority. This was just the kind of anathema that Bennett and his staff dreaded.

An issue that grew into a huge point of contention was the issue of nitrogen oxides or NOx tracking. Susan opined that BLM-Cheyenne and Pinedale deliberately ignored the issue of NOx tracking obligations dictated in the Pinedale Anticline Project Area Record of Decision (PAPA ROD) despite cautions not to do so. She even verified that the ROD contained language stipulating that when NOx reached a value of 700 tons per year (tpy), a new air impact analysis had to be initiated. The very year that figure actually happened was the same year that BLM stopped tracking NOx for the record.

At one point in our relationship, I asked why BLM maintained a position for an air quality scientist in Cheyenne if that person was not permitted to exert any policy influence. She explained that the position was to serve the function of reviewing potential air quality impacts from development projects and provide comments for the record but not necessarily to be acted upon. Adding to the absurdity, there were only four air quality scientists employed by BLM in *all of the U.S.* In Wyoming alone, that position was tasked to support 50 energy development projects that were currently underway or proposed.

Ultimately I asked why higher authority figures in BLM were so firm about fulfilling dubious directives from Washington D.C. Was it the promise of promotion? She explained there was a strong culture pervading the chain of command that only good news would be accepted and recognized.

Up until she left Cheyenne, I had many telephone conversations with her to learn what was happening there with regard to Sublette County. One topic of discussion developed around my speculation that our mountain ranges encircling the county on its east, west, and north were amplifying the buildup of drilling emissions impacting our air quality. I proposed to her the premise that perhaps temperatures were on the increase more than in regions beyond the mountain ranges in my area. I hypothesized this to be a result of the large number of drilling rigs dumping obvious and visible black exhaust plumes into the atmosphere but I had no empirical evidence at all to back me up.

My thinking had been clued by an article in a technical publication I was receiving from Lawrence Livermore Labs that discussed a possible link between regional warming and aerosols. These included carbon particulate soot from combustion sources.[16] I contacted the referenced researcher about my thoughts and was adroitly told only that my idea was "interesting."

Susan was also skeptical but open to anything so she passed my thought on to a highly experienced atmospheric scientist she knew in the employ of the U.S. Navy. Essentially, he shot me down on the grounds that there had been an ongoing drought in our area for a few years and among other influences, the drying out of the local earth surface could account for any perceived temperature increase. Realizing I had no leg to stand on with this idea, I dropped it and moved on.

As the months went by her superiors increasingly tasked her to review each drilling application from the operators and "make recommendations" as to their approval or disapproval. However, as more operators moved into Sublette County and flooded her office with drilling requests, she admitted to me that she was under eve-increasing and relentless pressure to simply stamp the applications with an approval and get them out the door. This became a source of great discouragement for her and eventually she couldn't take it anymore. She began to seek a transfer out of that regional office and ultimately succeeded.

A few years later I was attending a class on air quality modeling in Denver and found one of the attendees was a BLM guy who knew Susan. I inquired about her status and he said she was working in an analysis laboratory operated by BLM in Denver. She was much relieved because as he put it, "they did not treat her very well in the Cheyenne office."

A Dance Macabre

Around a year after Susan departed, I had occasion to chat with her replacement about an issue. I wanted to gauge what that person's attitude was and how he would follow in Susan's footsteps. He acknowledged familiarity with my name because of Susan's outgoing comments to him and also because of my reputation there. I received assurances that he and his superiors were staying on top of things and maintaining a close watch over developments in Sublette County. After another few years or less, he too was gone and replaced by yet another person.

This new replacement too informed me that she was familiar with my name by reputation but as I questioned her background, my heart sank. She stated she was a new hire to BLM and the spouse of a military member at the local air base. Our conversation further revealed her lack of experience as well as a serious lack of familiarity with the issues of air quality degradation in Sublette County because of gas development operations. This all became relevant as a result of a formal demand I placed upon the then most recent Pinedale BLM office administrator Shane Deforest.

One of the key sources of air emissions resulting from natural gas development identified in the Jonah and Anticline environmental impact assessments was gas compression facilities, also referred to as compressor stations. A detailed assessment of emissions species and volumes had been projected into the future from 2004 through 2011. These quantified emissions were tabulated in the form of existing emissions at the time of writing of the EIS and projected emissions by year 2011. The emissions species being analyzed were volatile organic compound (VOCs), hazardous air pollutants (HAPs), nitrogen oxides (NOx), and carbon monoxide (CO).

In December of 2011 I pulled together all the tables of these emissions declarations contained in several environmental impact assessment documents regarding the Jonah and Anticline. I wrote a letter to the Pinedale office demanding a review of the figures to ascertain if the assessments were in fact agreeing with the stated EIS figures.

My letter pointed out that Year 2011 was about to close and it was appropriate that the emissions history for Anticline and Jonah Field associated compressor stations be reviewed. This had to be done within the framework of the final approved EIS documents from which the ROD documents for those gas fields had been derived. I reminded the BLM that the EIS documents had stated specific compressor station emissions expectations for year 2011 but I was unaware of any concerted effort by BLM to evaluate the actual emissions results for that year.

Accordingly, I requested that such a focused review be formally implemented and findings (well supported by empirical measurement data rather than un-validated modeling) be published in the form of a status report available to the public. I asked that the review address the EIS projections and examine whether the actual emissions for 2011 fell below or above predictions. If they were above, explanations of why needed to be provided. If they fell below, explanations of why needed to be explained as well. In either circumstance, I wanted explanations to be substantive and free of vague catch-phrase assurances.[17]

I waited a month for a reply and when it seemed to not be forthcoming I called the Pinedale office. That was when I learned that my letter had been bucked up to the Cheyenne office. Again I waited into mid-January 2012 with no reply so I called the air quality representative in the Cheyenne BLM office. I wanted to know what was so hard about responding to a simple request involving data compiled by BLM and invoked by BLM in its EIS documents.

That was when I learned that yet another person had been employed in the position (and was the newest person I just described on the previous page). I asked what was going on regarding my request and I received a weak explanation to the effect that she had just hired on some weeks before and was still learning her duties. I sensed at once that she might be considering the idea of sending me a reply that would be superficial.

At that point I informed her of some of the details of my history with her office, her predecessors, and my own background and field research. I ended with the comment that if her superiors had failed to mention any of that, they had done her a disservice because I would not be lightly dismissed.

At last, in late January 2012 she crafted a reply which was then forwarded to the Pinedale BLM office. First, I was amazed because it was sent to me under DeForest's signature. I felt that Cheyenne had decided to throw DeForest under the bus by making him look like the author of the reply. Second, as I read through the letter I almost felt my jaw drop in amazement and fury.

The letter consisted of four pages of verbosity that only touched upon my specific question in one paragraph of comments. That paragraph stated:

> "*The emissions estimates presented in the Final EIS for the Jonah Infill Drilling Project are not specific expectations or ROD requirements since they are not representative of actual operations, and do*

> *not reflect the use or implementation of control technologies. The emission values presented in the Final EIS were extrapolated forward, as if no mitigation measures or controls would be used.... The actual emissions inventory and monitoring data submitted by operators to the WYDEQ-AQD for 2011 are used to assess compliance with DEQ-issued air permits and specific emission limits contained in those permits... The BLM continues to works[sic] cooperatively with the WDEQ-AQD to compile and assess annual emissions from the permitted compressor stations from the Jonah Infill Project area."*[18]

My response was immediate. I sat down and studied the reply line-by-line and crafted my analysis. This was more an exercise in immediate rebuttal to be used at some future opportunity in a public forum than intent to shoot back at Cheyenne. In truth, I knew I had been successfully marginalized yet again.

In that frame of mind, I put my conclusions on paper. The reply from BLM revealed the equivalent of a bait-and-switch action. When the Anticline EIS was open to public review, the agency presented it as a declaration of not-to-be-exceeded environmental impacts. I confirmed this understanding with Terry Svalberg. The purpose of the EIS from the BLM perspective was to advocate approval of the preferred option presented therein and, at least in part, that approval required public acceptance of the EIS declarations and assertions.

However, I and many other commenters warned that the EIS contained language that was far too non-specific and thereby left room for inventive interpretation in the event that project expectations failed to track as planned. In a nutshell, BLM created a hidden but implied escape exit.

The BLM reply to my letter invoked that escape exit by suddenly and conveniently redefining "emissions estimates" from EIS specific assurances into now reclassified non-specific "expectations." In the EIS, projected emissions were specifically identified as those that take place due to actual project development operations *including* unspecified but assured mitigation efforts. Now they were redefined as being none of this.

By the Rasping in My Lungs

Thus, BLM appeared to have engaged in development of the EIS and ROD documents strictly as a "pencil whipping" exercise calculated to buy off public acceptance while never intending to actually live up to the assurances contained in them. The agency could cynically but safely expect no further public scrutiny holding it to account. But just in case, the documents were crafted to provide enough vagueness to give cover in the event that reinterpretations such as those expressed in their response letter to me were to become necessary.

The reference to "actual emissions inventory submitted by operators ..." in the opening paragraph of the reply invoked what had become a much used phrase intended to imply empirical foundation. In truth, here has been no such thing from the very beginning of gas development in Sublette County. The phrase implied that some process was ongoing to actually measure the emissions. This was false. All emissions inventory compilation was and continued to be performed through the process of computer modeling. From the beginning, I had argued that none of the modeling had ever been validated by actual at-the-source instrumented measurements. Absent that, modeling was nothing more than wishful thinking displayed on a computer screen.

The last sentence in that lead paragraph of the BLM response was the sole reference to the fundamental question I originally asked which was for BLM to report on the accuracy of compressor station emissions predicted in the Jonah *and* Anticline EIS documents up through year 2011. That sentence only referenced the Jonah field and evaded the overall question with assurances that cooperative collaboration was ongoing with the operators and the state environmental regulators. In short, the entire letter was a non-responsive dismissal.

My concluding act on this matter was to send DeForest an email declaring, "... you have my condolences for having to be the one to sign the letter. You should have refused."

This entire sorry episode served well to illustrate that the BLM, created as a steward of public lands, had evolved into an agency of bureaucratic procedure. It had positioned itself behind a wall of impenetrable, labyrinthine administrative processes, rigged for exploitation by agency bureaucrats looking for an exit from regulatory responsibility. When it came to enforcement, try pointing to specific procedural violations and self-admitted decisions to disregard a seemingly legal obligation and you would be told you misunderstood or were misinformed.

Intrigues of the Inner Sanctum

Over the decade that I engaged in efforts to make the BLM and the operators clean up their act, there were two directors of the state BLM office. I never had direct interactions with either of them but I did have a useful contact in the form of a high school classmate Hank Castillon, who joined the Wyoming National Guard and over the course of his career, rose to the rank of brigadier general and state adjutant general. Following his retirement from active duty, he hired on with BLM and had his office in the Rock Springs field office down the road from Pinedale.

Whenever I was in Rock Springs, I always tried to stop by and visit with him. He was very forthright with me about the influence the state office was having on the management of environmental issues in both Sublette and Sweetwater Counties. On one of my visits, he informed me about the general attitude of state director Bob Bennett which helped clear up some of the recent statements to come out of that office through its public information office.

When I asked about Bennett's influence on his regional directors, Hank replied his influence was considerable. Bennett was a former lieutenant colonel who was still conditioned to take orders from superiors and execute them to the max. Furthermore, he imposed that requirement on his regional subordinates. Inside the BLM Bennett was referred to as "Drill it Bennett" and he had a close government relationship with Kathleen Clarke, the national director of BLM from 2001 through 2006. That must have worked well for him because Clarke had her own reputation for working to increase energy development on public lands in ways that favored the operators.[19]

This meshed well with the statements that had been coming out of Bennett's office in reply to growing public participation in the environmental impact comment process. At one point after a round of public comment his office was quoted in state news media as declaring that although he would continue to consider public comments, they had to be constructive and contain points of substance.

In the same breath, however, he dismissed voluminous negative public comment as having not been useful because it lacked specificity. Furthermore, he even dismissed written concerns from the State Governor by declaring merely that those concerns "… would be taken into consideration."[20]

My reaction was immediately one of suspicion. How did the BLM define "specificity?" I would learn to my great dismay some eight years later as a result of my compressor station challenge that it didn't matter. BLM would make even their written statements provisional and subject to convenient re-definition as required.

Hank also told me that his office was beginning to experience pressure to open public lands under its jurisdiction to drilling. In fact, the operators were becoming brazen about it. He told of one company that wanted leasing issues resolved immediately and threatened to go to the state's Washington D.C. delegation.

My ongoing conversations with Hank during this period were revealing. He informed me of a variety of events in the state BLM office of which only those inside the sanctum were aware. For instance, there was a regular telephone conference referred to as a "heads-up meeting" in which the latest rumors and information about environmentalist actions were shared for purposes of developing counter measures. This story agrees with what Susan Caplan told me and differs only about the detail of telephone versus personal consultations.

Also, despite strenuous public denials to the contrary, there were, in fact, quotas levied on each region regarding processing of drilling permit applications or "APDs." The regional offices were expected to process and approve minimum mandated quotas. This was proving problematic to Hank's regional office because its quota was higher than the actual number of applications being submitted. To fix this shortfall, the drilling companies were being encouraged to submit more APDs so the quota could be met.

This is backed up by a July 2005 press story quoting the Government Accountability Office as having charged that the Department of the Interior was too focused upon approving oil and gas drilling permits on public lands while failing to adequately oversee protection of the environment.[21]

In mid-2004, the national press reported that the BLM was on its way to issuing a record number of drilling permits on public land across the west. Specifically, 3,500 permits had been issued by June 25 of that year and the total number was expected to almost double by the end of the federal fiscal year.[22] Five western states accounted for 97 percent of the total reported. Not surprising to Hank and myself, Wyoming was at the top with 2,151 permits.[23]

Another July 2005 story revealed that across the west, approved APDs by BLM increased by more than a factor of three between 1999 and 2004. Those numbers grew from 1,803 to 6,399 and were craftily morphed by BLM into an excuse that they were overwhelming its staff to the extent that field inspections had become a lower priority. As a result, BLM was failing to meet its environmental protection responsibilities.[24]

In the face of all of this pressure, the Wilderness Society revealed additional notable facts in 2004. Although permitting to drill had gone up by 62 percent in 2004, the number of new wells actually drilled declined by 10 percent. Stated another way, throughout five Rocky Mountain States, 5,824 permits had been approved by BLM but only 2,489 had been drilled in 2004.[25]

This seemed to directly challenge the sense of urgency for approval of permits implied by the operators and the BLM. Nevertheless, the sense of urgency was underscored by a full court press to clear away the backlog of drilling permit applications. The obvious first conclusion was that the operators were acting within a highly favorable business environment afforded them by the Bush/Cheney Administration. They seemed determined to stockpile permits before any unfavorable changes in the political arena could intervene.

Interestingly, defense of the western states came from a senator from Connecticut. Senator Joe Lieberman who requested a GAO study, observed that the Bush Administration appeared to be stream-lining the drilling permit approval process for operators while showing little regard for environmental protection.[26]

This activity supported the concerns of many of us who wanted to see gas and oil development slowed. But our view ran contrary to that of Vice President Cheney's Energy Task Force. It was urging the BLM to speed the opening of federal lands to oil and gas leasing.[27]

The following month saw the Interior Department announce its intent to delay some new oil and gas activity on public lands until the effects on wildlife could be better determined. At first blush this looked like good news. But, there was a subtle catch. BLM state and regional managers could defer leasing if they were in the process of drafting a new resource management plan (RMP). These are plans that establish long-term land use policy.[28]

Pinedale BLM had indeed set about updating its RMP and did indeed consider placing a particular area off limits to development but the majority of relevant lands were well on their way to development already. Linda Baker of the Upper Green River Coalition highlighted this fact by pointing out that 75 percent of available public lands in our area were in fact already leased.[29]

Hank also warned that the state office had initiated a pilot project in three hot spots where special processing teams were in place for the purpose of expediting APDs and Pinedale was one. This involved the use of personnel paid by industry to process APDs. This practice was officially regarded to be unacceptable but there was no specific statutory prohibition. Up until recently, the practice had been successfully resisted at the regional level but pressure from above had swept that away.

Backing this up were press stories about the use of "volunteers" from the oil and gas industry working at BLM field offices. Their task was to help clear away environmental studies that were slowing the approval process for drilling applications. A press report in the Salt Lake Tribune revealed this was going on in the Vernal, Utah, BLM area management office but failed to include Pinedale, Wyoming.

The Vernal office was receiving its "help" through an oil and gas trade group called the Independent Petroleum Association of Mountain States which in turn funneled funds to a firm called SWCA Environmental Consulting. That funding was provided by a coalition of gas and oil companies that pooled their resources. This firm then hired consultants to "volunteer" for duty in the BLM field offices.[30]

Supposedly, these consultants were assigned to review projects under a series of safeguards which BLM maintained prevented them from learning if a project was associated with any company paying their salary. Their work was reviewed by BLM staff and they were not allowed to participate in final decisions.[31] This begs the question, however, if the same staff personnel were so overtaxed that they were unable to perform reviews of the project proposals coming across their desks, how was it less work-intensive to review the consultants' reviews of those projects? The likely implication is that such reviews were perfunctory.

I personally learned this was happening in the Pinedale office when I received a call from an individual assigned there who wanted to discuss my spectroscopic field research. He also wanted to explore a recent set of technical objections I had forwarded to the office as part of my public comments regarding local air quality degradation from gas development in the Jonah and Anticline.

I commented that I didn't recognize his name so he explained that he was an employee of the gas industry on loan to the BLM Pinedale office for the same purpose stated above about the Vernal office. He worked for one of the gas companies involved in drilling here and was on detached service to assist with permit analysis.

When I pushed back and argued that his industry was definitely having negative impacts, he replied that I should see the air in Houston. Annoyed with that, I shot back that Houston was not a relevant standard of reference against which to evaluate Sublette County. I argued that the more relevant metric was our pristine air quality before his industry had come into the neighborhood. That seemed to surprise him because he paused and conceded that I might be correct.

During this time frame, pressure from Washington D.C. to expedite energy development was becoming so extreme that anyone wanting to push drilling related agendas was welcomed to directly call Washington BLM department directors. Hank cited lobbyists, and individuals who were eager to do this and even told the story of a lobbyist in Casper who had called his regional office in Rock Springs. She demanded action on her issue but when advised that some time might be required, she proceeded to engage in threatening rants that her concern had to be addressed *immediately*.

At one point I asked if morale in the Agency was suffering from all of these issues. He confirmed that morale was indeed very low and there was a high fear factor over threats to jobs. He told of the deputy to Bennett being removed and assigned to Denver to oversee a "special" project that was in fact exile into obscurity. This was done because of his failure to execute Administration policy appropriately. Additionally, Wyoming had become such a politically contentious environmental battleground over energy development that the state BLM office was under orders to issue absolutely no public press releases until they had been vetted in Washington D.C.

Press reporting continued to corroborate much of what I had been told at the personal level. These reports illustrated a management attitude within BLM from the state level all the way down to field personnel that was dismissive of all input. Additionally, these reports illustrated a cynical determination by BLM regional directors to retreat behind denial and superficial assurances as a defense against any and all criticism.

In 2005 an uproar developed over BLM characterization of public comments directed to it in reply to calls for public comment about oil and gas drilling projects. The environmental group Wyoming Outdoor Council noted that the Rock Springs Rocket-Miner, had quoted an official in the Rock Springs BLM field office claiming that 68,500 comments received about a project were "unsubstantial, negative statements … with no suggestions of alternatives."[32]

Not unreasonably, this official noted that opinion comments like "I don't like this …" ranked as non-substantive. However, he went on to assert that public comments needed to speak to specific area issues by page number, line number, and statement content. Also, passages containing possible errors or insufficient explanation or analysis had to be cited.[33]

My own experience demonstrated that these arguments were in fact disingenuous. The environmental impact statements I responded to with pages of detailed comments and criticisms elicited one-line replies or less. In the Pinedale Anticline and Jonah final EIS documents, there was a large section acknowledging commenters by name and affiliation. Our comments were reference and distilled into one-sentence statements and fitted into tabular form. Next to these comments were BLM replies also in one to three standard sentence bullets.

In my particular case, I was credited with offering 43 substantive comments and treated to one reply in three variations: (1) *Thank you for your comment.* (2) *Thank you for your comment. The BLM believes the air quality modeling efforts, performed in cooperation with the WDEQ, EPA, NPS, and USFS, have been appropriate and comply with NEPA. The modeling used the best available information available at time of the analysis.* (3) *Thank you for your comment. The BLM believes that the data and analyses provided in the SDEIS and AQTSD are adequate for this impact assessment.*

Linda Baker reminded me about time constraints imposed upon us. BLM released its Finding of No Significant Impact Decision Record and Environmental Assessment for a Questar year-round drilling proposal modification in June 2008. A month later the public was notified that comments were being accepted until July 7th. That made available ten business days for interested public to read, digest, research, ask questions, and provide comments to BLM. This seemed at odds with Interior Secretary Gale Norton's 4 C's initiative of conservation through communication, consultation, and cooperation.

The Pinedale BLM Regional Office

From the beginning of the gas boom in Sublette County, the managers of the Pinedale field office demonstrated ongoing determination to advance the cause of the gas industry. This seemed to include equal determination to impede citizen intervention seeking even minimal control over what we believed to be an unwarranted rush to drill and produce.

This did not vary an inch during my decade of conflict with BLM state and regional factions despite the regular turnover of two directors at the state level and four directors in the Pinedale office. Indeed, in the middle years of this period, managers cycled through the office on their way to imminent retirement as if Washington D.C. saw the place as a dumping ground for end-of-career bureaucrats.

I experienced this behavior on a personal level in the course of my fieldwork and challenges to the way business was being done. On numerous occasions, low-level employees in the Pinedale office sidled up to me and begged me to keep doing what I was doing because they did not dare speak up.

Added to all of this, a wildlife biologist had recently quit the Pinedale office because of his dismay over the threat to sage grouse habitat. On another occasion, a recently retired wildlife biologist from the Pinedale office told me his contacts inside the office were informing him of weekly telephone calls from the office of Vice President Dick Cheney (as I related earlier), pressing for resolution of the then stalled environmental impact assessment approval for year round drilling on the Anticline.

Kabuki Theater

These events took place as a result of my activities as a citizen scientist in an informal capacity. My parallel formal involvement with the agency began in 2003 when the Pinedale field office began to reconstitute a collection of stakeholders from an earlier gas field endeavor into a new advisory group. This group was the Pinedale Anticline Working Group (PAWG) made up of local citizens, representatives from the operators, and representatives from the government land management agencies.

All were appointed for specific terms of service. Their charter was to review operators' plans for developing the Anticline and make non-binding recommendations to the Pinedale BLM field office administrator. These recommendations were to address best practices for protecting water, air, landscape, and wildlife in the region to be developed.

Then-Pinedale director Prill Mecham wrote in a public comment editorial that the Bureau of Land Management and the Department of the Interior chartered the Pinedale Anticline Working Group for the specific task of providing advice to the BLM on oil and gas development mitigation and monitoring in the Pinedale Anticline area south of Pinedale. The PAWG was to provide advice and recommendations to the BLM regarding (1) Setting goals and objectives for monitoring field development, (2) Drafting monitoring plans, (3) Evaluating mitigation measures contained in the Pinedale Anticline Record of Decision, and (4) Providing advice and recommendations to the BLM on monitoring and mitigation to better inform the BLM decision making process.[34]

The actions of the PAWG and its task groups, however, were constrained by what she and her Agency deemed Pinedale Anticline ROD and PAWG "legal constraints and necessary sideboards for its activities." The Federal Advisory Committee Act (FACA) provided rules on advisory committees. To the BLM this meant that the PAWG had to focus exclusively on mitigation and monitoring in the Pinedale Anticline.

This view proved vexing and impractical. The adjacent Jonah field was generating all the same problems already that we were to consider for the Anticline and yet we were routinely warned away from addressing the Jonah. Some of us insisted that there was no wall in the atmosphere separating the two; any issues and corrective actions we might put forward would be negated if the Jonah was ignored.

It should not be inferred, she stated, that BLM had was uninterested in the PAWG members' individual opinions with regard to other projects. Rather, she cited the collective expertise and insight offered by members regarding the Upper Green River Basin as a reason for the group's continued participation through the BLM's public involvement process. This input, she added, was "vitally important."

Mecham even invoked the process of "Adaptive Management," calling it an active part of the Pinedale Anticline Development. It was touted as a means to validate and/or adapt mitigation measures to ensure impacts were reduced to the greatest extent possible. It also was claimed by BLM to have been a key component in the Jonah Infill Drilling Project (JIDP).

Adaptive Management in its theoretical form is described in Appendix D of the JIDP proposal document but was applicable to all the gas development projects in the county. The goals and objectives were quite lofty:[35]

- Determine the effects of the [project] development on area resources
- Determine the effectiveness of mitigation measures contained in the project ROD
- Modify mitigation measures as appropriate to achieve the stated goals and objectives
- Assure that oil and gas-related BLM decisions regarding the

[project] are coordinated with non-oil-and-gas-related decisions such as grazing, and recreation as examples

- Provide rapid response to unnecessary and undue environmental degradation
- Validate predictive models used in the project environmental impact statement and revise those models/projections as needed based upon field observations and monitoring
- Accurately monitor and predict cumulative impacts through BLM maintenance of a Geographic Information System (GIS) for the [project] including all activities (natural gas, agricultural, recreational, etc.) on federal and non-federal lands and how they are affecting area resources
- Provide guidance for monitoring upon which the need to initiate consultation with the U.S. Fish and Wildlife Service will be determined

The reader should notice the use of vagaries in these criteria. "Effectiveness," "unnecessary," and "undue," are all lacking in empirical qualities and, open to interpretation by BLM managers. Furthermore, modeling and validation of the modeling have a nice empirical ring but in reality, when I pressed Pinedale managers on this, they retreated behind excuses that modeling was still an imprecise art.

Nevertheless, Ms. Mecham assured the public and the PAWG that it was a critical component of Adaptive Management. She insisted that the PAWG was an integral part of the BLM's management initiative to provide monitoring information and identify factors that would trigger changes in mitigation on the Pinedale Anticline.

With these statements, she invested a huge amount of personal capital in this concept because it became pivotal to winning over the local citizenry who were genuinely worried about what gas development would do to the land and wildlife. She personally asserted in run-up public meetings prior to development go-ahead that she had the power to judge when unacceptable impacts to the environment were happening and would shut down those activities generating the impacts until a solution could be developed. Several months later it became clear that this would never happen.

She summed up her public assurance campaign by noting that the BLM's goal was to "… create a cooperative environment based on mutual trust and understanding between everyone involved to enhance project mitigation and balance conflicting resource goals within the Pinedale Anticline Development Area." Yet again, she maintained that the BLM firmly supported the PAWG and its work. It didn't take long for those of us involved with the PAWG to conclude this was not true.

A major fight that developed between concerned citizens and the BLM had to do with year-round drilling on the Pinedale Anticline. For the first few years of drilling, that activity had been limited to summer and early fall months and then suspended through the winter due to the fact that the Anticline field was right in the middle of the wintering range for a large herd of mule deer.

Another reason the issue was so contentious was that the 2000 BLM Record of Decision (ROD) did not permit winter drilling *because* of the disruption to the wintering range of the mule deer. However, the operators subsequently petitioned BLM to allow winter operations citing various economic factors. Their successful lobbying resulted in re-issuing of the Record of Decision and it was signed in 2008. That new ROD allowed winter drilling but also included provisions the operators would have to meet. These included a five-year suspension of drilling on the flanks of the Anticline and the establishment of concentrated drilling areas within a six-mile circle to protect wintering herds.

The Questar Corporation which was one of the most active drillers on the Anticline, was the first to petition the BLM to allow it to perform year round drilling. It argued that impacts to wildlife and air quality would be shortened because drilling the field to completion could be accomplished in half the time that was otherwise possible.

At the Questar winter drilling proposal public open house, someone questioned Ms. Mecham as to what would prevent the other operators from rushing to the door with similar requests and receiving automatic approval. Her answer was vague. We suspected there would be no legal basis to deny one operator what had been given another. Sure enough, in the following weeks the other operators on the Anticline, Shell, Ultra, and Anschutze, followed suit and submitted their applications.

The aggregate potential from all operators' drilling would be 10,000 to 12,000 wells. Mecham was questioned sharply by some citizens as to the likely impact on the deer and sage grouse on the Anticline by such a rush to drill. She waxed eloquent about the adaptive management process and assured everyone in the room that when BLM detected negative impacts, it had the authority to suspend operator activities pending a solution.

A little over a year later at a PAWG meeting, specific and serious impacts were being revealed that had been on-going for some time. I challenged her directly, insisting she invoke her powers to suspend operations in keeping with assurances she had stated in the public open house prior to project go-ahead. I was unsurprised but, nonetheless, dismayed when she declared that "the adaptive management process was an uncertain mechanism and the criteria for invoking it were not well defined." So yet again, I received affirmation that the BLM was not to be trusted.

A quick aside is appropriate here regarding the BLM use of the "open house" concept. The Pinedale office began invoking this during the development of the Anticline. Prior to this, its public meetings had been conducted in a forum style in which the citizenry was invited to ask questions and express concerns. Apparently that proved too awkward because the Bureau transitioned to the open house concept.

It's utility to the Bureau was immediately revealed when some in attendance who started to ask questions were informed that such direct questioning of Bureau officials would not be allowed. Instead, they were directed to voice those questions at appropriate booths that had been set up around the room by the operators.

This was in fact a two-fold strategy to firewall BLM personnel from growing public resistance. First, it succeeded in "dividing and conquering" any reinforcement of public dissatisfaction that always developed as questions were asked and badly answered. Second, it forced the questioners who now had no direct means of challenge at their disposal to endure the operators' propaganda machines. This is not unique to BLM; in fact, I have seen other federal agencies such as FEMA adopt this strategy over the years.

An exception occurred in January 2008, however, when the Anticline project was in the final phases of EIS approval. The forum format was utilized by BLM ostensibly to take public comments but that turned out to be more of an industry testimonial fair. The meeting had been scheduled for a 6:00 PM start time but began an hour earlier, when the room started to fill almost exclusively with drilling industry representatives and workers.

It truly looked like drilling industry managers had put out the word that all employees were to pack the meeting. Whether this was fact or not, the effect was a shutout of local residents. By the start of meeting time the room was so packed with industry people that local residents being unable to even get through the door went home in disgust.

I was also annoyed by these events and a few days later obtained a copy of the attendance list from BLM. What I learned by scrutinizing this list was revealing. The list contained 230 names but many more were not recorded. Of that number, 78% were industry proponents from outside the county and even outside the state and 20% were local industry workers.

These cheerleaders for approval of the Anticline project were a mix of politicians and corporate captains and labor. Four people from the state Washington delegation were present as were the mayor and a member of the Chamber of Commerce from Rock Springs, a town 100 miles away. Additionally, the majority floor leader of the Wyoming state legislature was there as was a Wyoming Game and Fish Department official from the state capitol.

The record showed attendance and advocacy by persons from towns and cities hundreds of miles away. Drilling advocates came in from all over Wyoming: Cody, Afton, Thermopolis, Douglas, Lyman, Reliance, Casper, Green River, Riverton, Evanston, Glenrock, Cheyenne, and Greybull, and Rock Springs. Most represented was Rock Springs, a drilling support town, with a whopping 30 people filling the gallery. Finally, Colorado accounted for 18 attendees, and three came from Utah.

The operators too were well represented by 13 employees of the Questar Corp some of whom came all the way from Denver, CO.; 13 employees from Ultra Corp. some of whom came in from Denver and Englewood, CO., as well as Casper and Douglas, WY, a few hundred miles away; five employees of Rocky Mountain Shell Corp one of whom came in from Denver; 11 employees of Halliburton Corp some of whom came in from Denver and Broomfield, CO, and Afton, WY; and five employees of Schlumberger Corp who came in from Rock Springs, and Evanston, WY, as well as Denver and Greenwood, CO. Additionally, there were 110 people present from all the field service providers such as drilling support, trucking, etc.

A particularly vexing aspect of this bit of BLM theatrics was the fact that so many outsiders with no personal interest in local environmental issues were allowed to tilt the playing field. BLM could argue that its governing statutes required it to accept all commenters. In truth, the presence of so many persons representing industry from outside the county constituted interference by people who were transients and as such had no loyalty or concerns for the spectrum of issues that locals wanted to address.

The Pinedale Anticline Working Group (PAWG)

A depressing manifestation of BLM failures experienced in Sublette County took the form of the Pinedale Anticline Working Group (PAWG) which I introduced in the previous segment on the Pinedale BLM field office. The official description of the PAWG published by the BLM was as lofty as earlier-cited adaptive management goals.

Chartered under the Federal Advisory Committee Act (FACA), the PAWG was tasked with providing "... balanced recommendations to the BLM on the development and implementation of monitoring plans, mitigation strategies, and adaptive management pertinent to oil and gas activities in the Pinedale Anticline Project Area (PAPA)."

PAWG members were appointed by the Secretary of the Interior to serve two-year terms.[36] We were told that approval of membership actually did go all the way to that level. This initially proved a roadblock under Secretary of Interior Gale Norton, who was serving in the Bush Administration when the PAWG was first being constituted. However, her abrupt departure removed that impediment.

Interestingly, her exit carried echoes of the Harding-Fall scandal. Her critics described her as an administrator who engaged in regulatory rollbacks suppressing science, and colluding with industry. It was claimed that in 2005, half the Federal Fish and Wildlife Service biologist respondents to a survey admitted they had reversed or withdrawn scientific conclusions due to pressure through her office from industry lobbyists. Conversely, her industry support groups praised her for opening the West to energy development.[37]

Her close relationship with the energy industry became more open when the Federal Justice Department considered but ultimately declined to file charges against her in connection with questionable bidding issues by Royal Dutch Shell for oil shale tracts. The issue involved the fact that Shell won the leases from Interior after she had stepped down as the Secretary in March 2006 but before she accepted a position as a lawyer with Shell in December 2006. In a report by then acting inspector general of Interior, Mary Kendall, it was stated that Norton played a significant role in BLM's oil shale program while Secretary and noted that her participation should earn her a lifetime ban on communicating with the federal government regarding the program.

It was further alleged that BLM under her time of service appeared to give preferential treatment to Shell regarding two bids for larger than allowed parcels. This was then cleaned up by "someone" within the agency by altering Shells' acreage estimates to comply with the rule. Finally, it was noted that Shell submitted a complete bid document on the very same day that the Federal Register published the initial notice soliciting applications for leases.[38]

Additionally, Norton was potentially facing investigation for possible connections with Jack Abramoff, the lobbyist who had become infamous for involvement in an Interior scandal relating to Indian reservation tribal gaming. As heat continued to build on her, she resigned as Secretary.[39]

All of this discussion about Gale Norton is offered to establish the way the playing field was tilted against the citizens of Sublette County. Getting back to the PAWG, it had associated with it working groups tasked to address specific issues of concern. These included air quality, water quality, and cultural site protection (meaning archeological and historical sites).

It was becoming clear to me that BLM at all levels and epitomized by the Pinedale regional office, was the single most fundamental obstacle to combining practical natural gas recovery with achievable and affordable environmental protections. I observed the Pinedale and Cheyenne offices engage in what I came to characterize as "bureaucratic liturgical recitations of statutory scripture" as a defense against having to make decisions or suffer any interference with their determination to do things their own way. Indeed, Prill and her staff elevated doublespeak to an art form, consistently obstructing the PAWG with their self-protecting interpretations of such statutes, impeding citizens from achieving meaningful results and discouraging operators willing to try innovative ideas to reduce their environmental impacts.

One example had the impact of removing from PAWG and task group participation anyone who lived far from our meeting location. In somewhat perfunctory language, we were informed that monetary compensation for travel to PAWG meetings and task group meetings would be immediately terminated because it had been concluded that guidance had been "misunderstood" and in fact did not provide for such compensation.

In the beginning, Ms. Mecham sang the praises of the PAWG concept and assured us that BLM would not interfere nor could it interfere in the workings of the PAWG. However, within eight months' time, she was censoring PAWG agenda items if they addressed pending drilling project details over which BLM had yet to rule. Her assertion was that we had no jurisdiction over "pre-decisional" issues.

I found this attitude stupefying. Was it not appropriate to discuss concerns before they become decided? The fact is (and certainly with government bureaucrats) discussion after the decision is made becomes an exercise in moving the unmovable. Indeed, BLM had a strong history of refusing to alter a decision once it had been made and so, any discussion with BLM *after* a decision had been made was destined to become nothing more than talk radio without the radio. Discussion prior to a decision was the necessary thing to do.

Also early in the game, Ms. Mecham pointedly invoked the need for "effective recommendations" from the PAWG. This statement was almost insulting because in the first eight months of meetings I participated in, our task group on air quality issues submitted numerous recommendations but not one resulted in a reply from her offices. As this silence persisted we began to realize that BLM was lacking any mechanism to facilitate feedback or even acknowledge our input. It turned out there was not even a formal requirement that they do so. A red flag was up. The PAWG and its task groups were intended to be little more than public relations theater.

As my frustration grew, I increasingly took to writing public opinion pieces to the local newspapers and the Casper-Star Tribune which was the major state newspaper. As I explained earlier in the segment about the Pinedale BLM office, in 2005, Ms. Mecham wrote a guest editorial for the Star Tribune declaring a lofty-sounding view of the purpose of the PAWG. However, my experience with what was actually happening in the PAWG so contradicted her description, I simply had to challenge her head on.

I submitted my own opinion editorial to the Casper-Star Tribune. I explained that I was participating in the Air Quality Task Group as a scientist and engineer by training. I further explained I had been conducting independent air quality research in the Upper Green River Valley since 2003 when I first noticed the skies over my home in Daniel were becoming obscured by haze. I had, therefore, become qualified to offer several observations.

I recited numerous incidents that directly contradicted her assertions. I started by stating my fundamental conviction that the Cheyenne and Pinedale offices epitomized BLM in its state and regional incarnations. Together, both were a serious obstacle to environment protection and the quest to profitably recover gas reserves in Wyoming. It was possible to achieve both but flexible thinking was necessary by all parties. For example, BLM-Pinedale hindered EnCana's petition to attempt a drilling method used in the arctic to protect the ground surface under drilling pads. This will be covered later in the chapter on the operators.

I went on to assert that the agency avoided citizen pressure to change direction by reciting inventive interpretation of statutory and regulatory scripture. I pointed out that adaptive management was an empty promise and that recommendations by the task groups supporting the PAWG were being ignored. Finally, I cited her new tactic of blocking task group recommendations by invoking prohibitions on discussing pre-decisional issues.

I had done what I could with this editorial warning to readers. I tried to sound the alarm that the PAWG concept was shaping up to be a tactic for deceiving our community. I wanted readers to understand that the Agency was maneuvering to have them believe it was inviting public involvement even as it maneuvered behind the scenes with its bureaucratic skullduggery to negate that involvement.

Overall, this type of tactic seemed calculated to hold at bay any annoyances that might come from the PAWG and its task groups. A cynic could further draw the conclusion that from the outset, BLM launched the PAWG process with the ulterior motive of creating theater that gave the appearance of embracing stakeholder input but never accepting it or acting upon it. This was affirmed by an off the record warning expressed confidentially to me by a local BLM employee. That person advised that BLM never had any intention of altering its practices until *forced* to do so.

In the end, guidelines under which the task groups operated included a stipulation that BLM was under no obligation to accept any recommendations from the PAWG. That provision frequently appeared to be a BLM escape clause of choice.

Demise of the PAWG

In late 2012, after a long and dubious struggle, BLM exorcized the PAWG as a source of misery. The formal notice given by BLM-Pinedale was terse:

> *At the regular meeting of the Pinedale Anticline Working Group (PAWG) on Thursday, October 25th,* members voted to disband the PAWG, discontinue the meetings, and allow the charter to expire on August 3, 2014. They also included a formal request that a subcommittee be formed by the Wyoming Resource Advisory Council (RAC) to be available for local consultation. *The PAWG was created by the Bureau of Land Management (BLM) in 2004 to provide public input and recommendations on oil and natural gas field development activities in the Pinedale Anticline.*[40]

The editor of *Pinedale Online!* wrote a scathing commentary about this event that was right on target. "Any recent observer could have concluded that the

PAWG had died as a result of public complacency and so the public was to blame. The truth was far different."[41]

The commentary went on to correctly state that BLM-Pinedale had squeezed the life and energy out of a panel made up of mostly well-intentioned persons who tried to do the right thing in return for no tangible compensation whatsoever. Members had been appointed from the oil and gas industry, grazing, adjacent landowners, recreation, cultural resources, the environmental community, local governments, academia, and the public-at-large. These members donated hundreds of man-hours of substantial expertise towards the effort and it had all been wasted.

The editorial also correctly noted that even the operators had willingly agreed to many modifications to their activities. They invested in research that resulted in doing some things better and reducing negative impacts on various natural resources.[42] But then almost irrationally, BLM turned its attention to finding one reason after another to discourage and terminate this conduit of public input.

Initially, BLM insisted that because the PAWG members had a direct line of communication into "the ear of BLM" they were required to undergo a long vetting process. Recall my earlier reference to the review process at the Secretary of the Interior level. However, in 2010 without any such vetting, BLM-Pinedale elected to employ a costly private consultant from California to instruct it on how to recast the PAWG. This consultant then proceeded to declare that the PAWG and its many task groups had to change because they were not focusing on topics the BLM field manager wished to discuss.[43]

Among the growing list of topics not qualifying for discussion was air quality because the BLM again declared this not to be an area of its responsibility. This, along with socio-economic impacts, economic impacts, transportation, and water quality were one-by-one eliminated from eligibility on the grounds that other federal and state agencies had jurisdiction. Thus, BLM initiated a new plan for PAWG operations that eliminated all the task groups.

Eight task groups were informed they would be eliminated and with them would go hundreds of man-hours of detailed and hard won expertise in a myriad of environmental specializations. The air quality task group of which I had been a member was eliminated. Only one task group would remain but reformed as a responsibility of the PAWG membership itself.[44]

Adding to this purge, PAWG meetings were truncated to quarterly two-day meetings from the original monthly meetings. One result of this was a drying up of the volunteer pool because none could afford to skip their day jobs to participate. Somewhat disingenuously, the BLM managers then professed consternation over the dwindling willingness of the public to participate in either the meetings themselves or the PAWG process.

Thus, over the sunset years of the PAWG, willing and motivated volunteer citizens were shut out and qualifying topics for discussion increasingly narrowed to the very few deemed appropriate by the Pinedale BLM Field Manager. Meeting agendas were crafted to admit only superficial subject matter and field trips were used as a means to burn up members' time while simultaneously eliminating opportunity for discussion and evaluation. Eventually, useful exchange had become so diluted that on October 25, 2012, the Field Manager, Shane Deforest, announced no further need for the PAWG due to the absence of meaningful input.[45]

Adding insult to this injury, he went even further by alleging that the task groups had been in violation of law.[46] His stance on this was due to his personal interpretation that attempts to expand their scope of effort under their charters (at least as he saw it) somehow constituted illegal activity. This rang familiar to me as a long-time member of the Air Quality Task Group hearing constant warnings from BLM representative Caleb Hiner that we could not consider air quality impacts from the adjacent Jonah Field because we were charted only to address impacts from the Anticline Field.

Following a last exchange between Deforest and remaining PAWG members who challenged his sincerity in managing the PAWG process, the vote to disband was cast. Parting observations attributed to the closing minutes of life of the Group were acidic:

> "Well, that's one less meeting to attend. Mission accomplished BLM."
>
> "It took years, but you finally killed off the pesky PAWG as a meaningful public input process, and did it with the amazing skill to absolve yourself of any of the apparent responsibility for [its] demise."

> "The spin will no doubt be that the death of the PAWG is all the public's fault because it wasn't interested in participating anymore."

The editorial ended by observing: "As things move forward now with the many supposed ways the BLM says the public can still provide meaningful input into the process, it should be recognized that providing opportunities to speak is not the same as truly listening to what the public has to say."[47]

Now years later there are certain ironies to be recognized. Prill Mecham retired under a pall unrelated to gas and oil development and within a month was rumored to have gone to work for Shell Oil as a consultant. The next two managers were short lived and cycled through as placeholders pending their own retirements.

Deforest, the last Pinedale manager I dealt with, and who in his first PAWG meeting assured me that he had heard my warnings about how management would be crucial to PAWG success, is gone. Not only is he gone but he was removed for initiating a personnel dismissal action that was in violation of BLM due process. However, BLM then compounded its error in judgment by appointing the air quality task group's old nemesis Caleb Hiner as the newest Pinedale BLM Field Manager.

Looking Back

The overarching picture of gas development in Sublette County was that of BLM-Pinedale, backed up by its hierarchy all the way to the White House, pursuing a strategy of self-serving, narrow interpretation of guiding statutes. The objective seemed to be avoidance of the need to make decisions and avoidance of any management actions other than the movement of paper, i.e., approved drilling permits.

Furthermore, when challenged that BLM was not executing proper environmental enforcement oversight of public lands, its directors circled their wagons. They defensively quoted statutes containing enforcement powers but rarely, if ever, invoked them in a manner that resulted in actual, meaningful changes to the way business was being conducted in Sublette County.

Chapter 4

The U.S. Forest Service (USFS)

Beginnings

The U.S. Forest Service exhibited concern over the potential air quality damage to its forests in the Sublette County region of Wyoming well before the issue had appeared on the radar of any other agency or even the local citizenry.

In 1984 USFS scientists issued a cautionary report warning of events coming that would require national forest management attention. It noted that at the time of the report, oil and gas development was increasing rapidly due to discoveries of deposits in the Overthrust Belt, accompanied by increasing emphasis upon development of domestic reserves within the United States. The Department of Interior had estimated the Overthrust region harbored 65 trillion cubic feet of gas. As a result of validating discoveries in various locations within the Overthrust region, operators were in agreement that this was the hottest new area for drilling in the United States.[1]

Notably, the report worried that 80 miles of pristine Continental Divide mountainous territory was at stake. It was further noted that the Wilderness Act of 1964 gave the USFS responsibility to protect the national forest system from man-caused degradation but if faced with air pollution issues, action under the Act could probably only be taken *after* impacts had occurred.[2] This caveat would be well-demonstrated 20 years later.

The report went on to explain that an amendment to the Clean Air Act in 1977 established an air quality program called Prevention of Significant Deterioration (PSD). The intent was to maintain air quality in places where air was already cleaner than necessary to protect public health.

The elements that make up this program are: (1) designation of certain national parks and wilderness areas as Class I areas, (2) establishment of a permitting process aimed at certain new sources of air pollution, (3) establishment of incrementally small limits on the amount of sulfur dioxide (SO_2) and total suspended particulates (TSP) that can be added to Class I areas, (4) requirements on federal land managers to take affirmative action protecting air quality related values (AQRVs), and (5) establishment of a system allowing exceedance of Class I increments if the new source of pollution can be shown to federal land manager's satisfaction that no adverse impact will take place on AQRVs.

Items (4) and (5) would become the Achilles Heel of the entire process in the Sublette County region. Although the PSD program applied to both USFS and BLM, the latter would always play fast and loose with the regulation and USFS managers would hesitate because of a fear that more evidence would always be needed before they could act. Also, there is a difference between how USFS and BLM addresses the visibility AQRV. USFS applies a visibility threshold of 0.5 deciview reduction, defined as a change just perceptible to the human eye. BLM has no such threshold although when Forest Service lands are involved, a one deciview standard applies.

USFS feared that gas development in the region would alter the character of National Forests in the Wyoming and Wind River Ranges. The report noted that modeling air pollution levels in the mountains was problematic because of an absence of on-site meteorological data which in turn rendered modeling techniques unreliable by as much as plus and minus 100 percent.[3] When I came into the picture 20 years later, my own emphasis was aimed at the dismal modeling effort being done by state and federal regulators. I had no idea this report had preceded me.

In an attempt to anticipate these concerns, the report proceeded to describe a plan of action intended to supply USFS with data that would assist in future mitigation strategies. It set about identifying sensitive flora and fauna receptors, establishing baseline conditions, and designing a monitoring framework intended to detect any decrease in air quality.

In retrospect, USFS had looked into the future and saw what was coming. It also showed agency presence of mind aimed at crafting a plan designed to address the issues if and when they developed. Sadly, resources were always in short supply, especially for use in an as yet uncertain scenario so not a lot was done. Unfortunately, when that time arrived, political realities in Washington D.C. had filtered down to working levels with serious impeding consequences.

Another Agency Weighs In

The U.S. Department of Energy reinforced USFS concerns with a study it released in 2003. The study specifically warned about regional air quality. It cautioned that maintaining pristine air of the Rocky Mountain States posed a special challenge. It stated a fact well understood by regional residents already. Small changes in air quality can have noticeable effects on visibility. The report emphasized this to be especially true of the national parks and forests of Wyoming.[4]

It went on to acknowledge that the region was valued for its striking vistas and scenery. However, the ability to see these sights over long distances had degraded over time as air emissions had already increased. A number of sources were named including traffic, urban development, and industrial activities. Natural gas development was specifically named as an industrial example.[5]

The report further acknowledged that federal and state agencies as well as environmentalists had expressed concern about the contribution of natural gas drilling, production, and transportation. These were contributing to increases of particulates in the atmosphere, leading to lower visibility and regional haze. Of particular concern were emissions such as dust from service roads and nitrogen oxides from compressors. These can travel long distances and sometimes transform chemically to impair visibility.[6] A few years later, all of these most definitely came into play in Sublette County.

It was against this historical backdrop that I became directly involved with the Forest Service.

My First Day and Beyond

My introduction the USFS began as a result of my own action. In 2003 I decided that air quality was going to be the topic toward which I would direct my efforts. I had begun my initial field measurements of well completion flaring using my miniature optical spectroscope and I wanted the government land managers to be aware of what I was doing. Accordingly, I visited the Pinedale regional office of USFS and inquired if I could meet their air quality scientist who was located there.

To my surprise and pleasure I was immediately greeted by an air quality specialist named Terry Svalberg and his field operations guy Ted Porwoll. They sat down with me in a quiet meeting room and listened patiently as I presented my story. I started with an explanation of my science background in the Air Force that included time involved in atmospheric physics. This I went on, was associated with my primary function of working with a crew of technicians and scientists performing field measurements of infrared signatures from natural background sources. As result, I had become familiar with the science of optical spectroscopy as a tool to identify chemical species suspended in the atmosphere.

It,was a logical step to apply those skills to the task of identifying the chemicals that I was convinced came from the new drilling wave in the county and was impairing visibility. I explained that I hoped to build a body of evidence that characterized the emissions from gas development activities and establish the degree to which they might be contributing to the degrading clarity of our local air mass. I also mentioned that I had selected air quality as my subject of study in hopes I could leverage the Federal Clean Air Act which mandates protection of air quality in the nearby Class I region of the Bridger National Forest.

As I spoke, I felt relieved that I was sounding credible because they soon became quite interested in what I was telling them. They filled me in on what they had been observing thus far and briefly outlined their role in insuring protection of the Class I airshed. In later years they confessed that my first few minutes with them was a period when they were expecting something akin to amateur hour but quickly gave way to realization that here was a guy who just might know what he was talking about.

I look back on this event as my favorite moment in my air quality battles. My subsequent association with these two guys was an experience of genuine psychological and professional satisfaction. Our work together included our wonderful partnership on the Air Quality Task Group that was part of the BLM PAWG described earlier. I also traveled with Ted during his runs to the local IMPROVE measurement station. Here I was able to observe the process that USFS was applying to the task of monitoring impacts to mountain air quality from the nearby Jonah and Anticline gas fields.

A Voice in the USFS Administrative Wilderness

I often sat down with Terry in his office to engage in discussions about what he was experiencing as he interacted with WYDEQ, BLM, and EPA regarding the problems that developed in Sublette County. I teased him about his office resembling that of a university professor, shelves overflowing with papers and more stacks on the floor tilting over to the point of near collapse. His desk presented only enough empty space to accommodate a telephone and elbows. He had a tremendous workload on his plate due to his central involvement in assessing the impact of the drilling going into full tilt as well as similar projects in other parts of his region. The poor guy was the only air quality scientist in the intermountain region for many years and his tasking by his superiors was endless and crushing. Terry was a man with whom I would develop a close and effective relationship lasting to this day, followed closely by the same kind of relationship with Ted.

Over the years, Terry's dedication to excellent performance of his job description occasionally brought him into conflict with various individuals. In 2004 when Questar was pressing heavily upon the BLM to allow winter drilling, he let it be known that he was not pleased with the Questar environmental assessment showing that they were already emitting 1200 tons per year of NOx in excess of what the Anticline Environmental Impact Statement predicted. He pointedly reminded the BLM that its EIS had warned of 1000 tons per year of NOx being generated which would have a significant impact on visibility in the adjacent wilderness area. He opined that this battle was becoming never-ending.

Adding to this insult, he noted that the BLM had recently completed its Environmental Assessment, Finding of No Significant Impact, and the following formal Decision of Record for the Anticline winter drilling proposal, all of which justified project approval. He warned that a recently completed follow-up environmental assessment revealed that 1,895 tons per year of NOx emissions were being generated which greatly exceeded the original predicted 693 tons per year. That fact alone was supposed to trigger additional analysis before approval could be granted. As he put it, "Clearly, something is broken here."

Around the same time frame and during the environmental impact assessment phase for the Jonah Infill Project, a meeting of federal stakeholders was convened in Cheyenne that included EPA-Region 8 from Denver, state and regional BLM officials, and USFS officials. In that meeting BLM admitted that proposed drilling and well development in the Jonah Infill Development Project would seriously degrade regional atmospheric visibility.

BLM proposed to allow this to take place with the promise it would eventually implement emissions controls on the operators that would reverse the visibility loss over a period of ten years. This did not sit well with Terry; so he inquired of the Region 8 EPA representation if this was legal under provisions of the Federal Clean Air Act. He was essentially challenging the BLM notion that visibility impairment limits were something to be used up rather than a boundary not to be crossed. The EPA official verified that the idea would indeed be a violation of CAA law.

Word of this event got back to the state deputy director of BLM and he was unhappy. To correct this development, he instructed the USFS Capital City Coordinator to draft a letter of reprimand against Terry for signature by the regional forester, Harv Forsgren in Ogden, Utah.

Capital City Coordinators were explained to me as being USFS employees stationed in state capitols for the purpose of acting as liaison between USFS and other federal and state government environmental management agencies. The gist here is that a BLM director was ordering a sister agency to reprimand one of its employees and using an employee of that sister agency to do the dirty work. Furthermore, that USFS employee complied!

A sympathetic BLM employee in the state BLM office tipped USFS that this was in the works, whereupon Forsgren contacted the state BLM director and told him to knock it off. That ended the affair between the two agencies but not with the governor.

A short time later, Forsgren was in Cheyenne to attend government meetings and was called on the carpet by the governor in a closed-door scolding. Forsgren was subjected to a 30-minute thrashing because the governor perceived that Terry had precipitated a near-miss of his being publically rebuked for allowing environmental law to be violated in Wyoming. This story surprised me somewhat because the governor in question, Governor Freudenthal had been reasonably supportive of the need to observe environmental protection while still trying to support the oil and gas industry.

True to the common public perception that government employees are simply transferred elsewhere when they misbehave, the Capitol City Coordinator in question also landed on her feet. Shortly after the reprimand letter debacle, she went to work for BLM. Still later, she returned to the employ of the USFS and assumed the post of regional forest manager in northwest Nebraska.

My Own Voice in the Same Wilderness

In October 2004 the Pinedale office hosted a conference of northwestern states called the Western Regional Air Partnership Conference. It was a recurring event where USFS could review air quality issues and share information.

By then Terry and Ted had concluded I was a credible field researcher and wanted the other members to hear about my spectrometry technique. They invited me to be a guest presenter and allotted 25 minutes for the presentation plus 5 minutes for questions. Realizing this was my big opportunity to win

USFS interest, I put together a Power Point slide show that was packed. In my practice runs I was making my concluding comment right at 24 minutes and 30 seconds.

On conference day, I projected slide after slide with a running commentary. I presented photos of well completion flares, dehydrator stacks, and tank vapor combustors and the accompanying emissions spectra. As I proceeded, I pointedly admonished the WYDEQ attendees to "listen up" when I reached the section describing my discovery of sodium, potassium, and lithium in the flaring smoke columns. I did so because WYDEQ administrators were showing no interest.

When my time was done, I noticed the lead WYDEQ person scurrying over to the conference leader where I heard her almost desperately stating that the elements I detected in smoke plumes were not the cause of visible haze. I later told the conference leader I thought she was retreating behind administrative nuance. I believed she was invoking the Clean Air Act list of criteria pollutants that did not include these metals. Therefore, they could not be sources of visible haze.

As the conference leader and I talked further, he asked some questions that illustrated he had heard me and found my findings intriguing. Then to my supreme satisfaction he commented that he was amazed BLM had not been sued over its poor attention toward air quality. A few minutes later Ted Porwoll came over and complimented me, stating it had been satisfying to watch someone so well practiced in delivering a professional quality presentation.

Now fast forward to March 2009. Well into the period of gas development in the Jonah and Anticline I had my own encounter of sorts with Forsgren which illustrated how government stewards of the environment could add their own dysfunction to the mix. After several years of monitoring the decreased visibility in Sublette County I asked Terry to which official in his chain of command I should write a letter demanding a formal determination of visibility impairment in the nearby Bridger National Forest.

I was advised that Mr. Forsgren was the person. I crafted a letter to him presenting examples of my photographic and wind trend analytical evidence that I felt supported the need for the designation I sought. A few weeks later Forsgren bucked the letter down to Terry, tasking him with the responsibility of responding to me. Half-jokingly, I offered to write Terry's reply for him as a way to confirm the absurdity of this episode.

The letter I crafted to Forsgren cited several issues I felt constituted strong evidence for the action I requested.[7] I explained that I had worked closely with his air quality specialist in the Pinedale Ranger District, (the same Terry Svalberg) and suggested he might wish to seek Terry's thoughts about what I was about to describe. This information was what made Forsgren's tasking of Terry to write a reply seem so silly.

I went on to explain that as activity in the gas fields increased, I had observed a decline in atmospheric transparency such that by 2004 the changes had become severe. I thus began a personal field research effort using optical spectroscopy to identify pollutants in the air that might be traced to the gas fields. The result was a discovery not reported by anyone else, i.e., I found the elements sodium, potassium, and lithium being injected into the atmosphere by well completion flaring.

I continued by explaining that I had challenged BLM environmental impact assessments for the Jonah and Anticline fields on the basis of two areas of concern. First, I had studied wind fields in the local region for seven years. I had concluded that wind field data used by BLM in atmospheric modeling was highly suspect because they were too dated and too general in nature.

Furthermore, wind patterns in the area were far more complex and I felt my findings showed that. I believed I had demonstrated that they travel toward the Class I areas at least 40% of any year (45% in some years), thus transporting gas field emissions in that direction particularly in the winter. Consequently I argued, BLM estimates of NOx and VOC pollutant precursors to visible haze were being carried to those areas in larger quantities than BLM was acknowledging.

My second concern involved my challenges to BLM and WYDEQ on the grounds that their so-called "actual estimated inventory" of NOx and VOC volumes was seriously flawed. I noted that I had assembled my own inventory of 3000 wells using state records on the Jonah and Anticline fields and discovered a number of bookkeeping discrepancies. Also, I warned of possible weaknesses in the quantification of these pollutant emissions. Specifically, atmosphere modeling of emissions volumes were being run with NOx and VOC inputs that were themselves modeling results provided by the operators.

Operator computer models were being used to estimate volumes of emissions coming from storage tanks, dehydrators, and flaring stacks, none of which had in my view been validated for our altitude and meteorological regime. As a result of this possible shortcoming, I believed that estimated volumes of these emissions were inaccurate by a factor of two to a factor of ten.

What I didn't say was that Terry and I had arrived at this assessment together. The consequence of this was obvious decay in visibility toward the Wind River Mountains and the Wyoming Range. A likely additional impact was increasing acidification of the alpine lakes in both mountain ranges.

I then summed up by declaring what I wanted Forsgren to do. I stated my understanding that he was the USFS manager of our region with the authority to issue a Declaration of Visibility Impairment. I explained that I had been challenging WYDEQ for four years regarding its obligation to do a better job of protecting the Class I airsheds and I had carried my arguments to the officials of Region-8 EPA with signs of growing success.

I pointed out that our region was experiencing a new ozone exceedance history that was prompting the state (actually the Governor) to recommend that we be designated as a "non-attainment" area. I urged Forsgren to immediately issue a written statement declaring that visibility impairment was also underway in the Wind River Class I airshed.

I suggested that at a minimum such action would place the ball in the court of WYDEQ and force movement of its regulatory machinery in response. I further tried to encourage a favorable decision from Forsgren by noting that Region 8 EPA had been ill at ease with happenings in Sublette County; so action by him would likely be met with support from that agency.

I went on to state that enough scientific data had been amassed supporting an argument that visibility degradation was underway. Furthermore, it was unlikely to improve to the degree promised in EIS statements and Records Of Decision for the Jonah and Anticline. I drew this conclusion from my conversations with both Terry and Ted Porwoll who informed me of the actual evidence being collected in the region by the network of USFS air monitoring stations. In that context, I suggested my recent review of the Jonah ROD

By the Rasping in My Lungs

strongly indicated violation of half of the 13 air quality commitments it contained.

To his credit, Terry's reply to my letter for Forsgren was a skillful stroll on a political tightrope.[8] He thanked me for my concern and acknowledged USFS affirmative obligation under the Clean Air Act to prevent degradation of air quality related values that included visibility in the Class I wilderness area in question. Furthermore, the Forest Service "shared my concerns about decreasing visibility in the adjacent two mountain ranges."

Some weeks later we were together in a private setting and we talked about specific points made in the letter. I commented that Forsgren's assurances of a report soon to be released by USFS seemed without value. It was to present analyzed data regarding visibility impairment and increasing acidification of alpine lakes in the Wind River and Wyoming Ranges. So what?

I asked what good was yet another report about lake chemistry in connection with visibility impairment then underway. He reminded me that lake acidification was a direct collateral effect of lake surfaces coming in contact with air carrying gas field emissions. A time would come when the effects of gas development in the Jonah and Anticline would become an issue in a broader context than visibility impairment alone.

Next, his letter informed me that the existing IMPROVE monitoring station was not originally cited to measure regional haze so it did not exemplify visibility impairment being observed by citizens in the area. Accordingly, the USFS was working to obtain supportive data to prove actual impairment before so certifying.[9]

What Terry had to leave out of his ghosted reply in the name of Forsgren was that we had privately talked about this a great deal. The USFS had known for seven years that the IMPROVE site was not useful to our needs but, nevertheless, had been collecting the referenced supportive data for quite some time. The real problem was that Forsgren could not accept the fact that he had enough data with which to comfortably act and likely never would. We both recognized he was actually engaging in a delaying tactic.

The letter proceeded to address the fact that USFS was in the process of establishing a new IMPROVE site at nearby Boulder Lake to provide a better understanding of current conditions downwind from the gas fields. The ulti-

101

mate goal was stated to be validation of air quality modeling because modeling by WYDEQ and the operators had predicted high levels of visibility impairment. The new additional data could verify these predictions that then "… may be used to support certification of visibility impairment."[10] I noted this non-reassuring *"may be used …"* comment.

I asked Terry what return on investment for the Boulder monitoring station would be realized if its data only "may" be utilized to support certification of impairment? Furthermore, why had USFS alluded to its not understanding current conditions when at the working level, i.e., Terry's own realm, there was no doubt that gas field development was causative of our visibility impairment? And why had USFS been so slow to proceed with the Boulder IMPROVE installation if the need was apparently so acute?

He agreed that the rate of progress had been slow and legitimately noted that funding had been a severe impediment. We both were well aware too of the resistance that WYDEQ had been putting up because it felt this was a usurpation of its oversight authority. I talk in detail about this in the chapter about the Air Quality Task Group.

Also, we suspected WYDEQ was concerned that the data could prove embarrassing. USFS had in fact become so frustrated with BLM and WYDEQ stonewalling that it concluded it had to take independent action to defend its wilderness areas. Ultimately, Terry was hopeful that a critical mass of information from this new site as well as the other site would prove convincing enough to justify the certification I was seeking. However, that never happened.

Terry's reply next offered a fact that we had previously discussed with some chagrin. The decision-making authority for certifying visibility impairment had been taken away from the office of the regional forester in Ogden. As of January 2009, that authority had been relocated into the office of the Undersecretary of Agriculture, USFS's parent agency.[11] The impact of this was to place the decision making power in Washington D.C. far from the actual field locations where problems actually exist. Ultimately, this would further confound efforts to award visibility impairment certification because of the additional layers of bureaucracy inserted between Washington D.C. and USFS regional offices.

The letter ended with the assurance that USFS would continue to monitor and analyze data to protect air quality values and take action when the data indicated such action was needed. My challenge was the same as what I hit BLM with for invoking the same assurance. How long would "monitoring and analyzing" of data continue before action would result? I noted that incessant study had been the gold standard for regulatory inaction since at least year 2003 when gas field visibility impacts began to be noticeable in the form of the "brown cloud." Terry was again sympathetic.

The letter's closing reassurance was one I recognized as Terry's personal effort to convey a coded message to me. It expressed "... appreciation for my independent work that was assisting agencies to better characterize air quality issues and provide full disclosure of the impacts of their decisions." My continued collaboration in this effort was welcomed. He was telling me to keep the pressure on.

Thus, he had skillfully satisfied the desire of his superiors to appear noncommittal while at the same time conveying in plain sight a warning that agency decisions were having negative impacts.

Terry and I agreed that we could only hope that his regular warnings to his superiors as well as growing anger by the local populace would ultimately achieve traction.

Battle Over the Anticline

The development proposal for the Pinedale Anticline gas field was a situation to which USFS quickly reacted with increased attention. This was so because of its observations of what the Jonah field emissions were starting to reveal about their impact on air quality and alpine lake acidification. This attention took the form of extensive written comments for the record that were almost entirely the product of Terry's authorship and submitted under the signature of Harv Forsgren. They warned of serious potential added damage to air quality and threat to the Bridger National Forest in the nearby Wind River Mountain Range.

In February 2008, a very specific set of comments went to the State BLM director, Bob Bennett, whom I describe in the chapter on the BLM. In these comments, USFS (Terry) subtly signaled that USFS had full intentions to execute its obligations under the Organic Act and the Clean Air Act (CAA) mandating that it protect air quality related values (AQRVs) in its Class I wilderness areas.[12]

This comment document began immediately by citing concern over substantial increases in visibility impairment predicted to occur because of emissions from the PAPA project. Furthermore, concern was expressed about cumulative increases in visibility impairment from this project plus other additional regional project sources. Visibility modeling was cited for the Bridger, Fitzpatrick, Washakie, and Teton Wilderness Areas.

Emphasis for this point of view was added by referencing federal statutes. The Clean Air Act and the Wilderness Act were specifically invoked right down to CAA Section 169a, 42 and U.S.C. Section 7491 regarding the national visibility goal of preventing future impairment as well as remedying existing visibility impairment.

USFS then bored in on specifics it did not like. It noted that half the proposed development alternatives were not quantitatively analyzed and their impacts were not divulged. Therefore, SEIS proposals for air quality impact mitigation procedures could not be evaluated for effectiveness.

USFS was unsatisfied with the fact that unmitigated impacts from all development alternatives would generate visibility degradation beyond the 1.0 deciview impairment threshold in six nearby wilderness areas. USFS also took issue with the fact the BLM had performed modeling exercises that were not the same as those used by the Forest Service, the National Park Service, and the U.S. Fish and Wildlife Service. This resulted in BLM representing impacts with which the other agencies disagreed. Furthermore, those impacts had been originally stated in the initial SEIS Air Quality Technical Support Document but were omitted from the current version of the Technical Support Document.

USFS pointedly noted that while the SEIS reassuringly asserted that unspecified mitigation actions would produce zero days of visibility impairment beyond natural background levels, the modeling was less reassuring. In fact, both BLM and sister agency analyses were showing between 2 and 88 days per year of visibility reduction equal to or greater than the USFS threshold of 1.0 deciview.

USFS then presented a set of recommendations. First, it wanted the BLM ROD to require specific mitigation methods at the beginning of the PAPA project. The example of phased development was suggested. This by the way was also a common request from the local citizenry who wanted to see development pursued in a less frenetic manner. Next USFS wanted mitigation measures that were specific in nature and not constrained by industry notions of practicality and affordability. This echoed one of my own points of objection about which I had been hammering WYDEQ and BLM.

USFS next addressed another point I frequently demanded of BLM and WYDEQ. The BLM ROD needed to specify performance goals and end objectives to be accomplished within specific time frames. USFS and I realized that unless the operators had their feet held to the fire in some time-phased manner, achievement of visibility protection would simply continue to be a goal into the infinite future.

Then USFS stated a recommendation that again echoed my own demands of BLM and WYDEQ. It wanted a schedule of specific consequences to be applied against the operators that would incentivize them to actually accomplish effective and timely mitigation. Absent any pain for non-compliance, the operators would be able to simply shrug off any warnings.

USFS next offered a carrot by suggesting that a comprehensive monitoring network be set up around the area to measure mitigation accomplishment and reward those operators who met or exceeded goals. Conversely, if goals were not met, consequences had to be defined and applied. In truth though, I always worried that unless each specific rig emitter was being monitored, it would be impossible to actually identify an operator who was being good or bad. The mixing factor downwind would give them the cover to claim individual innocence.

The USFS comment document closed with the obligatory assurance that it looked forward to working with partners and by implication the BLM to achieve protection of its wilderness area resources.

This collection of concerns and recommendations presented to BLM was so familiar. Terry and I had talked about all of these in one form or another during the previous months and years. We both viewed with disappointment EPA failure to specify time frames and consequences for BLM and operators' noncompliance in meeting mandated corrective actions. We both felt that absence of performance criteria contributed materially to the difficulty we were all experiencing in persuading those two players to get serious about cleaning up their act.

It was with some satisfaction then that I became aware of the content of this comment input to BLM by the USFS. I couldn't prove it nor did I dare press Terry to admit it, but I, nevertheless, suspected that he had heard my concerns and deliberately incorporated them into the document sent to Director Bennett under Forsgren's signature.

Battle Over the Wyoming Range "44,000"

In the period between 2004 and 2012 USFS was in a position of having to multitask. By this I mean that it had a full agenda just dealing with the complexities of environmental impacts from the Jonah and Anticline gas fields. Unfortunately, there evolved simultaneous pressure from the drilling industry to develop the Wyoming Range on the west boundary of Sublette County.
In some ways this was a thornier problem because the landscape in question is national forest land over which USFS had more direct but not exclusive oversight. BLM was still a player. Furthermore, USFS was faced with having to address the likely cumulative impact to air quality in the Class I and Class II regions resulting from emissions coming out of the Jonah, Anticline, and any Wyoming Range gas developments.

Between March and August 2004 the USFS offered up 99 parcels totaling 175,000 acres to BLM for auction but then in September requested that BLM withhold these parcels. The reason was concerns expressed by the public and by Governor Freudenthal as well as U.S. Senator Craig Thomas. As a result, only 25 percent of those acres were finally submitted to BLM for auction.[13]

In April 2005 the supervisor of the Bridger-Teton National Forest, Carole "Kniffy" Hamilton had identified 38 lease parcels in the Wyoming Range to be forwarded to the BLM for public auction. These covered 44,600 acres that were part of a set-aside in the forest management plan for commodity resource development.[14] Hamilton's office cited decision criteria that included avoidance of roadless areas, compliance with forest management plan objectives, the Energy Security Act of 1980, the 2001 Forest Service Energy Implementation Plan and a variety of other unspecified federal guidance.[15]

Despite these efforts to walk a tightrope, the issue of "The 44,000" would become a heated subject of environmental combat that would drag on to the present time 13 years later.

The latest (early 2016) in a seemingly endless series of Forest Service environmental assessments contained the best summary of all that had taken place to date:[16]

- In March 1990 the original ROD for the Bridger-Teton Resource Management Plan designated 1.2 million acres in the Wyoming Range to be eligible for oil and gas development.

- Between 1990 and 1993, The Forest Service produced three environmental assessments that refined constraints for leasing within those Wyoming Range lands.

- In 2004, changes subsequent to approval of the forest management plan and issues cited by then-Governor Freudenthal forced a supplementary biological assessment and report. The result was a reduction of approved lands eligible for leasing to 175,000 acres but this was accompanied by a determination that the original 1990 environmental assessments of oil and gas impacts had been adequate.

- In 2005, the USFS authorized BLM to offer leases of 44,700 acres with stipulations attached. Also, BLM determined that National Environmental Policy Act (NEPA) processes had been adequate to support leasing.

- Between 2005 and 2006 all the authorized lands were auctioned and leases were pending award but environmental groups protested. BLM denied the protests and issued 12 leases but delayed another 23 and placed them in pending status.

- Later in 2006, appeals were filed with the Interior Board of Land Appeals which granted a stay of BLM decisions denying protests over the 12 approved leases. Grounds for the stay were that the leases had likely been issued in violation of NEPA based upon circumstances that included air quality concerns. This placed on hold issuance of the remaining 23 leases.

- In 2008, USFS issued a supplemental environmental analysis to address concerns over the original analysis and to deal with concerns resulting from the public involvement process. Resultant scoping revealed several significant issues requiring attention of which air quality was one.

- In 2009, BLM rejected bids of the three high bidders for the 23 leases. BLM Justification cited the Omnibus Public Lands Act indicating that Congress had identified a higher public use of these lands than oil and gas development.

- Later in 2009, high bidders appealed BLM's action to the Interior Board of Land Appeals challenging BLM's authority to reject their bids.

- At the beginning of 2010 one of the operators offered to settle its appeal by accepting its two lease parcels with a no-surface-occupancy stipulation. This would hopefully make the lease consistent with the Omnibus Act. Governor Freudenthal sent identical letters to the Secretaries of Agriculture and Interior signaling his support for such a settlement.

- During the following month in 2010, the USFS "Draft Supplemental Environmental Impact Statement for Oil and Gas Leasing in the Wyoming Range" was released for public review and comment. The preferred alternative was identified as no action, no leasing.

- Still later in 2010, the Interior Board of Land Appeals ruled in favor of two operators' challenges of the BLM decision to reject their high bids. The Board argued that the BLM had not provided a rational basis for its rejection of bids since the Omnibus Act did not prohibit issuing leases previously offered by the BLM.

- Late in 2010 and early in 2011, the *Final Supplemental Environmental Impact Statement for Oil and Gas Leasing in the Wyoming Range* and *Record Of Decision* was released to the public. The environmental analysis contained therein persuaded Forest Supervisor of the Bridger-Teton National Forest Jacque Buchanan to opt against authorizing BLM leasing of national forest lands in the Wyoming Range for oil and gas development.

- In February 2011, Sublette County Commissioners and three operators seeking leases appealed the Record Of Decision.

- In May 2011, Forest Supervisor Buchanan withdrew her decision against authorization of BLM lease award. Bridger-Teton managers assembled a new interdisciplinary team and began a new supplemental environmental impact statement.

It should not escape the reader's notice that two political entities, the Sublette County Commissioners and Governor Freudenthal opted to come down on the side of oil and gas development. This was always the agenda of the Commissioners and became increasingly so in the Governor's office during the period of his second term. The former were determined to deny that impacts to air quality were underway and the latter sought to walk an increasingly contentious path between operators' demands for drilling access and demands from the citizenry to stop gas related increases in air pollution.

Also, shortly after Jackie Buchanan assumed duties as the Forest Supervisor, I had an opportunity to talk with her when we happened to be attending an annual meeting of the Jackson Wyoming based Greater Yellowstone Coalition. I outlined my contacts and written comments to her predecessor Kniffy Hamilton and stressed the value of Terry Svalberg as her air quality advisor. I urged her to stand firm against what would become a strong headwind of pressure from the operators wanting to drill in the Wyoming Range.

She was gracious and listened to me politely despite the clamor of others around us wanting face time with her. When I finished, she assured me she would seek Terry's council and weigh all the arguments in a way that would be best for the national forest. However, as the months passed and the operators upped their pressure, she seemed to give in to it. Pressure from Washington D.C. added to the situation and after only a few years, she was reassigned to a higher post.

Year 2011 was not the end of the contest. In that year a detailed quantitative air model was designed and run which produced a final report in 2013. The upper Green River region had been designated (in 2010) an EPA "marginal non-attainment area" for ozone in accordance with the Clean Air Act. Also, a variety of issues unrelated to air quality but equally important came to bear upon Wyoming Range leasing that resulted in further delays.

The USFS's latest (April 2016) draft environmental impact statement document summed up the above points this way:[17]

> *There is a need to analyze new information that is relevant to environmental concerns and has a bearing on previous decisions associated with leasing these lands. There is a need to disclose the effects of reasonably foreseeable development and determine necessary stipulations to adequately mitigate potential resource impacts. The analysis is needed to inform the independent decisions of the Forest Service and the BLM regarding oil and gas leasing for the subject lands. The decisions of both agencies must be supported by an environmental analysis that adequately addresses the impacts of reasonably foreseeable development associated with oil and gas leasing.*
>
> *The purpose of this analysis is to evaluate new information and to correct deficiencies in previous analyses to ensure the potential effects are fully considered before a final decision is made as to whether leasing is appropriate on lands in the project lease parcels.*

Eight months later, the USFS ended the fight over the 44,000 with its no drilling decision and the USDA Undersecretary issued a public announcement making the decision final.

Battle Over the Wyoming Range Noble Basin

The Wyoming Range had been further confirmed as a drilling target by 2004 with the Eagle Prospect drilling project proposal. In September 2005 USFS released public notice of staking by Plains Exploration and Production Company (PXP) of Houston, Texas to perform wildcat well drilling in the Noble Basin of the Wyoming Range near the Hoback River south of Bondurant, Wyoming.

This project would start with three exploratory gas wells and a lot of new road construction within national forest lands.[18,19] This event launched what would become an eight year battle between local citizenry, USFS, and PXP and was coincident with but separate from the larger fight to stop drilling in the 44,000.

The details of the project immediately became unpopular because the main access route was to pass close by private lands supporting country homes and several active ranches. Part of this route was already in place in the form of paved and unpaved roads. However, the last several miles would require building new roadway through timber areas and along ridges that presented erosion and topographical difficulties. These and other details prompted me to send my first written comments to Administrator Hamilton.

I informed her that I had accumulated a solid year of well site optical spectroscopy measurements as the basis for my letter. Its purpose was to provide her with new scientific reasoning for disallowing any leasing and subsequent gas development in the Wyoming Range and upper Hoback region. I briefly outlined my technique and invited her to read a technical paper I had written and attached as supporting evidence. That paper was published in the Proceedings of the First Annual Coal Bed Natural Gas Research, Monitoring, and Applications Conference.

After giving her a hitchhiker's description of how optical spectroscopy works, I described my findings. I had observed and recorded spectral signatures of well completion flares and made a discovery that had, by my reckoning, not been reported in any such research up to that time. I had discovered that well flare plumes emitted a strong signature of the alkali metals sodium, potassium, and lithium.[20] There were other chemical emissions but these three were revealed most clearly in the spectra.

I explained that I had recently presented my conference paper to Terry Svalberg in the Pinedale field office and noted that he had shown keen interest. I went on to state that I had identified the source of these metals. That source was the family of fracking fluids being used in the Jonah and Anticline gas fields. I confirmed these findings with examples of fracking chemical supplier advertisements and technical presentations by fracking service providers. They named compounds containing these three metals and described their function in the underground region surrounding the borehole.

I then cited my photo-documented examples of heavy haze obscuration in the region five to six hours after my spectral observations of a flaring event. I cautioned that the situation would get much worse in the valleys of the Wyoming Mountains and the Hoback basin where the air could be more stagnant. These valleys and basin could act as a trap for flaring smokes particularly in winter when the lower atmosphere stratified due to cold temperatures.

I contended that resultant haze containing sodium, potassium, and lithium, as well as other components I had yet to identify would constitute an air quality threat to the forest and wildlife that could not be predicted. I further warned that these probable negative impacts might require considerable time to reveal themselves in the form of tree mortality and wildlife health decline. That delay was further justification for exercising extreme conservatism regarding leasing of those lands.

I urged her to take the admittedly difficult action in the current Bush Administration political climate to bar the Bridger-Teton national forest region from energy development. I asserted that exploration technology might not yet be sufficiently benign to prevent serious and irreversible damage to these mountain ranges.

I closed with my opinion that the gas would patiently remain in place until extraction technology became less damaging. Also, I provided a photo of a well flaring event on the edge of the forest in question as well as an accompanying spectrogram showing the strong signal of sodium and potassium and suggested that this scene multiplied by a thousand fold would not be a desirable outcome.

In February 2006, USFS held a public meeting to review the Noble Basin-Eagle Prospect project proposal.[21] Greg Clark, the Big Piney District Ranger, for the area in question was the master of ceremonies. Also the proposing operator, PXP was present. Actually, PXP was present in considerable force with several management level employees.

Clark presented about an hour's worth of commentary that frustrated us in the audience at an escalating rate. Essentially he engaged in a recitation of issues over which he claimed he had no control. He explained the process of development that would take place and listed a number of actions that the operator might take in the course of drilling on their parcel. Along the way, he ticked off those items he could not influence.

He could not stop the proposal because PXP had bought the parcel lease and had property rights to develop it. He would not be able to stipulate very much in the way of surface protections, water protections, or air quality protections. If questions were to arise about impacts, he would have to pass them up his chain of command.

Finally, after silently listening to this for as long as I could, I raised my hand and asked, "why are you standing up there? Clearly you have no authority so it seems to me your superiors should be standing there in your place." Without any noticeable hesitation, he shot back "I am the decider." That elicited disbelieving looks among audience attendees and a few giggles.

After Clark's formal comments, we all mingled with him and the PXP representatives. I pinned one of them down in a conversation that probed their readiness to perform drilling in a way that would not trash their parcel and pollute the air. I pressed for explanations of experience by the company in drilling in wilderness areas and what details they were planning to execute to accomplish safe drilling operations. I was a bit shocked to hear him claim extensive experience with such matters in offshore drilling operations.

I challenged him right away by warning that I doubted offshore experience would offer any relevance to issues in our national forest. That in turn seemed to shock him. He seemed to either disbelieve any disconnection between the two or he was amazed that I dared doubt the company's prowess in environmental protection. I was never able to ascertain which was correct.

The process of plodding through the NEPA protocols by USFS and PXP dragged on. In 2006 and 2007, USFS published a notice of public scoping in the Federal Register soliciting public comment. The notice discussed the intended Master Development Plan (MDP) and outlined the responsibility and authority structure that would be applied to development operations.

In particular, what I found worrisome was the stipulation in the scoping notice regarding division of authority.[22, 23] The Forest Service was to be the lead agency with the BLM and the State of Wyoming (make that WYDEQ and the WOGCC) acting in the capacity of cooperating agencies. The Big Piney District Ranger would make decisions on surface use issues while BLM would exercise approval authority over Applications to Drill (APDs) and subsurface mineral rights, design of the drilling program, and associated safety concerns.

By now I had become well versed in failures by BLM and WYDEQ to execute their stewardship responsibilities in accordance with federal directives. With this new gas development project proposal I saw a frightening situation. At that time, Wyoming was engaged in a battle over what is called a "split estate," whereby land-owners who possessed subsurface mineral rights exercised veto power over any restrictions that might develop from surface damage experienced by surface owners. For example, in a public meeting in Pinedale, a rancher described how drilling operations had destroyed his hay farm and he could do nothing to stop it because he did not own the mineral rights.

Thus, in this Eagle Prospect situation I felt there was a clear and present danger that BLM would have full authority and opportunity to override any USFS objections to surface damage that PXP might (would) inflict on the area. The division of authority essentially made BLM the more equal partner in a partnership of equals. This concerned me greatly because the BLM record was not particularly outstanding for its efforts to protect air quality.

With my subsequent set of comments to Ms. Hamilton in April 2006 I undertook the task of warning her in detail about the specifics of BLM behavior for which she should be on the lookout.

I explained that beginning with the Jonah gas field project and continuing through the Jonah Infill and on to what was the present Anticline development proposal, BLM had published a number of documents containing numerous requirements to be levied upon operators. These requirements were presented within a framework of assurances that gas development impacts to various environmental values would be monitored and continually addressed by adjusting those operator requirements as needed.

However, considerable time had since passed and experience had been accumulated to evaluate the effectiveness of those provisions. I then traced the history of those requirements and their apparent lack of effectiveness, starting with the Jonah and then citing the PAPA.

I began by invoking the Jonah Infill Record of Decision (ROD), dated March 2006. I quoted what I thought was certainly relevant to the Wyoming Range project: "Ongoing and future natural gas development projects in the region are contributing to observed changes in air quality and negatively impacting the nearby Class I wilderness airsheds." This signaled that changes to air quality and negative impacts to Class I airsheds had become a self-fulfilling prophecy.

I warned that the ROD invoked "… jointly developed performance-based mitigation requirements…designed to address potential adverse visibility impacts in … Class I areas." This included use of Tier II diesel engine technology for drill rigs, cooperative *tracking* of emissions, and *tracking* of numbers of wells, drill rigs, emissions, and compressor stations. Thus, WYDEQ would (cooperatively) merely *track* permitted emissions, operator submission of data on rigs, drilling days, horsepower, load factors, and emission factors *after* well drilling completions.

I tried to caution that the provisions for tracking simply conveyed the impression of an exercise in the accumulation of data for no enforcement purpose. There was little value in compiling such data without its primary purpose being explicit enforcement. Anything less could be viewed as an unstated plan for land managers to merely look back some day and point to what was done wrong.

As for reporting of such data after a well was completed, this could go on indefinitely with no useful return. Reports would only be useful if findings were turned around quickly enough to manage the next drilling cycle. This had to include further emissions reductions through improved engineering solutions.

I next warned it was inadequate for operators to submit annual operating plans that would merely *report* all emissions. The stated purpose was to demonstrate that potential impacts would fall below the 80% EPA-mandated reduction scenario. There was no stipulation as to consequences in the event of failure nor was there a clear definition of what form the demonstration should take, although modeling appeared to be the tool of preference.

That was borne out by the BLM assertion that it *might* run an air dispersion model to access impacts. This discretionary caveat offered too convenient an escape route. In fact, I argued a third independent party should be contracted to run the model for the specific purpose of impacts evaluation but only then in parallel with a field measurement program at the sources designed to validate the output from the model and vice versa.

I cautioned that BLM had signaled its intention to rely upon operators to come up with their own means for achieving emissions reductions. While this might seem reasonable given the dearth of expertise in BLM circles to specify solutions, there needed to be provisions citing specific emissions sources that had to be addressed and to what degree.

In particular I warned that dehydrators were multiplying literally by the hundreds and they could host as many as nine combustion stacks apiece. Each one of these stacks was a source of emissions pollution which when multiplied by thousands would not be a pretty picture.

I brought up adaptive management in the context of the specter of ozone exceedance. The language used by BLM invoked a cooperating agency committee approach to "... determine whether adaptive management would be needed to mitigate impacts." While that sounded reasonable, one cooperating agency, WYDEQ, continued to assert the need for a three-year period of study in accordance with EPA statutory guidelines.

However, I stated a reminder that the State Governor had signaled his interest in moving faster. In fact I argued the Federal Clean Air Act provided authority for more immediate action and adaptive management was needed immediately. This would slow further source permitting until assured solutions to the already existing problem were in hand.

I moved on to alert Ms. Hamilton to other elements revealed in Anticline emissions source mismanagement toward which she needed to be watchful.

BLM had by its own admission deliberately curtailed its management of emission sources. I quoted the environmental impact statement: "Restriction on the number of drill rigs present at any one time in the PAPA was not carried forward from the PAPA EIS to the PAPA ROD. BLM concluded that limiting the number of drill rigs…would be difficult to manage.[24] This revealed that BLM had no issue with dropping a commitment if it threatened to require more work on the part of state and district staff to keep up.

I submitted further evidence of retrenchment by BLM. Quoting the Anticline SEIS: "The number of drilling rigs operating on the PAPA has increased since issuance of that PAPA ROD.[25] There has been an increase in wells drilled and drilling rigs present each month during winter beginning in 2003-2004 due to the exceptions granted by BLM and the Decision Records for several limited winter drilling proposals."[26]

I suggested to Ms. Hamilton that all of this showed BLM had never been serious about executing its mandate to protect air quality or wildlife habitat but had instead followed a path designed to remove any obstacle that would block industry's gaining unimpeded access to the PAPA. I was suggesting the same could happen in the Noble Basin-Eagle Prospect project.

I then tried to caution her about the specter of NOx emissions. The Anticline SEIS listed mitigation measures that included "… emission reduction in NOx to 2005 levels within one year and an additional 80 percent reduction within 42 months."[27] I told her this was rooted in a flawed premise resulting from the closed-door meeting between several government agencies that declared 2005 as the baseline year.[28,29] I explained that this came about because of WYDEQ reluctance to establish air quality monitoring instrumentation four years earlier. As a result, 2005 became the de facto reference because it was the first year that had been fully monitored and recorded.

I reminded her that air quality impacts were well established by 2005. Proposals to go an additional 80% below the 2005 level were virtual. There were no protocols defined to force compliance nor were there penalties defined for failure to achieve the goal, let alone assess which operators might be at fault. Finally, the 80% reduction would be based upon *modeling* that I argued to be a flawed approach. Any modeling had to be accompanied by extensive field measurements of emissions at the sources for validation of such modeling.

My next caution was important and I emphasized that by quoting the PAPA SEIS:

> "Since issuance of the PAPA ROD (BLM 2000b), natural gas development in the PAPA has occurred at a pace greater than that analyzed in the PAPA DEIS. The PAPA ROD authorized the development of 700 producing wells or well pads ... The air quality impact analysis conducted in the PAPA DEIS assumed 700 producing wells and up to eight drilling rigs operating in the PAPA at any one time. As of December 2005, there were approximately 457 producing wells and over 26 drilling rigs operating in the PAPA ...The NOx emissions from all sources operating in the PAPA during year 2005 were estimated at 3,512.4 tpy which exceeds the 693.5 tpy analysis threshold specified in the PAPA ROD."[30]

I wanted to alert her to the fact that BLM had done two things here. First, it allowed well numbers to greatly exceed the planned number analyzed in the environmental impact statement. Second, it had created serious ambiguity in defining whether it had been planning for individual wells or well *pads* containing many individual wells.

The SEIS sentence contained a fundamental vagary to wit, "... 700 producing wells or well pads ...." That word "or" was the source of the ambiguity and presented a critical conundrum regarding air quality impacts. There is a stunning quantitative difference between 700 *wells* and 700 *pads*. The difference could be at least a factor of eight. She needed to be watchful for similar maneuvers with Eagle Prospect.

I reminded her that the SEIS contained the statement: "Based on the significance criteria, the PAPA DEIS (BLM, 1999a) stated that significant impacts to visual resources in the PAPA could occur for all alternatives except the No Action Exploration/ Development Scenario. Visual resources in localized areas have been significantly impacted, according to impact significance defined in the PAPA DEIS."[31] In other words, air quality was being degraded but BLM was pressing ahead anyway. I was hoping this would give her a good argument to slow the rate of development until sources of the impact could be brought under control.

My next goal was to convince her that BLM could not be relied upon to execute supposedly obligatory air quality impact analyses. I explained this by citing the SEIS Air Quality Impact Analysis Technical Support Document. Here it was admitted that a complete air quality impact analysis was not conducted in association with the Decision Record (BLM, 2004) that approved a Questar 2004 proposal for limited year-round drilling on the PAPA.

The same document went on to reveal that again a complete air quality impact analysis was not conducted pursuant to the Decision Record (BLM, 2005a) approval of a delay of the requirement for rig engines to be transitioned to Tier 2. Then when three more operators took advantage of the door opened by Questar by proposing their own PAPA year-round drilling program, BLM, for the third time, waved the requirement for an air quality impact analysis on the grounds that those operators had promised to implement as yet unproven mitigation methods.

I warned her that this demonstrated a clear BLM pattern of dismissal of air quality impact issues. Furthermore, I argued, BLM had elected to justify forgoing any meaningful air impact assessment. It did this on grounds that "commitments" by the operators constituted sufficient, albeit, faith-based oversight of its statutory obligations under the Clean Air Act.

I did my very best to give USFS an evidence based argument it could rely upon to guide its decision making process. However, several years would pass as the contest played out between citizens opposed to development, the operator pushing hard for development, and USFS caught in the middle.

• • • • •

The first public opportunity to comment specifically on the scoping document came in the form of a public announcement of deadline for comments effective April 28, 2008. I assembled a lengthy set of responses for Clark and also an official in the Forest Mineral Staff at the Bridger-Teton National Forest regional headquarters in Jackson, Wyoming. My comments to the Minerals Staff was a result of urging by Terry to insure that all the correct people would see what I had to say.[32]

I started by pointing out concerns over the scoping notice discussion of the MDP. It declared that once approved, it would constitute approval of all subsequent APDs. It also cited the host of "facilities" that would also receive automatic approval. This was accompanied by the claim that such an approach constituted a mechanism for "… better planning … as a whole …." However, the process as described appeared to present a wide-open door for a host of subsequent expansions unimpeded by any concerns over environmental issues.

Under the topic of "Proposed Action" was cursory discussion of the issue of cumulative effects early in the process. Also discussed was a lack of need for further analysis of these effects from subsequent development as long as they fell "within the parameters" of the Master Development Plan (MDP). I warned that the Jonah and Anticline developments had clearly demonstrated "planned, anticipated, and potential development facilities" were phrases that could easily be misused. They could open the door to endless waiver requests which BLM would likely grant. Thus, the concept of the MDP as laid out in the scoping notice presented the prospect of a blank check for the operator once the MDP was approved.

The scoping document next described the planned well count and spacing. It cited 40-acre well spacing but failed to clarify if this referred to down-hole spacing or surface spacing. This was crucial because in fact, the WOGCC had sole authority to grant higher density well spacing upon request from the operator. I commented that although this authority was outside USFS influence, it must anticipate and plan for the possibility. I brought this up because the WOGCC became somewhat notorious for its routine expansion of well spacing parameters in the Jonah and Anticline.

I then cautioned that the scoping notice did not contain specific references to air quality protection. I recommended as a condition of approval that the operator be required to establish and maintain at his cost an air quality monitoring station replicating those being administered by WYDEQ. I also recommended this be done for the life of the project through development and production, and that it be capable of continuous monitoring of ozone, NO_x,

SOx, and all meteorological parameters. I urged that sighting of this station should be in accordance with recommendations from WYDEQ, EPA, USFS air quality scientist Terry Svalberg, and myself.

I was seeking several objectives here. I wanted to insure that timely empirical data was collected to facilitate unambiguous assessment of impacts. Air quality, and visibility had to be monitored continuously. Potential toxic impacts to wildlife, plant species, and humans in low and high elevation habitats also needed to be monitored. I felt that such data would advance the state of knowledge regarding the poorly quantified realm of long-term effects upon flora and fauna. This would address what USFS had indicated to me privately was an area of serious uncertainty.

Finally, I commented that USFS had an opportunity to establish the beginnings of an empirical database regarding actual emissions from drill rig engines rather than modeled emissions. I urged that strong consideration be given to placing a requirement on the operator to implement an instrumented measurement program designed to collect empirical emission data on drilling rig engine performance.

I urged that such an effort span the entire envelope of load factors and encompass a statistically representative sampling of the rig emissions under those different load factors. A program like this, applied to even just one rig, would be extremely valuable but two or three would be priceless. Here I was attempting to get real information on how load factors actually influenced emissions history. I hoped to finalize the debate over that subject because it had been intense in the Air Quality Task Group that I discuss in a subsequent chapter.

As I stated earlier, Terry urged me to also write to the Minerals Staff for the Bridger-Teton National Forest. This I did at the same time. I addressed my comments to an official there whom Terry cited as the specific individual I should approach. I began by citing my earlier letter to Kniffy Hamilton and included it as an attachment. I then cited recent history of air quality degradation resulting from BLM's loose oversight of development in the Jonah and Anticline gas fields.

I explained that during the years since the first attempt to lease Wyoming Range lands, much cautionary data regarding air quality/visibility impacts to the Bridger-Teton airshed had been accumulated. Warnings were sounded and I was one of the most vocal but development had proceeded anyway. Now the Pinedale area was experiencing a series of ozone exceedance events.

These events were revealing the poor state of understanding about the dynamics of ozone behavior in our high desert environment. In fact, I had been first alerted by my BLM air quality contact in Cheyenne. Our ozone events were violating all the conventional wisdom about the subject. We were in unexplored territory of ozone science.

I argued that this history was important to USFS. Any further development of the region would exacerbate the problems. I warned that emissions from gas development sources that include ozone precursors could quite likely create situations in low areas and valleys that would present USFS with a mitigation nightmare.

I went on to explain that I had been engaged in a small project commissioned by local residents. It involved the placement of ozone detection badges from the Upper Green River Valley down to Bargerville and over to Big Piney. The project had been underway for only two of the six planned weeks of its initial phase but results were beginning to emerge. Very preliminary returns supported the probability that there were individual locations, primarily in low areas, where ozone was concentrating.

This would be in keeping with the facts. Ozone is heavier than air, VOCs, HAPs, and NOx emissions in the Jonah and Mesa regions were persistent, ozone exhibits a long lifetime in our winter temperatures (on the order of 12 hours), and stagnant air conditions in such locations were common.

These results appeared to contradict readings from WYDEQ monitoring stations at the time but that was suspect. The stations were few in number and their sensors were mounted high above ground. This likely resulted in non-sampling at ground level in possible high-risk areas. I suggested that these same conditions were everywhere in the Wyoming Range.

The Forest Service had recently published a report of ozone damage to forestlands of the northwest so I was sure Jackson based personnel were well aware of damage that could result. Add to this the specter of HAPs, VOCs and NOx creating unhealthy air conditions for birds, big game, and small game and there could develop a situation demanding remediation at tremendous expense.

Also, I noted that data was scanty regarding specific short and long-term impacts. Animal and bird populations would be thrust into the role of the preverbal "coal mine canary." Public outrage over this could become a public relations problem.

I then threw in a thought marginally connected to air quality. I opined that operations undertaken by drillers in the forest would necessarily result in destructive road networks resembling the maze present in the Jonah. Also, current prolonged draught conditions and risk of human caused fires from gas development and production operations would not be trivial. During the previous summer a forest fire in an adjoining state briefly threatened a gas field and I suggested that USFS did not want that to happen here.

I ended with a final caution. The Eagle Prospect project draft EIS had clearly revealed that in a partnership of "equals" between USFS and BLM, the latter agency had made it clear that in resolving any management disputes BLM would be more equal. I attached a summation of BLM failures to honor its own land management obligations and admissions confirming these failures. I cautioned that USFS could find itself a victim of such BLM behavior.

I thus maintained that the conclusion USFS needed to accept was this time an opportunity to prevent destructive development of forestlands before lease-out was within its grasp. Thus far, the legal rights of operators had blunted citizen opposition. They were allowed to develop a lease after its purchase as they saw fit but this situation could be eliminated under the current circumstance. USFS need only recognize that the political climate was due to change soon because of the upcoming presidential election and USFS could "just say no."

"No" was in fact declared eight years later.

Victories

I can close this chapter on an upbeat note. In 2012 there was a rare development that ended the Eagle Prospect saga. After heavy public resistance and the strongly contested string of environmental impact statements I have just described, the operator agreed to permanently abandon these gas leases in return for a public buy-out price of $8.75 million, due by December 31, 2012.[33] A coalition of conservation groups and citizens succeeded in raising the sum, thereby eliminating the threat of gas development in those particular lease parcels forever.

Furthermore, there was an equally satisfying outcome for the "44,000." In December 2016, USFS issued a public statement declaring the no leasing alternative in its most recent Draft Supplemental Environmental Impact Statement dated April 2016 remained the preferred alternative.[34] This decision was based upon more than 62,000 comments addressing the four available options. The decision was declared final and was signed by USDA Undersecretary Robert Bonnie 30 days later.[35]

I talked to Terry about this development in February 2017. I asked if the operators were mounting their usual storm of legal challenges and he said they had not done so yet and showed no signs they would. He also explained an added benefit derived from the USDA Undersecretary signing off on the decision and the timing. That official was at the top of the food chain so going over his head was not a promising option. Furthermore, the date of signing hopefully assured that the incoming Administration opponents of such decisions were blocked from easily overturning this one.

This rests on the fact that by signing, Undersecretary Bonnie finalized the rule, thereby placing it into a category normally requiring a complex process to revoke, or change it. This should entail months of due process involving review by the Agency, solicitation of public comment, review of comments, crafting a new rule that responds to the comments, and finally, dealing with likely litigation.[36]

Unfortunately the newly elected Republican-dominated Congress initiated efforts to legislate mass repeal of so called "midnight rules" signed during the Obama–to-Trump transition period; so finality of the Wyoming Range no-drilling decision may not yet be "in the bag."[37]

Finally, in a much earlier federal move in 2001 the Clinton Administration issued a rule prohibiting new roads in Forest Service administered lands.[38] The subsequent Bush Administration as well as a Wyoming legal challenge disputed USFS authority to do that. As a result, the Department of Agriculture announced in May 2005 that it was lifting the Clinton road rule. This opened up previously closed lands to oil and gas development that could further pit the Forest service and BLM against each other.

However, balance was restored when environmental groups counter-attacked with legal moves that resulted in a 2011 federal court ruling that upheld USFS authority to impose road restrictions. The decision was further supported in October, 2012, when the U.S. Supreme Court rejected a challenge to the Roadless Rule, thereby leaving it intact.[39]

Chapter 5

The Air Quality Task Group (AQTG)

I became involved at its inception with the Air Quality Task Group and over the course of three years learned the true meanings of frustration and bureaucratic stonewalling by BLM. I originally decided to participate in the AQTG for the specific purpose of ascertaining how serious the BLM would actually prove to be in regards to soliciting stakeholder input. I was skeptical and I would eventually be able to declare with authority that my skepticism had been valid.

The Air Quality Task group began in November of 2004 as a large roundtable meeting that included representatives from the Pinedale and Cheyenne BLM offices, the director of Wyoming Department of Environmental Quality (WYDEQ) John Corra, the U.S. Forest Service (USFS), representatives from at least two of the operators, and some citizens interested in membership. The BLM dutifully handed out a fistful of FACA guidance documents explaining the federal laws under which we were to operate, the conditions on how business would be recorded, qualifications to be met by applicants, and meeting protocols.[1]

The first need was for a chairperson and attention briefly focused on me. However, this all was new to me and I expressed reluctance to take on such a central role. Also, I feared the role would limit my ability to speak out publically. In due course, it was decided that Cara Keslar from the WYDEQ air quality division (WYDEQ-AQD) would act as a co-chair person. It was further agreed that the Forest Service representative, Terry Svalberg, would serve as a co-chair.

The AQTG Legacy

The AQTG started out with the best of high moral intentions intended to accomplish important goals by a dedicated and for the most part, constructive minded mix of participants. In the opening years, efforts of this group were focused, and determined. A lot of exploratory learning had to be experienced by all of us regarding federal requirements starting with the FACA and continuing with protocols dictated by BLM. The FACA hurdles were bad enough but the BLM requirements assumed the characteristics of a moving target.

Furthermore, WYDEQ inserted its own roadblocks through Cara who closely limited our activities in a manner that would not challenge WYDEQ's way of doing business. Also, in rigid pursuit of consensus she staunchly opposed allowing any dissenting opinions to go forward to the PAWG during her time as the task group leader.

Finally, the directors of WYDEQ and its Air quality division with whom we had to work were holding us at bay through administrative restrictions. Conversely, administrative restrictions on the operators seemed to be loosened in a manner that would cause less inconvenience for them in their rush to obtain drilling permits.

Despite the obstacles, the AQTG produced a comprehensive report in 2005 that was a prize of substantive science. It was also an informed recital of cause and effect regarding increasing air quality impacts locally and in the nearby Wind River mountain range. Drawing upon all of this good work, a series of important and directed recommendations rounded out the report.

It was with great frustration and outright anger that a year later we learned through an offhand comment from Pinedale BLM that our report had never been read. That immediately placed into suspicion our reason for existing. Had there been a genuine intent by BLM to involve public input or were we in fact political theater intended to only resemble public involvement?

As the years progressed, several of us increasingly questioned why we were having meetings at all. Little or no funding was being provided to even help us with travel. We were constantly being held back by our BLM liaison, Caleb Hiner, on grounds of charter restrictions. He seemed to be a classic example of what psychologist Stanley Milgram characterized in 1973 as a person who can assume a "state of agency."[2]

Paraphrasing Milgram's findings, this is a person with normal common sense suddenly shedding it the moment he crosses the threshold of his workplace. At that moment he assumes the identity of his bureaucracy and resists all perceived threats against it. And to add insult to our injury, we had requested him in the belief we would improve what had already been bad communications with BLM.

Slowly through attrition of our numbers as a result of discouragement, aging of members, and a growing sense we were being used for purposes other than what had been initially represented, our collective experience and corporate memory dissipated. The last meetings involved new members who evidenced personal agendas that only marginally coincided with the larger goal of reducing gas development impacts on the regional air mass.

As I have looked back and collected all the events surrounding the AQTG, I am amazed at how all the signs of BLM duplicity and WYDEQ turf protection were hiding in plain sight. Not until now after stringing them together in chronological order do the patterns emerge.

The Meetings

The following pages recount the proceedings of a long series of AQTG meetings. The events described have been extracted from the actual minutes that I and three other members wrote. In the course of my research into this part of my involvement with the issue of gas development impacts I had frequent conversation with Terry Svalberg and we combined our individual archives of meeting minute documents. In retrospect, I have reason to believe he and I now possess the most complete record of those minutes. I suspect this because when I found the public website where the BLM maintains those minutes, I was amazed to discover a total of only four documents and they only addressed meetings held in 2009.

If the reader thumbs through these pages, he might be inclined to suspect the content is too tedious to be endured. However, I urge patience and stamina because when you come out at the other end, you will have learned cautionary revelations about how the BLM and WYDEQ deflected the AQTG away from being an important influence on air quality assessment and impact mitigation and turned it into an impotent, ignored debate club.

2004

On November 3, 2004, the first official meeting of the AQTG convened.[3] I took on the secretarial task of keeping meeting minutes and this initial meeting resulted in a massive 20 pages of outline notes. They proved to be heavy with bureaucratic lists of ground rules, federal guidance requirements, AQTG task domains, consensus decision-making requirements, definitions of purpose, FACA familiarization, and the regulatory domain space of the Wyoming Department of Environmental Quality.

Attendees at this first meeting seemed eclectic. There were Susan Caplan and Terry who gave federal presence. Cara and her director of WYDEQ and his director of the Air Quality Division gave state presence. Kate Forsting who was an environmental chemistry specialist for a firm called Energy Labs gave an energy lobbying presence. Mike Golas from Questar Corp. and Jim Sewell from Rocky Mountain Shell Corp. gave operator presence. Jonathan Ratner from Wyoming Watersheds gave an environmental organization presence, while William Belveal and I gave a citizen presence.

My meeting minutes show that we listened to a BLM Public Relations representative drone on forever about the FACA. He recited part and subpart numbers of the code of federal regulations that discussed the FACA. Next, he recited the allowed task group objectives as determined by BLM. They involved assessing NOx emissions, developing an emissions monitoring plan, and filling administrative positions.

After that, we were tutored on the topic of consensus decision-making. I should have paid closer attention to this topic because it eventually revealed itself as a means of censorship by Cara in proceedings that would occur many months hence and by the PAWG chairpersons as well. The subject seemed innocent enough. It had to do with reasoned discussion protocols intended to resolve differences of opinion between members.

The fact that one hundred percent agreement would be mandatory should have triggered suspicion but I was still new at this. It in fact turned out to be a tool by which dissent was rejected and sidelined. Only one hundred percent consensus agreement was permitted to go forward to the PAWG in the form of assessments and recommendations. This consensus requirement was even invoked to disallow forwarding of minority opinions.

The core purpose of our activities was then laid out in the form of specific action items. We were to craft a plan for actual air emissions monitoring and methods. This became detailed to the level of measuring concentrations by mass per unit volume of air and deposition rates in both aquatic and terrestrial local environments. Emissions sources were to be identified by what pollutants they released and at what rates.

We were to recommend locations for monitoring sites, what state or federal agencies should perform the monitoring, and from whom funding should be requested. Lastly, we had to decide where our boundaries were. Would we include the Bridger Wilderness area? The Fontenelle area? The La Barge area? Some of these were active gas fields or federally protected regions, all of which reside in Sublette County. However, we were chartered under the Pinedale Anticline Project Area (PAPA) and that soon became a roadblock.

Our charter became the recurring point by which the Pinedale BLM office representative and the Shell representative routinely pulled the group back from meaningful actions regarding air quality impacts that respected no boundaries. We logically found that emissions from the PAPA and emissions from the Jonah field next door would not honor the administrative dividing line between them so we wanted to address both. However, the Shell representative continually opposed any such thinking and BLM adamantly objected that we were straying away from our tasking. As a result, we were always jerked back by the leash of the PAPA charter.

The WYDEQ-AQD air quality division chief then described his domain regarding its role in overseeing and regulating emissions that degrade air quality. He advised that he had received EPA approval of an implementation plan. This came as a result of EPA concluding that his agency had the resources, technical ability, and legal foundation to enforce the requirements of the Federal Clean Air Act. Time would demonstrate that all these assurances were dubious. In truth, it would seem that AQD did not have either the political will or technically competent staff to deal with the issues that grew almost geometrically with the spin-up in drilling that was to come.

The AQD monologue went on to state that development impacts would be evaluated through the maintenance of an emissions inventory as well as an inventory of sources. He was categorical that these inventories would support and confirm modeling to determine what might happen as development continued.

Looking back on this part of the meeting is annoying to say the least because emissions inventories became the focus of my battles with WYDEQ and BLM due of their failure to actually execute. As the reader will find in the science chapter of this book, I cite failures of the emissions inventory process because it never actually was implemented in an independent and field measured manner. Instead, it was compiled from model predictions offered in operator drilling permit applications. I will argue in the science chapter these were seriously in doubt.

Furthermore, The inventory of sources was never a specific entity as far as I could determine. All WYDEQ seemed to do was accumulate reams of paper in the form of permit application documents that were filed in rows of metal cabinets. These were in its Cheyenne offices where I spent scores of man-hours combing through them for well and emissions data.

AQD went on to declare it was the state's enforcer of the Federal Clean Air Act (CAA) and named the six specific pollutant species (called criteria pollutants) for which there were ambient standards of reference. The species named were nitrogen oxides (NOx), Sulfur dioxide (SO2), particulate matter, lead, ozone, and carbon monoxide (CO). These species were all that AQD would and could address.

I was to later discover the importance of this regulatory nuance when on one occasion I watched AQD confuse the County Board of Commissioners with one of its air quality monologues.[4] AQD's director was declining to address certain additional pollutants of concern to these commissioners. They did not understand why he was resisting so I felt compelled to intervene and coach AQD to do a better job of explaining the relevance of "criteria pollutants."

The AQTG meeting continued with tutorials on the subject of allowable incremental increases in criteria pollutants under the Federal Clean Air Act and other statutes. The premise here was that certain levels of the criteria pollutants already existed in the area. It would be necessary to empirically establish those levels and then determine the rate of incremental increase resulting directly from drilling operations.

The CAA stipulates that certain maximum thresholds of these pollutants cannot be surpassed so it was necessary to discover how much "space" existed between the existing levels and the maximum allowed. Then in theory, management efforts could be executed to hold down those new incremental increases so as not to exceed the maximum allowable.

Here again in retrospect, I was to discover the skillful tactics that the operators and regulators could invoke to stay within the law.[5] During the coming months I acquired clarity and realized a crucial aspect that would go missing, i.e., implementation of a timely measurement program would be crucial to setting pollutant baselines and also to clearly establish the rates of emission taking place on the Anticline.

It would soon come to pass that I would be hammering AQD publically for its delay in establishing its measurement sites on the Anticline. As a consequence of that delay, the official baseline values were not established until 2005 when they could have been and should have been fixed in 2000. By 2005, pollutant levels had already been seriously elevated by development operations in both the Jonah and the PAPA.

Next in the discussion was the topic of visibility impairment. Four years into the future this issue would become a major topic of conflict next to that of ozone. Susan Caplan confirmed that the Clean Air Act required states to protect visibility but there was ambiguity surrounding how much and by what regulatory methods. This would become a complicated issue because there were no previously established local visibility standards against which to measure what was to develop in our region

Here again time would reveal how these weaknesses would complicate any regulatory efforts. They would take the form of differing acceptable levels of impairment recognized by the BLM and the Forest Service and insistence by BLM and WYDEQ on modeling subsequent impairment rather than actually getting into the field and measuring it. This will be better discussed in the science chapter.

The AQD director went on to describe the state's plan to "manage air resources."He explained the practice of requiring drilling companies to obtain a permit to drill before they could proceed and their obligation to provide WYDEQ with data on their emissions via their own monitoring program. He further described the AQD in-house modeling capability and then described the basics of the WYDEQ compliance program.

A few years previous, I had walked the halls of WYDEQ talking to staff members about the subject of state response plans to address a possible nuclear materials spill on the interstate highway. I was regaled with the details of an intricate plan of action that existed only on paper and had never been actually tested in the field. Now I was hearing similar glowing descriptions of plans to regulate air quality threats in our county.

Time would reveal many caveats and exceptions as well as WYDEQ reliance upon voluntary compliance from the operators in the most crucial elements of those plans. Even worse, the time would come when one operator in particular would successfully persuade WYDEQ to adopt its proposal for essentially a cap and trade scheme which would weaken all the emissions control efforts that had come before.

The final relevant element of WDEQ's air resource management effort to be explained was its program to require operators to employ best available control technologies (BACT). This would become one of several WYDEQ firewalls against actually regulating the operators. WYDEQ staff lacked the technical ability or time to track BACTs in any detail so they instead relied upon the operators to make such technologies known and voluntarily employ them.

It should have been no surprise that the operators minimally complied and argued that the newest BACTs were too expensive. Instead they tweaked the old standard methodologies. In the science chapter, I will tell the story of a ground breaking dehydrator technology for which I facilitated field testing but WYDEQ killed through what I regarded as incompetence and bureaucratic "not discovered by us" stonewalling.

The remainder of the meeting was consumed by a variety of tutorials and presentations. One tutorial was in the form of a long discussion by the AQD director about a previous air quality evaluation project that had encompassed all of southwest Wyoming ten years before. That was followed by Cara's presentation of plans to expand her air quality monitoring station network (the network I previously cited as too little too late) and its place within the other monitoring networks operated by the Forest Service and a few operators.

Jim Sewell of Shell gave a presentation that sang praises of his company's own monitoring efforts and movement toward "green" well completion. This would become a buzz phrase for capturing fracking fluids coming back out of the borehole as opposed to burning the stream in the open atmosphere, a practice called well completion flaring.

The next presentation was my own in which I described my empirical field measurement project. I described how I had noticed declining solar ultraviolet light levels and speculated the cause was emissions from gas operations blocking them. I also described how I was measuring the presence of lithium, potassium, and sodium in well completion flares by means of optical spectroscopy.

Susan Caplan completed this part of the meeting by talking about her fledgling project to model the local atmosphere on a desktop computer. After her presentation all subsequent discussions were administrative in nature and this marathon meeting was finally adjourned.

I have recounted the details of this first meeting (perhaps too much) because they contain the essential foundation of what many of us initially believed was to be the roadmap ahead. We were very concerned over what was already a high drilling rate underway essentially without any serious attention by the agencies responsible for regulating the activity.[6] Thus, the reader needs to keep these details in mind because throughout this book, I recount events and decisions that violated nearly every assurance promised in this meeting.

R. Perry Walker

• • • • •

Our next meeting was convened three weeks later on November 30, 2004 and the number of people in attendance was much reduced. The same state, federal, and corporate representatives were there as well as the two of us who were representing the general citizenry.[7]

One of our first concerns was the lack of a BLM presence. We felt that any results we were to generate ultimately must go forward to the BLM Pinedale office through the PAWG. It seemed logical to have their presence in our meetings. Sadly, this soon yielded an outcome that fell into the realm of the old adage "be careful what you wish for."

In response to our request, a recently hired college graduate and former Pinedale high school graduate, Caleb Hiner was given the position. I knew this young man from when I was employed earlier as a substitute teacher and was told of his having graduated from the Uuniversity of New Mexico with public lands and wildlife management specialization. That made me comfortable with his appointment. However, as time progressed and we routinely grappled with thorny issues, he rapidly revealed his own institutional mindset.

He constantly intervened to protest that we were in violation of some charter limit or statutory restriction without offering any alternative solutions. He basically came to be a watchdog who almost regularly declared "you can't go there." Several years later I learned he had apparently managed to master the art of federal promotion ladder climbing. He had been appointed to the position of field manager of the Pinedale BLM field office replacing Shane Deforest.

This meeting was also when we were informed that members who had to travel more than 50 miles to attend would be eligible for per diem and travel expense reimbursement. This compensation would be provided by the BLM at the current approved government rates. As I described in previous pages, however; this benefit was terminated a few months later under the excuse that PAWG guidance from higher levels had been "misunderstood."

By the Rasping in My Lungs

It was noted that the EPA had signaled interest in participating in our meeting by assigning a representative from the Region 8 EPA headquarters in Denver, CO. That was immediately accepted as a good thing. Here again, however, expectations did not match deeds. I explained this detail in the section on EPA involvement with Anticline development events.

The major business of this meeting had to do with yet more federal environmental laws and tutorials for purposes of familiarization. Here again, I was so new to the business that I soon felt overwhelmed. However, only a year later I would be citing passages from some of the more directly relevant statutes such as the Clean Air Act and doing so with confidence.

The topics we reviewed started with the National Ambient Air Quality Standards or NAAQS and the analogous Wyoming Ambient Air Quality Standards or WAAQS. The Federal Clean Air Act requires EPA to set national ambient air quality standards for pollutants considered harmful to public health and the environment. Two types of national ambient air quality standards are defined.

First are primary standards which provide public health protection, including protecting the health of "sensitive" populations such as asthmatics, children, and the elderly. Secondary standards provide public welfare protection including protection against decreased visibility and damage to animals, crops, vegetation, and buildings. EPA has set National Ambient Air Quality Standards for the six criteria pollutants which I named earlier.[8]

Next were the PSD increments which stands for "Prevention of Significant Deterioration." This also would become a slippery subject that would offer ingenious operators opportunity to advance their field programs. Ultra Petroleum was one such example I discuss in the chapter on the operators.

Drawing from EPA explanatory materials, PSDs are incremental increases of criteria pollutants in the air from "... *existing sources in an area where attainment exists or cannot be classified with NAAQ. PSD's do not prevent sources from increasing emissions... their purpose is to protect public health, and welfare, preserve, protect, and enhance the air quality in national parks, national wilderness areas..., insure that economic growth will occur in a manner consistent with the preservation of existing clean air resources; and assure that any decision to permit increased air pollution in any area to which this section applies is made only after careful evaluation of all the consequences of such a decision and after adequate procedural opportunities for informed public participation in the decision making process."*[9] Reciting this aloud causes one to run out of breath.

The BLM and WYDEQ would invoke NAAQS and PSD's very heavily as justification to approve environmental impact assessments coming down the pike. While reviewing our history on the AQTG together, Terry laid a big surprise on me. As the BLM was trying to fast track the environmental impact assessment for the Anticline development project, he became concerned. He noted that BLM was applying PSD statistics within its assessment in a way that sidestepped protection of air quality.

His concern was due to the fact that the assessment would be the basis for the final EIS. After getting nowhere within regular channels he contacted the EPA regional office in Denver and warned that BLM seemed to be operating on the premise that it could as Terry put it, "screw up the atmosphere and then fix it later." EPA then stepped in and advised that was not an acceptable strategy.

This was the affair I related in the chapter about the USFS. Recall that the angry deputy director of BLM in the state office contacted a person in the Forest Service demanding a letter of reprimand against Terry. Such was the level of dysfunction existing between BLM, USFS, and WYDEQ, as the drilling projects around Pinedale moved forward. Now back to the events of the November 30, 2004, meeting.

We went on to talk about the National Environmental Policy Act or NEPA at some length because it was a central statute with which BLM had to comply. The key aspect of NEPA as far as our local gas development projects were concerned was the requirement that federal agencies include environmental considerations in their planning and decision-making through a systematic interdisciplinary approach. Specifically, all federal agencies are to prepare detailed statements assessing the environmental impact of and alternatives to major federal actions significantly affecting the environment.[10]

Our immediate discussion was about Questar's pending Year Round Drilling Proposal. Most of us around the table felt the proposal was well crafted but it was also felt that BLM had failed to perform the proper level of analysis. It was specifically noted that model runs cited in the Record of Decision depicted dubious cumulative impacts to visibility in several Class I areas.

Remember, Class I areas are defined as special areas of natural wonder and scenic beauty, such as national parks, national monuments, and wilderness areas, where air quality should be given special protection. These areas are subject to maximum limits on air quality degradation in the form of PSD increments and are more stringent than the NAAQS.[11] This was of importance to us because the Bridger Wilderness in the nearby Wind River Mountain Range is one of 156 Class I areas designated nationwide and we wanted that to be protected.[12]

To preamble the next paragraph, recall that an item called the deciview scale expresses diminishing contrast of a scene according to a complex logarithmic formula. In simple terms, 0.5 deciview is the minimal perceptible change in visibility to the human eye. The difference between 0.5 and 1.0 deciview can definitely be noticed when actually observed in the field.

The USFS reviewed the days of modeled cumulative impacts greater than 0.5 deciviews and determined that the cumulative impacts from the Pinedale Anticline Project, combined with other proposed projects in southwest Wyoming would significantly impact visibility in the Bridger Wilderness. This was one of several differences of policy between the many agencies involved that would routinely impede meaningful corrective action. In this case, the USFS used a standard of 0.5 deciview as its minimal perceptible change in visibility to the human eye but the BLM adhered to its 1.0 deciview standard.

Nevertheless, USFS determined that based upon PSD's and consideration of timing, magnitude and duration of the remaining projected cumulative visibility impacts, those impacts would be within an acceptable range.

This seemed a strange opinion due to inherent contradiction. USFS noted that the Anticline Record of Decision estimated NOx excluding the Ultra reductions would be 1693.62 tons/year and would constitute a significant visibility impact. On the other hand, the Questar Year Round Drilling Proposal Finding of No Significant Impact (FONSI) estimated NOx emissions at 1895.26 tons/year, creating no significant impact. This kind of debate continued through out the history of gas field proposals around Pinedale but BLM was rarely brought into check over its lenient attitudes.

Yet another revelation that would become routine had to do with the rig count on the Anticline. The Anticline EIS had been crafted on the basis of an estimated eight rigs drilling at any one time. However, during the summer of 2004, there were actually 32 rigs operating.

It turned out that BLM had badly estimated several classes of activity. It (1) failed to consider the cumulative effects of increased levels of development in the PAPA and the Jonah fields, (2) failed to consider the cumulative effects of increased levels of development in the Big Piney/La Barge field some 30-40 miles away, (3) failed to correctly estimate the total horsepower that drilling rigs would be operating under (this related to engine emission volumes), and (4) relied upon a FONSI level of analysis that was likely insufficient for a project on the scale of the Questar Year Round Drilling Proposal. Finally, documentation was lacking on how BLM arrived at the finding of no significant impact.

Ultimately, the task group concluded that WYDEQ-AQD had to make the PAPA and Jonah Fields top priorities for its State Wide Emissions Inventory after which the inventory should be expanded to the rest of Southwest Wyoming. The group agreed that an emissions inventory was needed before it could be determined how proposed new projects would affect air quality.

An associated issue came up regarding tracking of NOx emissions. In March 2000, a joint agreement had been crafted between USFS, WYDEQ-AQD, EPA, and BLM whereby BLM was to monitor and maintain a record of the presence and growth of NOx resulting from operations in the Jonah and Anticline gas fields. Inexplicably, BLM did not follow through with this obligation and as a result, the opportunity to establish baseline values for NOx early in the life of the two gas fields had been lost. For me, this was especially infuriating because those same agencies decided behind closed doors to set year 2005 as the baseline year for NOx and other emission species.

We ended our meeting by reviewing the elements of what we deemed to be a necessary and functional emissions monitoring plan. We realized that historical monitoring data should be included in the plan because there were a number of sites that had been operated by USFS for many years. It was felt this data could be invaluable in establishing the baseline situation prior to onset of drilling in the Jonah and Anticline.

We worried that USFS agreements with industry for funding these sites and the USFS Bulk Precipitation Monitoring sites would expire on December 31, 2004, which was fast approaching. Luckily, USFS has been able to piece together funding for these sites through September 2005, but beyond that, funding was uncertain.

We also agreed that visual scene monitoring under a process called the IMPROVE Protocol should be re-established at a location in the Wind River Mountains called Fortification Mountain. There were cameras available and the cost was considered to be minimal. The program had to be re-established though because it had been allowed to languish from lack of interest. However, now with visibility in that Class I region becoming a priority concern, it needed to be activated again but we had to determine the costs of re-establishing the program.

A quick explanation of IMPROVE is in order here.[13,14] When the 1977 Clean Air Act was created by Congress it was crafted to include emphasis on the issue of visual air quality in the Class I national parks and designated wilderness areas. Realizing that uniform haze impairs contrast and visual distance, legislation was included in the Act to prevent future impairment and remedy existing impairment in Class I areas.

To assist in the implementation of this legislation, the IMPROVE Program was initiated in 1985. It started a long term monitoring program designed to establish current visibility conditions, track changes, and establish the causes of monitored impairment. That visibility impairment came to be quantified through the development of the deciview scale.

A dominant theme of our meeting should be obvious……worries over funding. USFS had for years operated a number of monitoring sites that measured pollutant gases and particulates in the air as well as pollutant deposition in the region's alpine lakes. However, this always depended upon a tenuous funding stream that had to be squeezed from already stressed budgets. That stress resulted in USFS essentially having to beg from the operators on the basis that they would be helping clarify the state of the atmosphere in their own defense.

Despite all of our worries and unknowns, our meeting produced several meaningful recommendations for forwarding to BLM. They were:[15]

1.) Request more BLM involvement in the Air Quality Task Group.

2.) Ask for clarification from BLM on emission levels that were considered significant impacts in PAPA EIS, but identified as not significant in the Questar Winter Drilling Proposal FONSI.
3.) Require BLM to complete its NOx inventory for the PAPA and Jonah.
4.) Stress to WYDEQ-AQD that PAPA and Jonah be top priorities for state-wide emissions inventory and that the inventory be expanded from there to the rest of southwest Wyoming.
5.) Stress to BLM the importance of maintaining monitoring sites at South Pass and nearby Gypsum Creek; also maintain their long-term records as part of the monitoring plan for southwest Wyoming.
6.) Re-establish Scene Monitoring at Fortification Mountain and determine the cost of set up and operation of the site.

A few weeks later on December 14, 2004, co-chair Keslar sent all members an email containing her views on what had transpired in the November 30 meeting.[16] This would be our first indicator of her political correctness mindset that would become a form of WYDEQ censorship on the AQTG. Furthermore, this was the start of turf conflicts between WYDEQ, BLM, and to lesser degree, USFS.

First, she commented that AQD's statewide 2002 inventory effort has been underway for several months and the strategy for collecting data had already been established. That strategy was to collect data by source category (i.e., major stationary point sources, oil and gas operations, fire, etc.), instead of by region. The inventory was not yet final and, therefore, not available. Basically, she was saying we were late to the party and we had no say in the matter.

Then she addressed the group's intent to make BLM live up to its NOx tracking commitment in the Anticline and Jonah fields and surrounding vicinities. She explained that AQD had completed it's tasks specified in the Letter of Agreement for the year 2004 and had sent it to BLM the previous week. She then confirmed that, BLM had not fulfilled its responsibilities under the

Letter of Agreement since December of 2000. She further confirmed that tracking NOx was specifically addressed in the adaptive management appendix (Appendix C) of the Pinedale Anticline Record of Decision (ROD). Therefore, BLM's failure to complete this task for the previous three years would be an appropriate issue to discuss with the PAWG.

But then she waxed bureaucratic. She proceeded to warn that it was important to keep the Air Task Group's role in mind during discussion and formulation of action items. The Air Task Group was an advisory group to the Pinedale Anticline Working Group. The Air Task Group advised the PAWG on how to monitor the impacts of the Pinedale Anticline Project development on the air resource; the PAWG in-turn *recommended* actions to BLM. She was, thus, unhappy over the action item *requiring* BLM compliance with the NOx tracking commitment.

She declared that we could "request", "suggest", "encourage" or "strongly recommend", but we could not "require." She suggested instead that we, "strongly suggest to the PAWG that BLM reinstate NOx tracking under the Amended Letter of Agreement for Tracking Nitrogen Oxide Emissions as soon as possible."

I was quite annoyed by this attitude and shot back immediately with my own email to the members. I disagreed with her implied premise that because we were "only advisory" in nature, we must speak in whispering tones. By her own statement BLM had failed to execute and she stated clearly that, "BLM had not fulfilled their responsibilities under the Letter of Agreement since December of 2000." I then suggested there were ways for us to express ourselves in firm tones and yet avoid the appearance of making demands.

I also disagreed with an objection she expressed regarding our action item seeking clarification of the conflicting BLM statement about NOx levels as perhaps "... falling outside our purview...." Recall that this conflict resulted because BLM declared in the PAPA EIS that NOx emission levels would create "significant impacts" but then declared those emissions as being "not significant" in its subsequent FONSI approval document for the Questar winter drilling proposal.

I argued that her objection regarding our action seemed to imply that we had no cause to question BLM's conflicting statements. I countered that this should be a legitimate line of investigation because it went directly to the supposed purpose of the task group validating emissions threats to air quality.

I quoted her own statement that we were to "... advise the PAWG on how to monitor the impacts of the Pinedale Anticline Project development on the air resource." I then pointed out that to do so, we had a legitimate need to ascertain why BLM changed its view. That could drive our perception of what levels of primary pollutants qualify for concern in its management scheme and as a result drive our notional construct of an air monitoring system that would offer maximum utility.

She also expressed concern over our desire to take into account the other areas of gas development in the county that were contributing their NOx emissions to the regional airshed. She insisted that we must address the Jonah and Anticline impacts on the Bridger airshed in isolation from the surrounding additional gas fields in the immediate region.

I countered that this would be ineffective science because to understand Jonah/Anticline impacts on the airshed, we must take into account the large gas fields that were frequently upwind from us in the Kemmerer-La Barge-Big Piney complex. This might be possible on a limited basis using the EnCana weather monitoring station. While it was in the wrong place to sample the emission history of Jonah, it just might catch the flow of emissions riding on westerly and southwesterly winds from the gas fields on the western edge of the Sublette basin.

Given the tenor of Cara's comments, I sensed a need on our part for more definitive, justifying language to be part of each action item or finding. Drawing from the above exchange I offered the following suggestions:

> 1. Request more BLM involvement with the Air Quality Task Group. In order for the Task Group to execute its responsibilities, critical input from a knowledgeable BLM air quality expert is necessary. Such input would facilitate our understanding of constraints, limitations, and priorities BLM feels must be considered which might derive from influences of budget, manpower, or statutory guidance.
>
> 2. Seek a location with conference call capabilities for the next Air Quality Task Group meeting. This would mitigate any limiting influ-

ences on our ability to effectively address subject matter that the absence of certain members might render incomplete.

3. Ask for clarification from BLM on emission levels that were considered significant impacts in PAPA EIS, but identified as not significant in Questar Winter Drilling Proposal FONSI. In order for us to correctly advise the PAWG, we need to ascertain why BLM changed its view; that could drive our perception of what levels of primary pollutants qualify for concern in its management scheme and as a result drive our notional construct of an air monitoring system that will offer maximum utility.

4. Remind the PAWG that BLM has failed to fulfill its responsibilities under the Letter of Agreement since December of 2000 in which tracking NOx is specifically addressed in the adaptive management appendix (Appendix C) of the Pinedale Anticline Record of Decision (ROD). It is, therefore, imperative that the PAWG endorse in the strongest possible terms the need for BLM to immediately implement a NOx tracking program under the Amended Letter of Agreement for Tracking Nitrogen Oxide Emissions so that a baseline compilation of data can get underway before further gas development activity renders that baseline even further in arrears.

5. Stress to WYDEQ-AQD that the PAPA and Jonah be top priorities for a state-wide emissions inventory and that the inventory be expanded from there to the rest of southwest Wyoming. We recognize that there are other regions of Wyoming in which air quality concerns are growing and elevation of our region in priority may be difficult. However, we feel it can be legitimately argued that our close proximity to four designated Class I protection zones and key importance placed upon such regions in the National Clean Air Act justifies the request.

6. Stress to BLM the importance of maintaining NADP sites, (WY-97 South Pass, and WY-98 Gypsum Creek), and their long-term records as part of the monitoring plan for southwest Wyoming. Re-establish Scene Monitoring at Fortification Mountain. Determine the

cost of set up and operation of the site. With the exception of Fortification Mountain, these facilities are in locations that offer the advantage of providing already in-place elements of a potential Task Group notional air quality monitoring system, the continued operation of which would extend the accumulation of a rare long-term baseline data archive.

All of the points involving the BLM ultimately went nowhere in any form because the agency stood firm in its stonewalling about its obligations. I was starting to understand that federal law such as the Clean Air Act could be sidestepped if no higher authority would challenge a violator for non-compliance.

A side bar issue that was informally talked about in the meeting had to do with legality of participation by Terry as a USFS representative. Cara had sent us an email advising that this issue had been discussed during a conference call between her and the Pinedale BLM office as well as the other co-chairs of the various task groups.[17] The BLM specialist on FACA protocols had questioned the correctness of Terry's participation as a co-chair for our task group. It was apparently decided that the PAWG had jurisdiction over the matter and was responsible for defining operational procedures of the task groups; so the issue was referred to the PAWG for a determination.

When I saw this email, I was displeased to say the least. I immediately did my own fact checking and ultimately sent out my own email.[18] I explained that I had done my own research and as a result cautioned against over-interpreting ourselves out of the opportunity to involve the best people in our group. The full text of my response was as follows:

> *When Terry informed us of Cara Keslar's caution that he could not participate as a member of our task group other than as an advisor and facilitator, I have been concerned that we could be deprived of an important element of knowledgeable participation if her interpretation is correct. Accordingly, I have closely studied the content of the document titled United States Department of the Interior Bureau of Land Management, State of Wyoming Charter defining the PAWG and Federal Advisory Committee*

By the Rasping in My Lungs

Act, Subpart 1784, Sections 1784.0-1 through 1784.6-2 (both of which were hand-outs in our first meeting). Having done so, I suspect that Cara's comment as reported to the TG membership is incorrect.

Terry stated: - "Group Membership: Cara pointed out that in the Charter it states that Forest Service Employees are not to be members. Therefore, from now on I can not and will not weigh in on consensus decisions. I will be present as a technical advisor and facilitator. If we decide Ted should remain as note taker, he will also be shown as a technical advisor and note taker. Being shown as a technical advisor will allow us to participate in discussions where the public may not be allowed."

Admittedly, this quote is probably not a literal repeat of her comment but we can only go on what we have been told. I find no instance in the Charter in which specific reference to "Forest Service Employees" is made nor, to the best of my reading, can I recognize an indirect reference. Indeed, the only explicit reference to the task groups is in Section 13. "Task Group Meetings" and that does not address membership qualifications. If inferences are being drawn from the definition of PAWG membership, I submit that such inferences are incorrectly rigid. But if such a connection is to be insisted upon, I insist with equal determination that the PAWG member described as "a representative from a state wide or local environmental group" can be interpreted as someone from USFS on the grounds that:

Section 165(d) of the Clean Air Act confers on the U.S. Forest Service, as a Federal Land Manager, an affirmative responsibility to protect the air quality related values (including visibility) of any such lands within a class I area. 42 U.S.C. § 7475(d)(2)(B). In Wyoming, this responsibility extends to protecting the visibility of all mandatory Class I areas, including the Bridger, Popo Agie, Fitzpatrick, North Absoroka, Washakie and Teton Wilderness Areas.

Furthermore, there appears to be ample provisions in the Federal Advisory Committee Act, Subpart 1784 that justify and permit active membership by both Terry and Ted. In this document are the following statements:

Section 1784.0-2:
The objective of advisory committees...is to make available to the Interior and the Bureau of Land Management the expert counsel of ...knowledgeable citizens and <u>public officials</u> regarding both the <u>formulation of operating guidelines</u> and the preparation and execution of plans and programs for use and management of public lands, and the <u>environment</u>." (emphasis my own)

Section 1784.0-6:
... it is the policy of the Secretary to establish and employ committees representative of major citizens' interests...to advise....regarding policy formulation, decision making, attainment of program objectives, and achievement of improved program coordination and economies in the <u>management of public lands and resources</u>... (emphasis my own)

Section 1784.2:
Each advisory committee...shall be formed with the objective of providing representative counsel and advice about public land and <u>resource planning</u>, retention, management, and disposal. (emphasis my own)

Section 1784.6-1:
Members of general-purpose subgroups shall be representative of the interests of the following three groups (I named only the one I felt was most relevant): 2. Persons representing – (i) Nationally or regionally recognized environmental organizations.

USFS presence on our TG in the form of Terry and Ted can be defended on the grounds that they qualify as "public officials," and *"providers of expert council regarding both the formulation of operating guidelines and the preparation and execution of plans and programs for use and management of public lands, and the <u>environment.</u>"* Additionally, they are representative of *"major citizens' interests" (air quality in the Class I areas) and are uniquely positioned to advise regarding "policy formulation, decision making, attainment of program objectives, and achievement of improved program coordination and economies in the management of public lands and resources."* Finally, they arguably DO represent nationally or regionally recognized environmental organizations (the USFS) – please note that no definition of *"environmental organization"* is offered at all, let alone one that precludes interpretation of USFS as such, and it IS nationally recognized as a high profile steward of environmental resources.

I was much relieved and a little surprised when Mike Golas responded in support of my arguments. He wrote to me the following reply:[19]

Your comment "Let's not make the mistake of over-interpreting ourselves out of the opportunity to involve the best people in our group" reflects exactly my view of the task group opportunity and our charge to use the best available information and resources to work together toward the best solutions. Besides, who would challenge the design of the group? The oil and gas companies possibly but I have no inkling of that happening and, speaking for Questar, that approach would be contrary to our basic approach to making progress in the complexities of Pinedale development. Furthermore, I hope that these organizational discussions are not diluting our resources.

Now all that could be done was to wait for the PAWG decision to be handed down.

• • • • •

R. Perry Walker

2005

We next came together on January 4, 2005.[20] All the usual members were there plus members of the public. The latter consisted of persons from Duke Energy (three persons), Trout Unlimited, the local newspaper, a local environmentalist, and a Pinedale BLM observer.

Our big topic of discussion was the draft monitoring plan we were to produce. A flurry of scope related issues were placed on the table. Questar expressed reservations over any thoughts that the plan should extend beyond the PAPA and the Jonah field and suggested that any funding from industry might be a hard sell if we strayed beyond those two gas fields. Terry felt the plan needed to illustrate "the big picture" and then break out the elements falling expressly within the task group's charter. Jonathan Ratner opined that there might be a way for the group to establish our boundaries. He suggested assessing the regulatory needs and operating efficiency of existing monitoring projects in the region as a route to scientific defensibility of our own undertaking.

As usual, the issues of funding intervened. Questar suggested that we may need to concentrate on funding more than science and not allow existing programs to go unfunded while simultaneously creating a new program. Several members worried that outside technical expertise would be needed to assist in crafting a practical monitoring plan but how would that be paid for? How could we hope to engage in solicitation and managing funding grants, agreements, and other funds?

To address these issues, it was proposed that we create a matrix of current and historical monitoring programs to discover what data was available, what it addressed, and how we might use it. Our EPA member advised that his agency had numerous guidance documents available that discussed monitoring equipment, protocols, and sighting of equipment. They should be considered as direction for our monitoring program.

As we narrowed in on what we thought were emissions sources within our charter, i.e., those from the PAPA gas field, Mike Golas piped up with a complication. He seemed to issue a veiled threat that if individual committee members were going to claim there were no emission sources other than those from oil and gas development in the PAPA or nearby areas, then he would insist that our monitoring plan look farther afield at numerous, and rather obvious other sources. His point was that if we degenerated to finger pointing then there would be plenty of other emission sources we would have to consider in setting up a proper monitoring plan.

The problem with that view was that it ran contrary to what the BLM would tolerate. As time went on, the BLM and even our WYDEQ co-chair would become ever more restrictive as to what they would accept as falling within our charter. We would increasingly be constrained by BLM's tight grip on our leash and to the rigid definition of consensus agreement that the WYDEQ co-chair would come to impose on the group.

William Belveal, our public representative counterpart worried aloud about the visibility in the Upper Green River Valley and later questioned what WYDEQ was doing to regulate well completion flaring and the resulting emissions. Cara replied that her office was in the process of rule-making to address that issue.

This seemingly simple reply was a revealing point about WYDEQ and how it would be in the vapor trail, so to speak, behind industry for months to come. From the beginning WYDEQ was in a mode of catch-up. Operators were pushing hard on BLM to approve their drilling agenda. The result in the field was growth in drilling and production infrastructure that outstripped the glacial pace with which the regulators were accustomed to operating.

The last piece of information to be placed on the table came from Susan Caplan. She informed us that the PAPA Record of Decision had estimated emissions of NOx to be 693.5 tons per year but in fact they were just under 2000 tons per year. Furthermore, the total number of sources (wells under development) in the Jonah and Anticline were "much greater than what had been analyzed" in the BLM environmental impact assessments.

Another few years would have to go by before the importance of this admission could be realized and understood. Eventually in the PAPA EIS, BLM-Pinedale would admit that this low-ball result had been the product of its conclusion that maintaining oversight of the growth of well development posed too great a management burden, so they simply stopped trying.[21]

The meeting concluded with comments from visitors of the general public who were present. The visitor representing Trout Unlimited, pointed out that the 1999 PAPA ROD referenced emission levels that had been analyzed as part of the initial PAPA EIS. She recommended that those data be designated as the baseline for future reference. In retrospect, this comment seems prescient because it preceded my own outrage in 2005 when I learned that the federal and state regulators had elected to designate that year as the baseline year for future emissions assessments.

• • • • •

On January 25, 2005, we sat down for our next meeting in an effort to make some progress on the issues and problems that had been highlighted three weeks earlier.[22] A big item was the request for BLM to re-activate NOx tracking but that proved a dead end. Susan Caplan from the state BLM office advised that no funding was available for this nor was there any date offered for when money would be available; additionally, BLM had not even decided if it wanted to re-activate NOx tracking.

Our second big item was our search for funding. We had asked the PAWG to request funding from appropriate entities. Terry had already written to the director of the air quality department of WYDEQ requesting such assistance. To our surprise, that director, Dan Olsen, agreed to release $32,000 for the purpose of maintaining operation of existing monitoring sites. That sum would keep them running until September 30, 2005.

Another big item was Terry's continued presence on the task group. In an almost "Oh, by the way" manner, it was announced that he had been cleared to continue with us. The justification was that this had been decided by the PAWG which possessed the authority to decide such matters. Privately, I was informed that in fact, the BLM specialist on FACA was wrong with his restrictive interpretation. The rule he was invoking applied *only* to the PAWG membership and not the individual task groups. Thus, USFS active participation in our group had been saved.

When we grappled with the ongoing effort to craft our monitoring plan, William Belveal again brought up the issue of well flaring and the growing discernable effect that was having on visibility. Cara replied that flaring was monitored and permitted on a company-by-company basis. In my view, "monitored" was the WYDEQ euphemism for doing nothing.

WYDEQ was simply issuing permits for emissions volumes declared by the operators and filing those permits in a cabinet. Essentially, the process was implying that paper on file solved the problem. This book keeping-instead-of-action attitude would seem far too lenient and ineffective as future events and my own field research on chemicals contained in well flaring smoke plumes would establish.

We next proceeded to consider and accept or modify a number of sections in the draft monitoring plan. In the previous meeting we had agreed that both the Jonah and the Anticline fields would be the focus of our attention and recommendations because they were side by side and we just couldn't credibly address one and ignore the other. Accordingly, here, we formally named the region of our treatment as the PAPA/Jonah II Area. The numeral II derived from the fact that the Jonah development area had recently been approved for additional drilling under a plan approved by BLM and titled "Jonah Infill" or Jonah II. What we all continued to ignore was that we had been chartered under the Anticline development project. This would be effectively used against us by BLM.[23]

• • • • •

On March 18, 2005, a member of the PAWG asked us to pull together a ranking of priorities for the recommendations by the Air Quality Task Group so it could request funding through a Wyoming Natural Resource Grant. Because of the quick response time needed for this information and the lack of a scheduled meeting, the task group was polled via email asking us to forward the priorities as we saw them. The result weighted the items in the direction of those mentioned most often and was not a result of explicit deliberative discussions within the task group. Nevertheless, it did capture the essence of concerns and direction we wanted addressed.

Our list presented details by priority and sub-priority with estimated cost. We admitted that many of the items were likely too expensive to finance given the money available to all of the task groups. We had to hope the PAWG would wisely prioritize and make decisions based on project costs, money available and honest assessment of the importance of task groups' individual endeavors:

Priority List[24]

Priority 1

1. A Provide funding to continue existing monitoring that is lacking funding.

   1. National Atmospheric Deposition (NADP) Sites (funding would expire 10/05)
      a. Gypsum Creek NADP site - $20,500/yr
      b. South Pass NADP site - $20,000/yr

   2. IMPROVE Transmissometer (funding would expire 10/05) Estimated cost - $25,500/yr

   3. Bulk Deposition Monitoring (funding would expire 10/05)
      a. Hobbs Lake site - $48,382/yr
      b. Black Joe Lake site - $44,341/yr

1. B Re-establish Scene Monitor at Fortification Mountain (new monitor) Estimated cost - $15,800 for first year

Priority 2

   A. Finance independent analysis of all existing data to determine baseline and trends. It is estimated this will cost $25,000 to complete.

   B. Buy one or 2 monitors to be placed in other areas. The monitor type is not specific, but this could be included in the vote to re-install the Scene monitor (camera) at Fortification Mountain.

The list served to warn that some existing measurement sites were nearing the end of their funded lifetime. We wanted them to continue operating because they were well placed. They could capture any degradation in important environmental metrics in the down-wind alpine regions located in the all-important Class I Bridger Wilderness of the Wind River Mountain Range.

Beyond that, we wanted to insure that existing data (which we suspected there was a lot of) would be put to use primarily to establish an atmospheric and alpine baseline. Such a baseline would establish the pre-drilling environment. Lastly, we wanted to enlarge the monitoring network to cover the Fortification Mountain area due to its position downwind from the Jonah and Anticline fields plus add one more site elsewhere.

• • • • •

On April 7, 2005, when we next met, we focused primarily on air quality measurement data that existed or was needed and their monitoring site origins.[25]

We sat through long presentations from government staff and government consultants about what was and was not available in the way of data. This included a discussion of data available from the Bridger Forest IMPROVE site that proved very useful. The data was of high quality despite complicating factors such as multiple source regimes contributing to the pollution of the airshed and changes in data processing protocols that rendered "apple-to-apple" comparisons problematic.

Aerosol trends were showing an increase in nitrates in the Bridger area from 1999 through the most recent year of analysis, 2003. Data for 2004 was preliminary but seemed to indicate the early part of that year was the highest so far. It was noted that nitrates were on the rise nationwide but sulfur was holding steady after having been on a rise in earlier years. It was also noted that potassium and sodium were "difficult to read" and trends were uncertain. In retrospect, this observation was unfortunate in light of my own field spectroscopy work showing heavy emissions of those two chemicals from well completion flaring.

I learned from Ted Porwoll that this IMPROVE data was outsourced to a lab at Lawrence Livermore Laboratories for analysis, and involved a process that took a year and a half to turn into a final report. This was in my view not good because the gas guys were not letting any grass grow under their feet in their push to drill out the fields before anything could get in their way.

Our discussions then moved to the region's airshed and impacts from energy development. The representatives from EPA and USFS offered observations that were fairly direct. Questar's winter drilling proposal was the first analysis to show a definite increase in NOx due primarily to oil and gas development. The proximity of the Jonah Infill to the Class I Bridger Wilderness was creating more significant visibility impacts and yet initial Jonah Infill modeling failed to include Questar's Winter Drilling estimated NOx increases. This was why EPA had mandated that BLM perform remodeling of the Jonah Infill. Furthermore, EPA was becoming concerned over developing high levels of VOCs and ozone.

The good news was that particulate from construction and increased traffic did not show additional visibility impacts to Class I Areas. To reinforce that trend, EPA and WYDEQ would continue to encourage BLM to reduce particulate emissions.

Our first really in-depth discussion of modeling then ensued and looking back, I can see that had I been as versed in the practice as I would become a year hence, I would have strenuously called out the numerous red flags that were hoisted on that day.

Joe Delwiche, our EPA presence noted that the BLM preferred alternative for the Jonah Infill was not modeled. EPA wanted to see the preferred alternative modeled with more complete and current inputs. He reminded us that NEPA required a proposed project be analyzed for a range of alternatives. They should address scenarios covering varying levels of all-out development with their attendant environmental consequences. Also, a maximally conservative approach had to be analyzed that practically speaking, might elevate environment over efficient project execution.

In between these extremes should be a preferred alternative that would strike a balance between the two. This was an aspect of the EIS process that in my mind became a mere check-the-box exercise by BLM because it opted for a preferred alternative in both the Jonah infill and the Anticline projects which differed only marginally from the all-out option.

We went on to discover additional reasons for doubting the efficacy of modeling. Terry noted that it seemed modeling had understated NOx emissions because use of lower emission drilling rigs was assumed due to their use of "Tier 1 and Tier 2" diesel engines. This was an EPA mandated phase-in requirement because of their reduced emissions.[26] However, neither BLM nor WYDEQ were requiring the phase-in of those engine types.

Now Shell's representative stepped in. He asserted that these engines were indeed being phased in. T-1's were in fact in the PAPA and both T-1's and T-2's were in the Jonah. Shell specifically was operating three T-1 engines on one rig in the PAPA as early as 2004. He also speculated that because of unique drilling requirements in the Jonah, there "probably" were more in operation.

The rest of us had to take him at his word because we had little means of accessing exact information about operator activities. This became a class of debate that went on more than once due in part to such specific operator held information always being hard to come by.

One rather strange subject was introduced into our discussion about NOx. It was noted that the Jim Bridger Power Plant east of Rock Springs some 100 miles to the south had recently begun installing low NOx burners with the expectation of a reduced NOx level in the range of 8000 tons per year. This would be good for the region east of that plant because the prevailing winds traveled in that direction. Only in rare cases when the winds tracked northwestward from there and over us would this event have any value to our Class I air shed.

A similar item barely mentioned in AQTG meetings had to do with the Naughton Power Plant outside Kemmerer 100 miles to our southwest. A few years earlier Ultra Petroleum had bankrolled an upgrade to that facility to reduce its NOx emissions. This is described in more detail in my chapter about the operators.

• • • • •

Our next meeting took place on June 16, 2005.[27] This was when we were advised that travel cost reimbursement was disallowed. A BLM representative stated that "closer study of regulations" applying to travel expense reimbursement for task group members revealed that we were in fact not eligible. Furthermore, they had determined that only two positions on the PAWG were eligible for such reimbursement. However, no explanation beyond this perfunctory statement was offered. That was the bad news.

The good news was that we were not subject to the federal directive requiring meetings to be advertised in the federal register 45 days in advance. Instead, we were only required to advertise in local media two weeks in advance. The 45-day rule had been quite cumbersome because we all had constant difficulty looking so far ahead and choosing dates when all of us could reliably be available.

When we reviewed the minutes from the previous meeting, Mike Golas complained that the details were, in his opinion, insufficiently recorded. This produced a round of suggestions as to how minutes could be better crafted, including the use of appendices, inclusion of prepared summations from presenters, and focused review discussions of presented materials. Ultimately, it was agreed to leave the previous minutes as they were and draft future versions according to a template suggested earlier by Svalberg. Golas' complaint would come back to haunt him when in a later meeting he complained that the minutes were *too* detailed.

There was a brief discussion about how to fill the need for a note taker that revealed more about the agenda of the operators. When we learned that our current note taker would be dropping out due to health issues, I suggested that members of the former Wagon Wheel Committee would be possible candidates. This group had formed up three decades earlier to successfully oppose the federal government's plan to conduct an experiment to stimulate gas development in Sublette County by means of an underground nuclear explosion.

They had conveyed to me an interest in being involved with our group but Golas and Sewell instantly vetoed that idea stating that it would be undesirable in their view to have former opponents to gas development become active in our group. This was yet another confirmation that these guys were in our group to conduct rear guard actions.

I was then given the floor to describe my project by which I was facilitating a field test between EnCana Corp. and a small engineering company that had invented a ground-breaking dehydration technology they called the "Quantum Leap Dehydrator." The technology reduced emissions from tons per year to a few tens of pounds per year. I tell this story in detail in the chapter about the operators.

Sewell and Golas seemed unimpressed and proceeded to describe their own company initiatives, which were really only minor adjustments to long-used methods. These included consolidation of field support facilities to serve multiple wells and off-pad pipe transport of waste water. Sewell went on to negatively opine that improving one aspect of production could lead to degradation of another.

We then moved to the thorny issue of money. Susan Caplan announced that WYDEQ and BLM were recognizing a need to cooperate in the arena of funding; however, the source of such funding remained to be determined.

Jonathan Ratner was quick on the mark. He recognized the next logical step to be a search for funding to pay for analysis of current data and to implement a comprehensive monitoring plan. He also cited a need for long term funding to support current monitor stations in operation. He wished to see an end to a crisis management approach to funding and wanted to see establishment of a long term (years) program of funding stability. His characterization of the issue being one of regular crisis management was quite accurate.

With his trademark ability to cut right to the chase, William Belveal bluntly asked why we were holding meetings and making recommendations if no agency present had any funding to support our ideas nor would likely obtain funding in the foreseeable future. No one responded and after a few moments of awkward silence, the next subject was quickly opened for discussion.

That subject was the creation of a monitoring plan. Two federally recognized monitoring systems WARMS, and IMPROVE were explained as were their trade-offs, costs, limitations, operating protocols, and utility for attacking the measurement of particular species of interest to us. Other details discussed included quality of data from each. Susan Caplan offered to perform a comparison of sample data from both instruments. Then Jonathan stated he would prefer to see funding aimed toward regulatory-based measurements and Susan felt the State would agree with that preference. Joe Delwiche then opined that another IMPROVE site under WYDEQ oversight seemed to be a best option.

Jim Sewell suggested it would be valuable for a summit meeting of top data analysts from all appropriate agencies to take place. The goal would be for them to present views in support of a goal to define a comprehensive and reliable air quality monitoring system that would basically fall under a definition of network design.

To this end, Terry asked Joe to poll his EPA monitoring experts for willingness to attend such a meeting. They would be asked to address the viability of our monitoring ideas and to supply their own ideas. This led to the determination by the group that it would be necessary for us to craft a set of questions to be addressed that would mimic a straw-man scope of work. Jim and Susan agreed to compile all of our ideas into such a scope of work designed to evaluate the existing monitoring network and make recommendations for a new improved network.

Toward the end of our meeting, we asked Delwiche how our counterpart air quality task group in the Powder River gas development region in northern Wyoming differed from us and what lessons we might take from it. He described its operating environment and commented that if it were better than us in any way he would have to characterize that as doing better in obtaining direct involvement from BLM officials. He recommended that we try harder to get such participation from the Pinedale BLM Office.[28]

One final story that came out of this meeting needs to be told.

Mike Golas informed us that WYDEQ was scheduled to hold hearings toward the end of the month to notify operators of intent to impose green completion requirements on well completions. These requirements would go into effect 30 days later. He further took the opportunity to toot his company's horn by claiming that Questar had been implementing green completion for some time. Be that as it may, this WYDEQ action was long overdue and a direct result of my hammering WYDEQ over an incident that took place on an Anschutz well pad many months previous.

The incident in question was what was referred to as an "upset." Details were sketchy beyond a cursory explanation from WYDEQ that an emergency

By the Rasping in My Lungs

event took place requiring an immediate venting of the well's pressure into the atmosphere. The result was thick haze across the region from a black mushroom cloud. I talk more about this in the chapter about the enviros and the media.

• • • • •

Discussions in our July 21, 2005, meeting delved deeply into the regional air monitoring system, its gaps, and new monitor needs.[29]

Terry began by reviewing monitoring being performed by the U.S. Forest Service in terms of monitor locations and data coming from them. Already, the acid reducing capacity of alpine lakes in the Wind River Range was in decline. This is an important metric because it signals the health of these lakes regarding their capacity to absorb acidic rain runoff and neutralize it into a less caustic presence. At a minimum, failure to neutralize the acid content poses problems for fish there and downstream.

A recurring point of contention between regulators and industry and the local citizenry would have to do with well density. The operators' started out with regulatory allowance to develop the fields on the basis of 20-acre spacing or "down-hole density." This addressed the number of wells they could drill on a given lease parcel. As the wells continued to come in with high volume yields, the operators pressed regulators for higher density and that became a 10-acre standard or double the surface drilling density. Terry spoke to this. He reported that the well count was increasing along the crest of the Anticline plateau at a rate that seemed to be going in the direction of the 10-acre spacing density and USFS was becoming concerned over the impact on air quality this could produce.

The results from the air quality monitors being operated by WYDEQ in the Jonah and Anticline environments were showing that particulate matter (PM) and NOx levels were still typical for the area. Nevertheless, NOx monitoring was showing fairly high levels comparable to an urban area. In addition, there had been several ozone exceedances above the EPA maximum allowable eight-hour level of 85 parts per billion (the currently stipulated limit for that time). This was noted as a possible indicator of a problem. As a result, precursor volatile organic compound (VOC) monitoring was determined to be necessary at these high ozone sites even though VOC monitoring was understood to be expensive.

It was further decided that ozone monitoring in Pinedale should become a priority. An upwind ozone monitor at or near Big Piney was deemed to be necessary but recent ozone data should be verified (by WYDEQ) before too much effort was expended. Because of the public health issues related to these exceedances, the data should be analyzed quickly and if advisable, a news release be issued.

In fact, ozone would indeed become a major issue that would ultimately force the State Governor to submit to local citizen anger and recommend to EPA that the region be declared in "non-attainment." This would establish that Sublette County was in violation of EPA limit stipulations, thereby triggering stringent corrective actions on all sources of ozone precursor emissions that in fact were primarily the gas fields.

Lastly, we decided that a sulfur dioxide (SO2) monitor in the area was needed, especially for modeling purposes. An IMPROVE site in the Big Sandy area was recommended due to its down-wind proximity to the Jonah and Anticline. Furthermore, we felt that monitors to the west of the gas fields were needed primarily because that was upwind from the fields. We wanted to be able to compare the emissions content of the airshed before and after an air mass had passed through the gas fields. A new site called the South Daniel site was placed for this purpose but the $250,000 cost and power issues limited the placement of additional sites like the one at Daniel.

Finally, because of the lag time in getting an IMPROVE site at the Big Sandy site, a camera facility was suggested as an interim measure. Industry would be approached for funding of future monitoring. Long-term lake monitoring at a Wind River lake called Lazy Boy was suggested because of its sensitivity and resultant willingness by the USFS to possibly fund.

• • • • •

By the Rasping in My Lungs

Our July 28, 2005, meeting served as an indicator that we were not even on BLM's radar screen when Terry reported that the Sublette Examiner had just carried a story regarding 2006 funding by the BLM.[30] The news story announced an allocation of $1 million by BLM of which two thirds would go to a project in northern Wyoming and the remaining one third would come to Sublette County. None of this remaining money would be directed toward air quality issues. Clearly, 2006 funding was in a bad state for our AQTG.

Susan Caplan attended by phone and at one point brought another state BLM official named Ken Peacock on the line to address the funding allocation decision. We complained about having to read the news as opposed to being given a courtesy notice by BLM. Mr. Peacock could only opine that this "must have been an oversight."[31]

Joe Delwiche from EPA then pressed Mr. Peacock for his feeling about the likelihood that we could get a BLM representative from the Pinedale office to regularly sit in on our meetings. Delwiche tried to illustrate why we felt we needed a regional representative in attendance by invoking the issue of NOx thresholds as a trigger for adaptive management actions we might wish to recommend to BLM.

Peacock recited objections rooted in a missive from the Director of the Department of Interior that established a protocol and intimated that no deviation was possible. He also invoked BLM interpretation of FACA guidelines but Delwiche challenged him on that, arguing that the mere fact of FACA not mentioning an option should not of itself be taken to mean that the option is forbidden. Seeing that he had been trumped, Peacock then asserted that FACA requires us to forward our issues through the established chain of reporting (PAWG, BLM regional office). However, he confessed there was no one at the regional level who had the power to assist us in the manner we were suggesting anyway.

Delwiche next attempted to address his belief that on the basis of practicality we were necessarily moving beyond our charter because of the inseparability of Jonah from the Anticline or Big Piney from the Anticline. Our problems were not just air quality related but went to management related concerns. He was trying to break the restriction upon us of only addressing the Anticline air impacts.

He cited limits expressed in documents like a Record of Decision and our latitude to press for management actions necessary to address exceedance situations. For instance, could we recommend a threshold that might trigger an adaptive management reaction in such unregulated areas as visibility degradation? Can we define a level of deterioration to be grounds for required corrective actions?

Peacock responded that BLM had no air quality regulatory powers and he again cited FACA as the overarching guideline applicable to us. Finally, he retreated behind the argument that BLM simply could not attend every task group meeting due to insufficient staff.

Probably out of sheer exhaustion and certainly out of exasperation, we were forced to concede that we would be allowed no latitude in communicating with BLM other than through the PAWG. Nevertheless, I was amazed and gratified to finally witness the EPA representative standing up to BLM.[32]

Moving on, we were further informed that the scoping notice for Shell's proposed winter drilling would close in a week for public comment and that it contained no provisions for air quality impacts modeling. Furthermore, any extension of the deadline for comment was not contemplated by BLM even though it coincided with the projected release date of the re-accomplished Jonah Infill air quality analysis. Terry observed that this would be unfortunate due to the fact that the impacts from the latter combined with the undetermined impacts from the former could be important.

Our final piece of bad news was that a provision in the new Federal Energy Act calling for fees on the energy industry for application processing was on hold. This was particularly galling to me because of what I had privately learned from Susan about the burden of permit processing she was struggling under.

I next took the floor to report on progress with my dehydrator project. I gave a quick verbal update regarding EnCana and the inventors of the Quantum Leap Dehydrator. I explained that the two companies had set in motion a project to fabricate and test two skid mounted retrofit packages that would be plumbed into existing dehy units and tested through the winter.

I took this opportunity to describe my growing perception that a few of the operators in the area had been indeed showing responsible attitudes toward environmental impact mitigation but were running into administrative road-

By the Rasping in My Lungs

blocks with BLM. I described my efforts to enlist the Haub School of Environment and Natural Resources at the University of Wyoming as well as the Governor's State Planning Board to circumvent this trend.

My concept was to facilitate a meeting on the neutral ground of the University where operators would present their best ideas for environmental protection methods they wished to test. The highest BLM official possible would be in attendance and the meeting would end with a small committee designating the most promising ideas to be fast tracked through BLM administrative hurdles into field-testing. In general, the idea won support by the other members and Jonathan Ratner posed the question as to what form of action item we could initiate to make this happen. It was recommended that the PAWG contact the director of WYDEQ and request that he facilitate such a meeting.[33]

Terry gave a point-by-point rundown of the results of an Interagency Roundtable Monitoring Assessment (IRMA) meeting held in Cheyenne recently. Of the key points he discussed the following stood out:

- DEQ was planning ambient air quality monitoring at Wamsutter[34] (120 miles southeast of the Anticline)
- SO2 monitoring was contemplated for the Wind River Reservation
- Preliminary ozone values on the Anticline had spiked on several occasions
- It had been concluded that there was a monitoring gap for hazardous air pollutants (HAPs)
- Lazy Boy Lake would be added to the lake deposition measurement project
- There was a gap in air quality monitoring at the north end of the Wind River Mountain Range[35]

Our WYDEQ and EPA members briefly discussed the first quarter monitoring results that revealed a few days of ozone exceedance. However, these exceedance events took place in February and later times when solar ultraviolet levels are low thus rendering these results curious and suspect.[36] As a result, both agencies chose to defer any conclusions pending more results from subsequent calendar quarters.

Joe Delwich proceeded to wax eloquent about the extensive EPA database that existed regarding air quality measurements. As I listened, I became suspicious and commented that consequences resulting from exceedance situations he was describing sounded primarily administrative and bookkeeping in nature. I wanted to know when actual engineering responses would kick in. He conceded my question was a good one and engaged in a lengthy description of process steps that lead to that end. He opined that our group could be important in influencing adaptive management steps through the PAWG to prevent a situation from developing that would actually necessitate such engineering solutions.

He felt that we needed to become involved in defining levels of importance for incremental increases in NOx, VOC's, ozone, etc., that should trigger an adaptive management response. However, he also cautioned that we needed to clarify that our charter even allowed such involvement. This was the fly in the ointment as we were to learn from the resistance BLM would eventually apply. That came in the form of tight limits on what it would allow us to do under its notion of our charter provisions.

We subsequently moved on to the issue of VOC and CO2 emissions monitoring. There was some general discussion about the complexity and cost of VOC analysis. We learned that there were over 100 species making up the category of VOC's and each would cost from $100 to $200 to be monitored. Mike Golas clarified that there are VOC's associated with gas field operations and then there are "all others."[37]

Our discussion of CO2 was brief because it was only a sidebar consideration. Nevertheless, we were stunned to learn from the WYDEQ that the Shute Creek facility, a very large gas processing plant some 100 miles south, emitted 7 million tons of CO2 per year. From this moment on, when I collected statistics from WYDEQ drilling permit documents, I paid closer attention to the declaration of CO2 emissions for each well.

The remainder of the meeting was consumed by conversations about expanding our reach to include other gas fields in the county, prioritizing yet again the issues we felt should go forward to the PAWG, and finding funding and deciding how to apply it.

The final piece of business to close out the meeting was the revelation by Terry that although we would be forwarding these recommendations to BLM through the PAWG, our success to date was dismal. Thus far, BLM had not responded to a single recommendation we had sent forward as a result of our many previous meetings. We began to realize that this was due at least in part to the absence of any formal mechanism or deadline requirement obligating BLM to respond. Some of us began to realize that our conversations with BLM were going to be one way at best and insincere theater at worst.

• • • • •

Despite the depressing revelations of our previous meeting, on August 9, 2005, some of us assembled for the sole purpose of finalizing our formal list of concerns and recommendations to the PAWG.[38]

Our first four recommendations related to needs of the Air Quality Task Group for information and financing, while the remaining four were more administrative in nature:

1. *Once again, we request the BLM to complete the NOx Tracking Report for 2005. The Task group feels we need this important piece of information to determine if current conditions are within those modeled in the Pinedale Anticline EIS. The Task Group requested this information in January, and we were told the BLM was working on finalizing the report. To date we have not seen a completed report.*

2. *We request a more formalized response mechanism for BLM to address PAWG and TG recommendations in a clear, concise and timely manner. While the processes for making recommendations*

from the TGs to the PAWG and from the PAWG to the BLM are relatively clear, it seems that there is no mechanism in place to communicate back from the BLM to the PAWG and then the TGs; for example, all of the Task Groups submitted recommendations for funding in February and April. However, no notification of how the money would be allocated was delivered to the Task Groups. This information became known to the TGs only when it was published in the newspaper in July. It seems that there is a need for better communication.

3. *The Air Quality Task Group would like to stress the urgency of securing funds for existing Forest Service monitoring sites (Bulk Deposition, NADP and Transmissometer) for FY 2006 which starts October 1, 2005. If this monitoring is to continue, agreements will need to be initiated in early September to be in place on October 1. We request that the PAWG convey this urgent message to the BLM.*

● ● ● ● ●

Our next meeting date was on October 18, 2005.[39] That gathering was probably when real disagreements about modeling of NOx emission volumes started to evolve. The BLM decision to approve the Jonah Infill Development Project (JIDP), referred to as Jonah II in previous meeting notes was on the verge of taking place and some of us were sensing the start-up of a steamroller.

Mike Golas asked Joe Delwiche for a summary of EPA comments on the JIDP. Joe proceeded to explain that EPA comments hinged on the BLM preferred alternative stated in a BLM letter to EPA. In the letter, BLM withdrew all development options except for an 80% emissions reduction scenario. Thus, EPA only responded to that option identified in the environmental assessment as Option EC-1.[40] EPA assumed mitigations would go into effect in the early stages of development and that EPA comments had been flowing to

BLM for many months. Jonathan Ratner asked if the 80% scenario still allowed 3-5 days per year of exceedance and it apparently did.

Susan Caplan advised the group of a high level meeting to occur in Cheyenne for the purpose of drafting mitigation procedures for the JIDP and Jonathan asked how a project could be approved if air quality bounds were to be exceeded. Susan cautioned that the projections were *model predictions* only rather than measured events. This was our first clear signal that BLM decisional actions would be based on dubious modeling in place of actual field based empirical sampling.

Joe commented that the 80% reduction scenario assumed a high emission situation. Mike then stated his concern that the 80% scenario assumed definite negative impacts which mandated a requirement for the modeling effort to be of impeccably high quality. Nevertheless he conceded that folks like Jonathan and the public had no other information on which to make judgments. His comment was almost certainly defensive; he was worried that impact modeling would bite the operators.

Susan then explained at length the concept of deciviews and their connection to definitions of limits regarding visual impairment. This confirmed that BLM and USFS recognized different deciview levels to define visibility impairment as had been explained to us in the November 30, 2004 AQTG meeting

Jim Sewell then jumped in to caution Susan about Shell's concern over modelers' use of what it believed to be excessive emission inventory values but Delwiche replied that more than one modeler was achieving the same results. I then offered my observations regarding field operations in that I spent time with Questar observing drill operations and fracking operations.

I had learned that diesel power plants were being operated with specific strategies designed to deliver maximum horsepower only when needed for as long as needed chiefly as a fuel conservation measure. I suggested this was likely a subtle and probably little realized practice within the modeling community. It was necessary for modelers to connect better with field operations. They might need to fine tune their understanding of actual operating times for power plants so they could better quantify NOx emissions from these sources.

R. Perry Walker

Mike Golas seconded my comments and suggested the public was reacting on the basis of information coming only from government agencies charged with oversight of gas development. If those overseers were using bad data or applying incorrect factors, this would only serve to aggravate the public's perceptions and feed the emotional component of the public debate.

Terry expressed concern that BLM began the modeling process some time back with numerical input data that very possibly had yet to be updated and that may have been applicable to one field but not another. Should this be true, the result would be a perpetuation of an invalid database justifying stronger consideration of comments received.

This is the appropriate place to explain an added issue that went directly to the subject of modeling NOx emissions. In the JIDP ROD Appendix A, one of the criteria cited was load factor.[41] During my time on the Air Quality Task Group, there were heated arguments between Susan Caplan and the operators as to what the load factor really was. The issue was important because the degree and duration the drill bit is under heavy cutting load directly tasks the engines. That in turn causes engines to run more or less heavily which dictates their exhaust emission history.

Although both sides of this argument were certain they were right in their position, neither side could provide definitive evidence that their position was in fact correct. My own concerns included doubt as to whether they ever considered the issue of actual engine combustion performance at our altitudes. This last point became one of my recurring criticisms in my public comment replies to gas field environmental impact analysis documents.

These conversations thus ushered in much future debate and extreme skepticism on my part. I came to reject the efficacy of modeling as it was being pursued and its reliability in predicting effects that were becoming increasingly obvious to even a casual observer.

We next reviewed, *yet again*, the final version of our formal recommendations going forward to the PAWG. We discussed each item and worked out a basic agreement as to how they should be rewritten to satisfy everyone. Thus, rewritten suggestions of the three most contentious topics were hammered out. This exercise elevated our concern over failed NOx tracking by BLM and the resulting exchange revealed a growing mistrust toward BLM by most of us.

By the Rasping in My Lungs

We pressed Susan to describe the status of the NOx report with her Cheyenne superiors and she commented that it was stuck in the minerals department. Terry reinforced the purpose of the reporting effort as being a yearly tracking of NOx evolution. Our timely access to the report was the key to its value. She promised to press the minerals folks to release the reports for 2004 and 2005.

Sewell and Golas offered that the operators had been providing Jonah and Anticline 2004 NOx statistics to WYDEQ for a long time and they should not be difficult to pull together. Joe Delwiche seemed to reinforce that by noting WYDEQ was on the record requiring resumption of NOx tracking by BLM. Furthermore, EPA was interested in knowing the emissions from even perceived small sources such a well completions.

This led to a discussion between Susan and Joe about revising the applicable letter of agreement in a manner that would move the reporting responsibility from BLM-Rock Springs to BLM-Cheyenne. I was surprised and perplexed to discover at this moment that the Rock Springs BLM field office in Sweetwater County would be tasked to generate a report on events happening in another jurisdiction.

She explained the history of the letter and its having come into existence as far back as two prior state-level BLM directors. We asked if it was time for the current director to be pressed for his commitment toward honoring the content of the letter. Joe cautioned that bringing it up in any context of revision could risk its being weakened or ignored despite its being an element of the JIDP ROD. This triggered much conversation about the wisdom of revision. Jonathan worried that revision could result in outright suspension while Joe suggested we simply beg BLM to live up to its existing commitments outlined in the current letter. Susan offered that such begging had been tried without effect.

Terry summed up the discussion with the suggestion that we not ask for a new letter but instead ask BLM to follow up on the NOx report and add VOC tracking to the effort. Sewell voiced doubt as to the value of VOC tracking because the operators were reporting this already but Joe expressed his opinion that there was indeed value in such an added request.

At last, we decided on a set of action recommendations to be forwarded to BLM through the PAWG: (1) Regarding NOx tracking, we opted to let

WYDEQ press BLM for compliance but that we would submit to the PAWG a draft letter for forwarding to the BLM state director requesting compliance; (2) Monitoring ozone would be classed as a low priority for us for the time being but we wished to review state collected data on precursor VOCs; (3) we had intended to press the PAWG for monitoring funding but since the money was now gone, the issue was dropped.

We closed out the morning session before breaking for lunch with a final rehash of the two of the more contentious items which were our desire to have a permanent BLM member attending our meetings and expanding the our scope to take in the Jonah field. Some debate followed which questioned the need for elevating the first issue. I reminded everyone of the frustrating debate in the last meeting between Joe Delwiche and Mr. Peacock of BLM-Cheyenne over this issue.

Others wanted us to at least ask BLM-Pinedale if it concurred with Cheyenne and to request BLM visitor presence. As an alternative, we could settle upon any BLM staffer able to answer management decision questions as they developed in our meetings. Perhaps indicating his own resignation, Joe recommended we simply accept BLM's interpretation of the FACA and focus on our charter.

We then returned to debating the logic of our group expanding its scope beyond the Anticline. Some of us felt we should not treat the PAPA in isolation from adjacent fields. Others strongly objected to such an expansion. Joe Delwiche commented that the impacts we were addressing would be influenced by many other sources so, if we confined ourselves to PAPA only, we would fail to take those sources into account. Also, we were addressing the central issue of the Wind River Mountain Class I air shed which was a far-field monitoring concern justifying a broader scope.

Added to this, there were indications that BLM was thinking of creating a new non-FACA air quality group attached to the JIDP. We were concerned this would be a duplication of effort.[42] As usual, Sewell strongly argued that expansion of our scope was wrong and that what we accomplished within our narrow charter would likely be the example any duplicative group would emulate. We finally had to drop the item due to a lack of consensus.[43]

After breaking for lunch, we reconvened to hear about experimental technologies being tested in pursuit of lowered impacts on the air shed by gas development. We first heard a presentation from EnCana about the Quantum Leap Dehydrator. This was the same project I had brokered between EnCana and Engineering Concept, LLC of Farmington New Mexico. I had arranged this presentation so the task group could hear particulars of the project from an actual gas guy. I was hoping Sewell and Golas might be a little more open to the concept if it came from one of their own.

The EnCana speaker gave a short talk on the Quantum Leap Dehydrator explaining that it was to be tested during the next few months. This technology had been tested earlier by EPA in Colorado and found to be exceptional in its reduction of dehy emissions of VOC's and NOx. He advised that two units would be delivered around mid-November for testing through the winter.

He cautioned that his company did not know if it would embrace the unit but was interested enough to give it a thorough testing during the most extreme season of our year. He commented that the unit would use hydraulic apparatus rather than electric (valves, etc.) and would utilize a 6 horsepower internal combustion engine for prime power.[44] It would drive a high reliability motor unit developed by York for air conditioning application in remote locations.

He admitted the company's lack of information regarding the dehy's performance due to its having been in the field in small numbers thus far but recognized its promise as described in EPA test records. I also pointed out its potential for conserving wellhead gas that could be sent to market because I wanted Sewell and Golas to see the economic advantages that would offset the higher per-unit price.

Sewell next proceeded to advise us on the status of drill rigs in the area. This was a topic of intense interest because of the higher population of those machines than BLM had originally modeled. Three of the most active operators, Shell, Ultra, and Anschutz were to field six rigs under an approved application for winter drilling on the Anticline. These rigs would utilize a total of 20 diesel engines carrying advanced NOx control technology.

The engines would be rated in excess of 1000 horsepower and the operators committed to having them meet or exceed EPA Tier II standards. Four rigs (14 engines) would employ catalytic techniques and two rigs (6 engines) would employ bi-fuel (hybrid natural gas/diesel) techniques. Test protocols had been developed and submitted to WYDEQ and EPA for approval. The operators also committed to reporting results in a formal manner to those agencies. Further characteristics associated with these methods were offered as follows:

Catalytic Method

Advantages

- 75-90% NOx reduction
- Proven reduction method
- Urea based agent as NH_3 source
- Converts NOx to N_2 and H_2O

Disadvantages

- Possible excess NH_2 in exhaust stream to atmosphere; NH_3 is monitored for this Urea supply
- Requires 2-3 truck deliveries/month; This is still much lower truck traffic and less NOx

Cost is about $100k per engine

Bi-Fuel Method

Advantages

- Simple concept requiring only a fuel controller
- Uses existing engines
- 30-50% reduction in NOx

Disadvantages

- Requires source of Nat Gas fuel
- Increases in CO result

Cost is about $20k per engine

Sewell went on to claim that his calculations showed NOx emissions would drop from 978 tons/yr to 350 tons/yr as a result of applying these methods. Time would show, however, that the catalytic method would drag on for quite some time before reaching successful operation.

Additionally, these operators would accept the BLM prohibition on well completions during winter months, would bus workers to and from the work sites (thereby reducing vehicle emissions), would continue funding of the Boulder weather monitoring station, and would report results to the PAWG.

We ended with the Questar representative advising us that the company had agreed to provide funding to WYDEQ in support of air quality monitoring. The sum under consideration would be $80,000 per year for five years allocated toward monitoring operator responsibilities as stipulated in the Anticline ROD. A Questar condition for the funding was that $50,000 must be applied toward employing an additional air quality compliance inspector.

This meeting generated our restated list of concerns and recommendations to BLM-Pinedale through the PAWG. However, political correctness kicked in on the part of WYDEQ and the draft list became the sanitized version that follows:

Draft Version

1. *Once again, we request the BLM to complete the NOx Tracking Report for 2005. The Task group feels we need this important piece of information to determine if current conditions are within those modeled in the Pinedale Anticline EIS. The Task Group requested this information in January, and we were told that the BLM was working on finalizing the report. To date we have not seen a completed report.*

2. *Because of high ozone events monitored by ambient air monitors at Jonah and Boulder monitoring stations in the first quarter of 2005, we request the BLM to initiate an annual VOC Tracking Report similar to the NOx Tracking Report cited above. VOCs are a pre-cursor to ozone development so knowledge of the relative availability of VOCs in the air will allow us to better understand the process of ozone formation in this area. The State DEQ currently provides information on permitted VOCs with their portion of the NOx Tracking Re-*

port to the BLM. The BLM would just have to make small additions to cover non-permitted sources such as traffic and drill rig emissions to account for all VOC sources in the area to complete the report.

3. We request that un-obligated BLM Monitoring Funds ($50,000), be used toward an evaluation of air quality monitoring architecture analysis by independent scientists. This request was developed as a result of discussions at the July 21, 2005 meeting in Cheyenne where Federal Land Managers and Regulators discussed monitoring in the Pinedale area. It was felt by the group that there would be some value in an independent analysis of existing and proposed monitoring analysis of data gaps as well as integrating under one Quality Assurance Plan all of the existing monitoring protocols. The group felt this would add credibility to the overall monitoring programs. Also, this project would include the development of a long-term exit strategy for the monitoring. The AQTG thought this proposal was warranted. This contract might be accomplished through existing BLM agreements or possibly through coordination with the State DEQ. (Draft Statement of work is available.)

4. We request that the PAWG facilitate an industry meeting at the Ruckelshaus Institute (UW) through WYODEQ, and BLM for the purpose of developing open communication between operators and regulators that will develop new best management practices. This recommendation goes beyond just air quality, and may encompass all of the task groups. Many of the natural gas operators in the Pinedale Anticline (and Jonah too) area are being very innovative in finding ways to reduce emissions and increase efficiencies. While adjacent operators may be aware of what is being done, we have been told there is not a mechanism to openly discuss these things. We feel this may be a great mechanism for open dialogue between operators, regulators, Federal Land Managers and the general public.

5. We request a more formalized response mechanism for BLM to address PAWG and TG recommendations in a clear, concise and timely manner. While the processes for making recommendations, from the TGs to the PAWG and from the PAWG to the BLM are relatively clear, it seems there is no mechanism in place to communicate back from the BLM to the PAWG and then the TGs.

For example, all of the Task Groups submitted recommendations for funding in February and April. However, no notification of how the money would be allocated was delivered to the Task Groups. This information became known to the TGs only when it was printed in the newspaper in July. It seems that there is a need for better communication.

6. *We request presence of BLM decision maker at our meetings to better understand our complex subject matter, meeting dynamics, and to provide some consistent messages and updates to the BLM management. The AQTG feels that we are doing a great job in our discussions and developing our recommendations; however, we feel that this information is not getting to BLM managers in a timely manner. We feel the BLM management could directly benefit from a better understanding of the complex and far reaching problems related to AQ they are facing.*

7. *We recommend the creation of a unified PAWG (Pinedale Area Working Group) for the BLM Pinedale Area Office. New NEPA analysis for the Jonah Infill Project proposes a group similar to the PAWG and Task Groups to oversee adaptive management in that project area. The AQTG feels the result is a fragmentation of the skill base currently focused on the PAPA. This is in part because there will be duplication of administrative processes (more meetings, more reports ...). It seems to be much more logical to develop one AQTG for the Area Office with the charge of looking at the entire air quality picture across the Field Office. Incorporation of this type of adaptive management group in the Revision of the Resource Management Plan would be appropriate.*

8. *The Air Quality Task Group would like to stress the urgency of securing funds for existing Forest Service monitoring sites (Bulk Deposition, NADP and Transmissometer) for FY 2006 which starts October 1, 2005. If this monitoring is to continue, agreements will need to be initiated in early September to be in place on October 1. We request that the PAWG convey this urgent message to the BLM.*

Final "Sanitized" Version

1. There were no changes here.
2. This version eliminated all references to high ozone events in 2005, initiation of a VOC tracking report, our justification for the report, recognition of WYDEQ possession of VOC statistics, and BLM administrative adjustments to assess non-permitted sources. Instead, it was all replaced with cajolery of BLM over its workload and an entreaty that BLM respond to our recommendations and request submissions.
3. This version eliminated all references to monitoring funds, the Cheyenne meeting, independent monitoring architecture analysis, independent data analysis, a single quality assurance plan, and a long-term monitoring exit strategy. Instead, it was all replaced with an attempt to cajole BLM into granting us a dedicated staff member to provide us with administrative guidance.

The remaining body of recommendations contained in Items 4 through 8 did not survive the review process.

Our task group submitted an update document to the PAWG, dated October 25, 2005 which included the sanitized list of concerns and recommendations.[45] We requested that these recommendations be moved forward to the BLM in a timely manner and we tried to give them emphasis by noting that they had been formulated as a result of Air Quality Task Group meetings on July 28th and October 18th, as well as an Interagency Roundtable meeting between the BLM, DEQ, EPA and FS in Cheyenne on July 21, 2005. Here federal land managers and regulators had discussed existing and planned air quality monitoring for southwest Wyoming.

Next the never-ending issue of funding was addressed. We highlighted gaps in current monitoring as well as the most critical unfunded portions of the current air quality monitoring network in the Pinedale area. These included USFS monitoring of the Gypsum Creek and South Pass NADP sites, the Bulk Deposition monitoring at Hobbs and Black Joe Lakes and the IMPROVE transmissometer located at Pinedale.

We learned that on Sept. 9, 2005, WYDEQ and the operators had met in Cheyenne to discuss funding for current and proposed new air quality monitoring in southwest Wyoming including the monitoring done in the Pinedale area by WYDEQ, operators, and the Forest Service. All parties agreed it was important to continue existing monitoring while also adding new monitors where a need existed. WYDEQ was said to be working out the details with operators over funding for the next two to three years. Also, WYDEQ and USFS agreed to look for longer term funding mechanisms that would continue air quality monitoring in the area.

At the Cheyenne meeting, Exxon Mobil volunteered to resume funding of the Gypsum Creek NADP site and the Hobbs Lake Bulk Deposition which they had been supporting for the previous two decades as a condition of approval of their Wyoming air quality permits. This was done even though those permit conditions no longer applied. WYDEQ also came forward and committed to financing the rest of the USFS monitoring (for the near term) once details were worked out with the operators.

Thus, all USFS monitoring was assured without any break. Additionally, the proposal given to the operators included the installation of a camera at Fortification Mountain to digitally record visibility conditions daily. This was considered an important observation location to gauge downwind effects from the Jonah and Anticline.

This set of funding agreements was one of those rare occasions when the big players actually seemed to come together constructively. It was due in no small measure to the looming threat in the minds of the operators that to do otherwise would risk limiting their option for conducting drilling business as usual. Furthermore, WYDEQ directors were under increasing pressure from the citizens of Sublette County to show they were addressing the problems. In sum, the combination of public pressure, negotiation, conversation, and dogged perseverance by the Air Quality Task Group seemed to be yielding positive results.

• • • • •

Our last meeting of 2005 took place on December 1.[46] We started with an approved motion to appoint me as a new co-chair in part because I had been producing meeting minutes for quite some time. Also, I volunteered to fill the void left by Cara who was often redirected by her own WYDEQ responsibilities. Furthermore, the presence of a Pinedale BLM staffer was recognized as fulfillment of our request. William Belveal immediately inquired of that person if there was any improvement in communications back to the task groups from BLM and the answer was "not yet."

Mike Golas described the recent creation of a joint operators group that was focusing on the Jonah and Anticline gas fields. This was an industry initiative and was expected to operate under the title of "Joint Operators Group for Air Quality Management." Almost every operator in the two fields was expected to participate under a charter that targeted all operations in a way that would reduce emissions and lead to improvement of best practices and improved air quality. The management approach to the issue was to include development of a comprehensive understanding of how current practices impacted air quality, development of a unified team approach to address the issue, sharing of information between operators, and assembling it into a useful format.

Mike thought there would be an initial assessment by the group around late 2006 as to how successful it had been. He further explained that a compilation of capital and maintenance costs for the next six years had been pulled together. This encompassed such diverse topics as cataloging present and expected well site location and emission modeling software. The cost figure was $7.8 million for which industry desired a 50/50 split with state government. Furthermore, a methodology was being finalized by which industry would seek to divide its share of costs between its most significant members. These significant operators included Exon-Mobile, Shell, Ultra, British Petroleum, Questar, EOG, Anschutz, Chevron, Gates Petroleum, and Devon Oil.

Dan Olsen, the DEQ Air Quality Division director, next addressed our group. He explained that he did not like relying solely upon computer models without their results being validated and he considered monitoring stations to be the necessary tools for such validation. His desire was to deploy a network of stations designed to diagnose the effect that gas and oil development was having on the local air mass. He noted that already there were bulk deposition and NADP measurement devices in the field measuring air pollutants. However, there were no direct measurement methods in the field to directly capture primary and secondary species creating those pollutants being detected by the bulk deposition and NADP apparatus.

His belief was that both measurement regimes were necessary to assemble a complete picture and his desire was to set up monitoring for things like VOCs, HAPs, and formaldehyde. He cautioned that some of this would be difficult and it would be necessary to establish reliable methods and protocols. Also, he declared that there would likely be more utilization of cameras and transmissometers to better quantify visibility impairment.

He described the state budget process and stated that February 2006 would be the budget session period for fiscal years 2007 and 2008 which started in July of each year. He observed that the Governor was very supportive of the WYDEQ proposed budget that would be submitted to the legislature. Included in that budget request was a requirement for 22 additional staff.

This tracked well with my own efforts at lobbying WYDEQ for more personnel because they were being overwhelmed by the operator's full court press to drill. Unfortunately as subsequent events would reveal, all I succeeded in doing was to help WYDEQ fill its office space with inside-the-box thinkers who impeded progress toward effective regulation.

Dan then summarized the present situation. He noted that incremental consumption standards (PSDs) were not at risk yet. WYDEQ had considered and acted upon the issue of open pit well completion flaring. It was studying drill rig engine emissions. It was seeking to catalog emission sources and create an inventory out to a radius of 300 kilometers from the Wind River Mountain Range. Finally, it was looking at other upwind states' contributions to our air pollution.

All of these activities were directed toward the goal of comparing the present state of the atmosphere with a baseline established in 1988. Ultimately, his hope was to conduct annual updates of the inventory and use that to analyze the resultant incremental impact – all to validate the capability of computational models so they would tell the correct story.

This is a good point to insert a personal historical comment. All of this assurance from Olsen was late in coming and almost certainly the result of public pressure, not the least of which came from me in the form of a letter-to-editors campaign. I had been hammering Dan and his boss as far back as early 2004 over their refusal to get into Sublette County in force with atmospheric monitoring stations.

The fact is they had been dragging their feet because the cost of these stations was high and they simply would not believe that gas development was going to proceed in a manner that would impact local air quality. At one point, the relationship between us became so strained that my contact in the Governor's office asked me to tone down my criticism because, as she put it, "Those two lock up whenever your name comes up in our staff meetings."

In a question and answer session, Sewell asked what would be the process by which data from the monitoring stations would be analyzed. Dan explained that the operators would "write a check" to WYDEQ and WYDEQ would obtain consultant services to quality control the data. Analysis would be either in-house or outsourced. The exact option was yet to be decided. Jonathan Ratner asked if $7.8 million was adequate and Dan offered that the figure was his best guess given available information. He considered this at least a good starting point and subject to refinement as better understanding developed about what would constitute adequate monitoring and analysis.

Jonathan also questioned Dan about the method applied to determining the architecture of the monitoring network. He specifically asked if there was a document describing the analysis process that created the network. Actually, this was an excellent question intended to ascertain if there was any kind of structured approach underpinning how the network would be designed.

Dan responded that there was not. The network architecture evolved from a kind of roundtable discussion rooted in ambient air quality guidelines as well as a quality control approach to instrumentation selection by "knowledgeable people." He indicated that the process was essentially experienced-based. Joe Delwiche jumped in to assure us there was a network review process in effect and later Dan observed that an annual review of the architecture would be conducted.

At this point, I inquired if WYDEQ would consider a request to fund an aircraft based air quality measurement program out of the University of Wyoming. I cited my recent conversations about the topic with Harold

Bergman and Derek Montague who were professors involved with environmental issues, air quality in particular. They had informed me that UW was operating a medium-sized aircraft containing a suite of very sophisticated monitoring instruments.

I reminded Dan of the use of that aircraft in a southwest Wyoming air measurement project in the mid -'90s of which he had been a part. I also advised that Montague had cited the existence of raw airborne data collected by the aircraft in the Sublette region in 1996. That data constituted an unexploited baseline data source because it had never been analyzed. Dan replied that he would talk to his boss, John Corra, about the aircraft and he suggested that perhaps there might be a possibility of a grant to some graduate student to analyze the raw data I was citing.

Again a personal historical comment is in order. The aircraft I was so keen on bringing into the picture was never tapped. I tried also to talk some of the operators into funding it but they would have nothing to do with the idea. Eventually though, WYDEQ yielded to the practicality of air measurements and hired a local private pilot (who happened to be another high school classmate) to fly a few instruments over the gas fields as part of their effort to obtain a basic three dimensional picture of what was going on above the gas fields.

When we returned from our lunch break, we yet again dug into the issue of the missing BLM NOx report. Our group still had not received it nor had anyone else. Susan Caplan, was asked about this. She responded that staff personnel needed to produce it were in short supply and that it would be helpful if one of our group would care to work on it. She advised that there was missing information about drill rig horsepower, number of drilling days, and rig engine load factors. She also worried about the magnitude of effort needed to contact the large number of operators from whom these parameters had to be collected.

Jim Sewell then reminded her that WYDEQ received this information from operators as a matter of normal reporting and would continue receiving it on an annual basis. He further explained that WYDEQ was collecting information on both permitted and non-permitted sources examples of which were contained in Dan Olsen's PSD consumption analysis presentation to the public in Pinedale last spring. Jim advised that industry had submitted a report for 2004 and anticipated continuation of the requirement well into 2005. Tony Hoyt, our newest WYDEQ member offered to query his Cheyenne office for clarifications. Joe Delwiche then suggested to Susan that she craft an outline describing how to proceed with collection of the desired parameters and how they would be utilized.

This exchange serves to illustrate one of the endless problems we faced. The problem here was compartmentalization and little cooperation. WYDEQ was collecting data from operators but seemed to be hoarding it. BLM was collecting information but what they did with it only they seemed to know. When one tried to undertake an action that the other thought was poaching within its jurisdiction, heels dug in and behind the scenes feuding ensued.

Mike Golas observed during this meeting session that the State's technical role had appeared to be somewhat undefined in recent months and he worried that our limiting relationship with the PAWG could cause us to bog down.

Late in the afternoon session there was a discussion that seemed to suggest some of us didn't believe what we had been hearing earlier. Ratner revisited his concern that the method of design for the monitoring network being assembled remained an unknown quantity. Ted Porwoll from USFS voiced his concern that visibility monitoring was in short supply and thus was failing to validate computer modeling which had been the basic assessment method of choice within agencies tasked to evaluate visibility degradation.

Delwiche seemed to support the skepticism by observing that a "loose end" existed in the form of little if any formal assessments of monitoring network design. Thus on the one hand, we were agreeing with Olsen's view that monitoring stations were needed to validate modeling but on the other hand we were unconvinced that Dan's stated reliance upon "round table discussion," instrument selection by "experienced people," and unspecified "field experience" would result in a successful approach to building a monitoring network.

We ended the day with yet another idea for prying the missing NOx report from BLM. This time, at the suggestion of Joe Delwiche, the NOx tracking report by BLM would again be requested. However, in this iteration it would be requested within the context of the local operators' reported emission data to the State that Jim Sewell had earlier alerted us to. Perhaps this might facilitate BLM's execution of that task.

2006

Our next meeting took place on January 11, 2006.[47] This would be the year that our meetings would become ever more frustrating. My own growing conviction was that this entire exercise in public outreach and empowerment by BLM had always been intended to be little more than theater and optics. The real agenda had been to deflect and delay all public interference with its determination to facilitate drill-out of the gas fields in fulfillment of the operators' long publically stated schedule.

Adding to our woes in 2006 would be the new representative, Caleb Hiner whom the Pinedale BLM office had assigned to us in fulfillment of our request for a BLM administrative guide in our meetings.

We began with a review of funding being provided by WYDEQ, USFS, and the operators for monitoring programs. Mike Golas gave us an update on the industry working groups' efforts to secure some of that funding. We were aware they had settled on a $7.8 million package that would be split between Industry and the State of Wyoming. There would be five core contributors from industry which would provide $3.7 million to the effort, (EnCana, Questar, Shell, Ultra and BP), and the industry working group was continuing to solicit funding from other industry sources. The $7.8 million package would cover monitoring from 2005 – 2010. Furthermore, the Forest Service was making known its desire to have one agreement with the State rather than several agreements with industry to pay for Forest Service monitoring programs.

We next tackled the timing of our Annual Report to the PAWG. The task group looked at a schedule of PAWG meetings for 2006 and agreed the best time to present the report to the PAWG would be at its May 10, 2006 meeting. This would allow for all of the previous year's data to be received and analyzed as well as providing the task group with enough time to assemble and edit the report. The PAWG was planning eight meetings for the year. The task group agreed that giving an update on AQTG activities at each of those meetings would be a good way to remain visible to the PAWG.

Yet again, we discussed the increasingly infamous NOx tracking report which had been promised by BLM. We were assured that Susan Caplan and another state level BLM staffer were working on locating information for the report but wanted WYDEQ and oil and gas representatives to help. WYDEQ was to provide actual emission data collected for 2004 from industry. However, this was on hold because WYDEQ was currently dominated by a need to purge its computer system of a virus infection.

Looking ahead, a similar report might be requested for 2005. Until then, BLM advised that it would look at the data to determine if there were deficiencies requiring correction. This was said to be relevant for future reports in hopes it would facilitate their preparation. However, by this time, we were all skeptical that BLM would ever live up to its obligation.

Terry next presented notes from his attendance at the July 21, 2005 Interagency Roundtable meeting. There had been discussion of the adequacy of the current air quality monitoring network in southwest Wyoming. The group looked at possible data and monitoring gaps in the current system and considered proposals for new or additional monitoring equipment in the future.

A WYDEQ contractor representative for monitoring services sitting in on our meeting stated that current monitors were sited using approved siting criteria but which he did not specify. The task group agreed that an independent review of the network and data should be proposed to the PAWG and included as a need in our annual report.

Susan Caplan stated that she had spoken to people interested in doing such a review, but BLM funding was probably not possible. Nevertheless, she committed to sending the BLM draft statement of work to the task group for review. With this document plus similar documentation resulting from a like initiative regarding a Powder River Basin Monitoring Network, our task group hoped to craft our own guidance recommendations. Terry offered to kick start this project and the WYDEQ contractor representative offered to provide monitoring history of the Upper Green River Valley to help with the effort.

We finished our day with summary discussions on how to proceed with our second annual report to the PAWG. Details about format, data to be included, and assignment of sections to be drafted by our membership were worked out. We closed by setting dates for our next meetings on February 16, 2006, March 23, 2006, and May 4, 2006.

By the Rasping in My Lungs

· · · · ·

Although a snowstorm made attendance difficult, we nevertheless met on February 16, 2006.[48] Despite the weather, nine individuals from the public and the task group were able to show up, with four members tuning in by telephone. We immediately learned that Cara Keslar was dropping out as our co-chair but would continue as a general member from WYDEQ.

Our review of funding for air quality monitoring activities revealed that WYDEQ had issued a memorandum of understanding (MOU) for funding to a select group of operators and that the State might require some kind of interim agreement to fund Forest Service NADP activities. The companies receiving the MOU were some of those identified earlier (EnCana, Questar, Shell, Ultra, and British Petroleum). They were expecting to chip in $3.9 million of the $7.8 million monitoring package that had been formulated. As part of this, an additional field compliance inspector was to be hired by WYDEQ. However, that inspector would be stationed in Casper a day's drive away so effectiveness was suspect in our minds.

We learned that special testing was underway on 20 drill rig engines in an effort to validate new techniques being evaluated for NOx emissions reductions. Testing under engine high and low load factors was to be performed. At the moment, two bi-fuel engines, running on a mix of diesel and natural gas were being tested. Although they were not EPA Tier II, they supposedly were demonstrating greatly reduced NOx emissions. After these, Shell's SCR (Selective Catalytic Reduction) engines would be tested. SCR was reported to work well with diesel/electric engines and was expected to reduce NOx emissions by 80-85%. Questar was expecting to be operating its first full Tier II rig beginning in May of the current year and all six Questar year-round rigs would be the Tier II type by late 2006.

Turning to the task group's monitoring report for the PAWG, we agreed to include an executive summary in the document to help the PAWG and others interpret the report. Terry showed some graphs of Forest Service data and advised that the Forest Service had enlisted two people to do a statistical analysis of its data for the report. The task group accepted several specific items regarding the contents and presentation of the monitoring report. They included adopting a standard visual range to characterize visibility impairment, focusing our analyses on sulfates, nitrates, particulate matter, and ozone, emphasizing alpine lake acidification, and maximizing the use of statistics and graphs to convey the science we were offering.

While these seemingly mundane details may appear to be a lot of boring bureaucratic meanderings, in fact they were the inescapable elements that were essential. We had to develop a sound, professional document because we expected BLM to use it as its guide. Accordingly, we labored long and hard to be relevant, productive, and scientific. Above all else we felt that our work would be accepted and applied by BLM and WYDEQ only if we strictly attended to these details.

Martin Hestmark of EPA Region 8 next gave a brief presentation of EPA's Oil and Gas Initiative and answered questions from the group. This initiative was touted to be EPA's approach to being proactive. It started in year 2000 when EPA Region 8 began a dialog with several states, BLM, and the US Forest Service concerning oil and gas development issues in the west. EPA "saw an opportunity" to work with operators and the states to develop technical specifications for present and future development and ambient air monitoring. This was also regarded as an opportune time to resolve past compliance issues and help streamline industry/regulator interaction. An action plan had been signed by the participants in September 2003 intended to avoid NAAQS or SIP problems requiring regulatory solutions.

This all sounded marvelous. The premier federal environment protector had apparently looked into the future from year 2000 and seen the dysfunction that would develop. However, federal doublespeak was at work here. Although the Region 8 EPA staffers whom I came to know were in fact sincere, politics and money being wielded by the operators would dominate. It would develop that the reference to industry/regulator interaction would in truth turn into (in the case of WYDEQ) the operators telling the regulators what they wanted to see done, followed by the regulators falling in line.[49]

Next came our ritualistic discussion of the BLM NOx report. Susan Caplan updated the task group on BLMs progress and informed us that BLM had located figures for 1987 through 2005. Then we were told these figures had been produced by taking the well counts from each year and multiplying them by an average emission factor of 13 tons/per well of NOx for the PAPA. This factor had been worked up by WYDEQ but was not explained.

Statistics for the Jonah and Anticline drilling versus well completions had been broken out and initial estimates were showing NOx emissions of 2,200 tons for 2004 and 3,100 tons for 2005. However, these emissions factors had in turn been derived from an *estimated* drilling emissions study of PSD consumption performed by WYDEQ. Some of us were incredulous!

We pressed for details about what was specifically stated in the amended NOx Tracking Agreement between BLM and WYDEQ regarding the degree of rigor they were supposed to apply to their efforts. Several of us felt betrayed because it was instantly clear that this level of analysis was inadequate. This was to become my first realization that both WYDEQ and BLM would continually invoke dubious desk-bound studies in place of more meaningful field observation. In fact, this approach became a central theme of my increasingly critical written and public vocal criticisms of BLM's near total reliance upon modeling based studies that in turn relied upon highly questionable data.

All we could do with the remainder of the meeting was schedule a few more meetings so they could be entered into the Federal Register. We determined that a meeting for May 4, 2006, had to be dedicated to finalizing our Monitoring Report that would be presented to the PAWG shortly thereafter. However, there is no record of more meetings until January 2007.

• • • • •

2007

When we reconvened on January 25, 2007, yet new membership changes had to be processed.[50] Cara Keslar from WYDEQ would again be taking over the position of chairperson but because she was working out of Cheyenne, the rest of us would be tasked more heavily to perform support activities. Our membership was dwindling; so we discussed how we should go about recruiting new people to fill the existing two vacancies as well as a third that was pending.

By June 2007 I was reaching my limits of tolerance with WYDEQ-AQD and ready to quit the AQTG. In a letter to its Director Dave Finley having to do with his recent memorandum to operators, I laid into him for his staff's obstructionist behavior over the Quantum Leap Dehydrator project and other examples of bunker attitude. Besides lambasting his Casper staffer who was key in their killing the project I addressed Keslar's performance on the AQTG:

> "You may as well know one more fact. I may quit the AQ task group in part because of Cara Keslar. I bitterly regret my original comments urging her to remain in a co-chairmanship role in the Air Quality Task Group. I soon discovered the truth of the adage 'beware what you wish for' because it soon became apparent that she lacks the seasoning to handle authority. This was the opinion of three of the six of us as well as some folks in Cheyenne who have worked with her.
>
> She manifested this in her approach to incorporating dissenting opinions, her presentations to the PAWG, and her intent to unilaterally 'reclassify' me as a representative of environmentalists because of the paltry $3000 I received from them toward my ozone monitor project. I guarantee that this would have led to a serious conflict had I not exited. Before you dismiss my criticisms of your staff, remember, I was in government service for 20 years and supervised a large force of civil servants, so I know the symptoms when I see them.

2009

The record of meeting minutes becomes quite sparse from here on. The next available meeting minutes are the result of a gathering on April 7, 2009, but they seemed rather cryptic and only made sense if you had been there. The title page of those minutes stated that this was the first meeting since fall

2007.[51] Of the 20 persons present, six were with the natural gas and electrical power companies, six were with government land management agencies, two were environmental group representatives, five were local citizens, and one was a local representative to the Wyoming legislature.

The primary topic of discussion was the growing ozone problem resulting from emissions from the Jonah and Anticline gas fields. There was considerable give and take about what could be done to address this, who should and indeed who would, and finally, how to deal with the constraints that seemed to bog the process down.

It began with a statement from a WYDEQ staffer that it would be recommending Sublette County to EPA for ozone non-attainment status under the national 8-hour standard EPA guideline. After a five-month review process by EPA, it would release a public notice of intent to accept the proposal after which there would be a 30-day public comment period. If all went according to schedule, EPA would issue its final decision and define the bounds of the non-attainment area. This milestone would be implemented eleven months hence on March 12, 2010. It was further explained that our county would likely be designated as "moderate to marginal" for ozone health risk.

It was again pointed out that our ozone problem challenged established science because what had been believed to be a primarily urban phenomenon was apparently not valid. WYDEQ claimed to be in discussions about this with EPA but it left something out. WYDEQ was attempting to "restart the clock" by challenging the environmental conditions that had created a previous ozone exceedance event. I refer to the protocol by which such an event can lead to non-attainment determination. It must be supported by low and high ozone concentration measurements during a consecutive three-year period. If the three-year cycle is broken the clock is restarted. I discuss this in detail in the chapter on WYDEQ.

The conversation went on to again reveal the bureaucratic nature of rulemaking. A new ozone exposure rule would take up to nine months to implement. WYDEQ was claiming to be in discussions about the technicalities of non-attainment designation while at the same time EPA was in the throes of unspecified legal challenges about the subject. We were warned that these challenges could delay the agency's response to the non-attainment designation for our region.

In an almost chicken-and-egg dilemma, it was pointed out that a health threat had to be established before appropriate regulations could be crafted.

In frustration I asked what WYDEQ's administrative procedure actually was for crafting regulations but the response was circular. Wyoming statutes dictate how to proceed with rule-making, which takes seven to nine months. However, the new Administration could choose to maintain, modify, or reconsider rule making, thus altering that time line.

Moving on, Jonathan Ratner hit a nail on its head. He questioned the utility of WYDEQ's claim that a new ozone study would be forthcoming by March 31, 2009. With that in hand, the agency would supposedly be better able to perform in an advisory role to the AQTG. He asked bluntly how this would be any better than the last time. He was referring to 2007 when similar actions had been promised but failed to deliver. He reminded us that that last relationship with BLM had been dysfunctional and he wanted to know what had changed.

Now the Pinedale BLM field office manager Chuck Otto who had chosen to sit in piped up that the reason for the prior failure was a lack of guidance from BLM higher headquarters. Furthermore, he recalled events leading to a mass resignation of PAWG members over the way business was being conducted which eliminated a quorum presence in that group. As a result, no formal actions could be taken. Now with a newly constituted PAWG, the main purpose for existence would be oversight of PAPA monitoring.

Unmoved, Jonathan and I continued to bore in. He pointed out that our first formal report of recommendations had not been acted upon by BLM. I coldly added that it wasn't until a full year after it had been submitted that Pinedale BLM managers bothered to read it. Furthermore, I reminded him that our second formal report had not been read for *six months* after submission.

With a modicum of squirming, Otto assured us that he was determined to make the PAWG process succeed. He then added his assurances that advisory groups like ours had been valuable to other offices and he would personally make certain that our future submissions were not neglected.

The PAWG member in the room who was representing one of two big environmental groups in the State then made a pretty unfounded statement. She asserted that Otto had made a difference by increasing the level of trust between the PAWG and BLM. This was unfounded because she had not been a member of the PAWG long enough to render such a determination and I had to wonder what her agenda might actually be.

Ratner wanted BLM to immediately review and reply to our reports of recommendations and Jim Sewell suggested that we review them first with an eye toward selecting those we would wish to forward yet again to BLM. This sounded to my ear like another operator delaying tactic so I felt compelled to point out that our task group was in need of better balance because there was too much industry maneuvering at work within our membership.

In a test of Otto's assurances of support, Jonathan raised the issue of travel compensation for our members who come long distances. He again recommended that BLM assist with the expenses. However, without a pause, Otto declared that regulations only provided such assistance to PAWG members. Clearly he had learned from his predecessor Prill Mecham.

In the discussion on what our focus should be, I tried to change thinking about the artificial separation between the Jonah and Anticline being imposed upon us by BLM. I observed that it seemed a sensible action that we drop this distinction because there were no barriers between them as far as the atmosphere was concerned. One of the Forest Service representatives reminded us that there was a time in the past when our group had discussed mitigation actions applicable to both fields. He speculated that there could be a useful outcome if only we could be allowed to address air quality outside the immediate Anticline area.

Others tried to probe into possible alternate tactics that would enable wider application of our deliberations and recommendations. Otto dug in by continuing to emphasize his jurisdictional limitations and insisting that we were there for purposes of giving advice and information regarding air quality concerns, not jurisdictional issues. The Forest service guy countered that we were pretty limited even in that regard if we could only talk about the Anticline.

Jim Sewell quickly supported Otto's distinction although his motivation was likely more business as usual in nature. A second Shell attendee backed him up by invoking our duties as defined within the charter limiting us to Anticline emission sources. Jim added some praise for past field data reviews by our group and wanted that extended to drilling currently underway. He tried to sweeten the pot by suggesting that the other operators could provide technical data on their drilling operations. In truth though, he could only speak for Shell and that was even suspect

The meeting ground to a halt over opinions about the causes of our ozone. One of the citizen attendees observed that our ozone exceedances were "low level" which complicated scientific understanding and prevented determination of appropriate solutions. I countered that while EnCana had recently bragged to WYDEQ regulators about its emissions reductions, at the same time visibility was declining and ozone was increasing. This implied a lack of understanding about cause and effect and the effects which were *not* low level. Ratner underscored this by observing our pristine air quality back in the year 2000 was now only a memory.

In a state of frustrated exhaustion we closed this meeting of the AQTG and agreed to a telephone meeting on April 30, 2009.

• • • • •

The April 30, 2009, call-in meeting was perfunctory in nature.[52] Five of us were present and five of us attended by telephone. Three new members were introduced and no one volunteered to fill the chairperson's position. We decided that the 2005 Air Quality Report was likely outdated by now, thus justifying a revisit of the original recommendations and placing them on hold for the time being. It was further decided that the old report should be reviewed with attention placed on suggestions as to how the document could be improved and updated to reflect the current state of air quality issues associated with the Anticline. New recommendations to the PAWG would be crafted accordingly.

The meeting actually seemed to generate more uncertainty than substance and in so doing, signaled the declining morale, energy, and relevance in the minds of many of the long time members. We wondered who would be willing to accept the burden of chairperson? How would we be able to take on the heavy lifting involved with editing the 2005 report or rewriting it entirely?

Who would be willing and qualified to take the lead on this? Terry and Cara had filled this role the first time but were no longer routine members.

What were BLM's current expectations and requirements for our task group? Both seemed no clearer now than when the task group first came into existence. Finally, there was a legitimate concern that the current members, newest, and original, lacked the technical expertise needed to engage in continual updating of our annual report with current technical field developments and data.

It was becoming apparent that the years of stonewalling by BLM and WYDEQ, aided by glacial movement of EPA, had drained the task group of its will, energy, and dedication to purpose.

• • • • •

The last AQTG meeting to be found in the BLM public record convened on June 4, 2009.[53] I was no longer involved with the group because I had given up any hope that we would ever be heeded by the PAWG and BLM. A reading of the proceedings of this meeting seemed to confirm my gloomy assessment.

The turnover in membership had eliminated the corporate memory of all our work from previous years. The new members were thrashing around because of their novice limitations about understanding the clash of scientific and bureaucratic processes that was the driving influence on the group's endeavors. This was illustrated by the general view synopsized in the minutes. They revealed a flawed assumption that our original 2005 Air Quality Report to the PAWG had been itself flawed because of questions about its significance and public usefulness.

Discussions began with queries about the PAWG and it's current operating guidelines. New PAWG member and Wilderness Society representative Stephanie Kessler was in the room. She stated that a document had been crafted in the previous month defining its role, its processes, and its duties. She then seemed to use that as a wedge to advocate for the AQTG to do the same for itself and included specifics about how future reports should look.

This had all been well addressed in our first organizational meetings back in 2004 but apparently, BLM and certainly the PAWG had lost its historical memory. Evidently, this individual was going to fix that by requiring administrative redundancy.

Appearing deaf to all of the above, the BLM representative Lauren McKeever put the group on notice that the next annual report was due to BLM in three months. However, reality was unavoidable, so she admitted that would likely be insufficient time for the inexperience members to craft the document and suggested they decide on an acceptable new deadline.

This dilemma was minimized a bit later when Ms. Kessler observed that provisions in the Anticline ROD essentially made it necessary for the AQTG to delay beyond the upcoming deadline because of time needed for information collection, analysis, and dissemination. This issue of deadline surfaced several times during the meeting and ultimately produced agreement that January 2010 was the most achievable option.

When the necessity of extending the report deadline for these activities was questioned, Ms. McKeever countered that more time was going to be needed specifically to analyze visibility impacts. This led to exchanges about just what the report must contain.

A big topic of discussion was emissions from the PAPA because Kelly Bott, the WYDEQ representative noted that a major emissions inventory was due in a month and would be a key indicator of whether or not operators had met their visibility thresholds. Furthermore, ozone modeling results were due at the same time as the AQTG report deadline.

Additionally, reports were due from the operators on performance of their liquid gathering systems designed to reduce emissions. All of this would weigh heavily upon WYDEQ consideration of consequences of non-compliance with the offset credit scheme it had recently put in place.

These references to emissions inventories and offset credits trigger unpleasant memories. A few months earlier WYDEQ had embraced a self-serving proposal from EnCana to allow its NO_x emissions in return for its "good work" at reducing VOCs. Once approved for one operator, the others clamored for and received the same consideration. Thus, a major precursor to visual haze remained insufficiently regulated. There is more on this in the chapters about WYDEQ and the operators.

Returning to meeting proceedings, the question was posed as to data resources available for use in any report update effort. Terry advised that there was a lot of 2005 data but that was now obsolete. He recommended three options: (1) re-accomplish the 2005 report with new information, (2) craft a report supplemental to the 2005 original version, or (3) present the 2005 report appended with bibliography links to new data sets. However, Kessler chimed in to warn that the new PAWG guidelines she had just cited forbade any use of data older than one year.

McKeever tried to clarify this restriction by assuring that the PAWG had elected to grandfather older report guidelines and the AQTG still retained authority to select any structure it preferred. WYDEQ's Kelly Bott then offered that McKeever was correct and previous guidelines still constituted a valuable basis for AQTG purposes.

This entire exchange seemed to have been done in code whereby references to "data" were being grafted onto the word "guidelines." One had to wonder if this was an attempt to transform sound but perhaps unwelcome field measurement data into malleable currency offering less definitive and less threatening possibilities.

Another exchange between Bott and Kessler demonstrated the reliance of federal and state overseers upon the operators for compliance information. Kessler asked who was responsible for visibility milestones and Bott replied that the operators were required to show federal and state agencies they had met all visibility milestones stipulated in the PAPA Environmental Impact Statement.

Kessler asked if the PAWG had further requirements not stated in the PAPA Record of Decision (ROD). The question was odd because the PAWG had no authority to create any requirements. Adding to this oddity, Bott responded there were some in the PAPA Air Quality Technical Support Document (TSD) and stated the page and table numbers.

This truly seemed a non sequitur because the TSD had nothing to do with the PAWG or any wishes it might have held. The purpose of the TSD was to report air quality impact modeling results for the various development alternatives of natural gas development on the PAPA.

Then McKeever exhibited the BLM preference for administrative paperwork by recommending that WYDEQ include a summary cover letter attached to all upcoming WYDEQ reports and communications. Bott was happy to comply.

Next came discussions that seemed to underscore a lack of direction and weak understanding of the entire relationship involving BLM, the PAWG, and the AQTG. Kessler asked if WYDEQ was responsible for some of the deadline-driven inputs. Bott replied that her agency would provide data sets but would not engage in interpreting it. The question was then asked why the PAWG would be viewing that data (or even be qualified to do so) and who was the intended customer for the AQTG's report to the PAWG in the first place?

Bott responded that the AQTG was to compile the available data for the PAPA, and other locations into a single source document thereby providing an important convenience for public use and information gathering. Svalberg tried to explain the technical nature of the AQTG report project by noting that the end result would enable useful trend analyses presumably in search of common emissions sources and their impacts. This would be difficult stuff for general public consumption.

Kessler revealed further ignorance of AQTG purpose and history when she insisted that more had to be done than what had been the case with the 2005 Air Quality Report. One had to wonder if she even looked at the report to see its detailed technical content of extensive scientific data and resultant specific recommendations.

This aspect of the proceedings seemed to expose a sad inevitability. If a second report were ever to be crafted and released to the PAWG, it was likely destined to serve as little more than an exercise in administrative information compilation. It would be merely another collection of pages that BLM and WYDEQ could point to as evidence of a citizen-operator-regulatory partnership even though it would result in no tangible influence on air quality oversight.

Now came a round of discussions about who should do what regarding air monitoring. Kessler opined that the PAWG should be a clearing house for air monitoring information and act as the connection between the public and the air quality impacts mitigation plan (presumably she was referring to provisions in the PAPA ROD). In that role she advocated for the PAWG to serve as the conduit for notifications to the public about whether air quality standards were being met.

Now an attorney in the meeting (the minutes do not explain what his role was) spoke up to caution that the recognized goal of the PAWG was to provide BLM ideas for reducing pollution. In so doing, he seemed to be cautioning that the AQTG and the PAWG were operating under charters that did not include talking directly to the public. Also, no regulatory powers were included. Svalberg seemed to feel that some regulatory latitude was indeed conferred on the PAWG but ultimately this entire conversation seemed to simply grind to a halt.

Finally an important point was raised when Kessler asked what would be the consequences if air quality benchmarks were to be violated. She correctly worried that a new round of modeling might be the result although she incorrectly assumed the PAWG would be the entity to request that. McKeever suggested that the PAWG should create a summary of all air monitoring activities, include valuations of compliance with air quality goals, and guide the public to locations where more information could be found. This appeared to be an attempt to cast the role of the PAWG, and by connection the AQTG, in the role of guides toward information repositories.

At this point the conversation seemed to take a curious U-turn back to the subject of the 2005 report. It seemed as though a different report was being discussed because there was a lot of praise for it. However, Kessler continued her criticism by arguing that the report required a "redo" because of much more new data and newer more challenging issues. Furthermore, she complained that in its original form it offered no usefulness to the public and should not be duplicated in any form. Again, she demonstrated either misunderstanding of the original intent, which was to influence BLM's air quality oversight, or she was more interested in pursuing the politics of public image.

On the other hand, WYDEQ's Jennifer Frazier commented that the 2005 report had been a good product because it had indeed accomplished the very points I described above that McKeever had cited a need for. Ratner added his opinion that the report which he had helped craft had been of such quality that an update with current data would suffice. Ratner also advised that he had recommended a three-page summary for public consumption because of the full report's 86 pages of content but that was not done.

Echoing Ratner's desire for a report summary, Frazer advised that the next report needed to be more user friendly. Kessler commented that this time the report must not be filed away as had happened to the 2005 version. McKeever promised to do her best to prevent that. So, lots of good intentions were voiced but action looked to be in short supply.

The group then moved on to an attempt at agreeing on a format to be used in preparing the report and again, parochial defensive self-interests assumed dominance. McKeever requested consensus on a format made up of a summary and reference guide. Secondly, she wanted a separate report document listing all emissions-related activities taking place on the PAPA including an explanation of detected trends. Frazier immediately went on the WYDEQ defensive by declaring that the AQTG had no business engaging in data trend analysis. McKeever then backtracked by circularly stating she only meant a report on analysis of trends and not original trend analysis.

This really looked like an attempt to thread the needle in such a way that WYDEQ would not feel threatened and that its turf was not being invaded. Ratner spoke up and seemed to imply that analysis was needed because simply providing the public with links to other information sites would illuminate nothing; the public needed help in understanding fundamental meaning.

Bott weighed in with further WYDEQ firewalling by claiming she was good with the idea of revealing trend data but the report must not draw conclusions in either a negative or positive manner. This was almost stupefying because it essentially mandated that the report be no more than a compilation of numbers, and activities thus becoming only a tour guide.

Next came a silly argument about report length. Someone apparently suggested the report be only three pages long and McKeever warned that would be too brief. Kessler piped up with the suggestion that ten pages might be acceptable but under no circumstances should it stretch to 200 pages. On the other hand, Ratner felt that ten pages was too long. As I sat back after reviewing this I laughed aloud over the preposterous image of everyone debating

By the Rasping in My Lungs

page count. What happened to the idea that quality and content should have been the driver dictating document size? This was yet another signal that the original purpose of the AQTG had been lost amid the competing currents of politics and science.

Yet again the group circled the mulberry bush about its purpose and still insisted on muddying the conversation with similar issues about the PAWG. Kessler insisted on addressing the PAWG role that she argued should be one of providing the "big picture" to the public. Again the group argued over its role with field data but one member worried they were unqualified to draw conclusions. Svalberg argued that was appropriate in part because the BLM and WYDEQ were not stepping up to that role.

One of the members asked about any sources of funding (that nagging detail again) and Jennifer Frazier responded that the Pinedale Anticline Field Office was taking proposals for projects as candidates for funding. The deadline for application was only eight days hence so that was out of the question but a second round of applications could be submitted for the next application closing some months later.

Later in the meeting she again raised the topic and stated that the AQTG could qualify for more than $15,000 in funding assistance. She further believed that this was a useful action to follow because she felt certain the operators had begun to accept the growing level of concern by the citizens of Sublette County. She also warned that these funds were intended to be applied toward mitigation.

In a replay of all the years of meeting before, the group mired down in this search for sources of funding for its report. Bott noted the PAPA ROD stated that funding from the Pinedale Anticline Project Office (PAPO) could be applied to research intended for public access. The PAPO was tasked to distribute operator contributed funding for the purpose of supporting approved air quality mitigation projects. Kessler confirmed this by citing another page in the ROD.

Bott also advised that the PAWG could request funding for contract assistance in preparing an air quality report. Finally, McKeever asked for consensus on crafting a summary report and submission of an application for funding from the PAPO. Kessler offered to draft the grant application to support the report effort. However, she needed to know what skill sets were needed to engage in data related sections of the report because that would affect the amount requested

It was suggested that a writer with background in air quality would be important. Also, the contribution of a data analyst yet to be hired by WYDEQ would be important. With these thoughts, a ballpark sum of $5,000 to $15,000 was deemed the likely spread. Further refinement would be necessary and because he had been instrumental in preparation of the 2005 report, Svalberg was asked to help estimate the number of labor hours likely needed to accomplish the task of this new report preparation.

What followed was more machinations about what statistics would be useful, how they should be processed, how they should be presented and with whom in mind. Also, more angst was expressed over what types of monitors needed to be included and whether they should be air monitors or lake acidity monitors or some of both.

A lot of chatter then ensued about what and how many data sources would be incorporated, the title to be used that would reflect the report's content, and the language needed in the application that would conform to PAPO guideline requirements. The issue of recent ozone findings was raised in the context of it being added content to this report or a separate report. Kessler stated her desire to write a grant request for only one all-inclusive report.

Yet again the group verbally milled around in a discussion over what should be included in the report. Svalberg drew on his air quality science experience by warning the group that it had to be explicit in selecting air quality impact types because that in turn drove monitor selection and the kinds of field data that would have to be collected. He was asked about the usefulness of graphs of field data to which he replied they could be key to revealing trends and linking those trends to particular oil and gas development activities nearby.

One of the group members then recommended that the AQTG identify gaps in field data, settle on a limited objective for the report, and decide how to fill any identified gaps in the field data. The purpose for the latter would be crafting recommendations to appropriate groups about how to correct those data gaps.

McKeever wondered if mitigation actions could be levied upon operators on the PAPA by this or any other group but no one seemed to know the answer. This was a further illustration of the uninformed nature of this newest incarnation of the AQTG. In previous years, quite heated debates had taken place about BLM and WYDEQ authority to do just that. Some argued that BLM had ample authority under the Clean Air Act and its obligation to safeguard public lands under its care.

Others argued that BLM had no authority to involve itself in issues of air quality oversight and occasional behind the scenes conflicts between WYDEQ and BLM over the subject attested to this. Finally, when I challenged a BLM speaker in a public forum about this level of enforcement, I was flatly told that BLM had no authority to impose any requirements on the operators.

Yet again, further observations were in abundance about obligations to understand any data that would be incorporated, how to process it, and what it should look like to be most useful to the public. Finally, Frazier and Kessler committed to craft the grant application and the meeting was adjourned after agreement to reconvene on July 16, 2009.

No record of such a meeting could be found, creating the question of whether it ever took place.

Through a Glass Darkly

Several aspects of this last meeting in the public record stand out. First of all, it is surprising that so much time in a meeting of the AQTG was consumed with debates about the role of the PAWG over which the AQTG had no control. This was an inappropriate place to argue about the role and functionality of the PAWG. The AQTG was subordinate to the PAWG and the latter frequently made that plain to us in the earlier years when we tried to move it in directions intended to repair the decaying air quality situation downwind from the Jonah and the Anticline gas fields.

Second, it was very clear that the corporate memory of the task group had been seriously dissipated by years of turnover in our membership. That was a result of the near obstinate resistance we experienced from a succession of directors of the Pinedale BLM field office as well as from the state BLM office. Several of us including myself saw no value in expending our time for a cause that neither the BLM nor WYDEQ shared; so we exited from the group.

Toward the end, newer members appear to have been more interested in creating a public image of proactivity. They displayed little understanding or perhaps actual interest in doing the heavy lifting necessary to bring about real regulatory oversight.

In particular, there seemed to be a reluctance on the part of those newest members from BLM and WYDEQ to allow actions by the group that ventured too close to actual interpretation of the field data. That was a crippling paradox. On the one hand they professed to see a crucial need for measurement data but on the other hand wanted to content themselves with merely throwing it out to the public as if to say "see, we are monitoring."

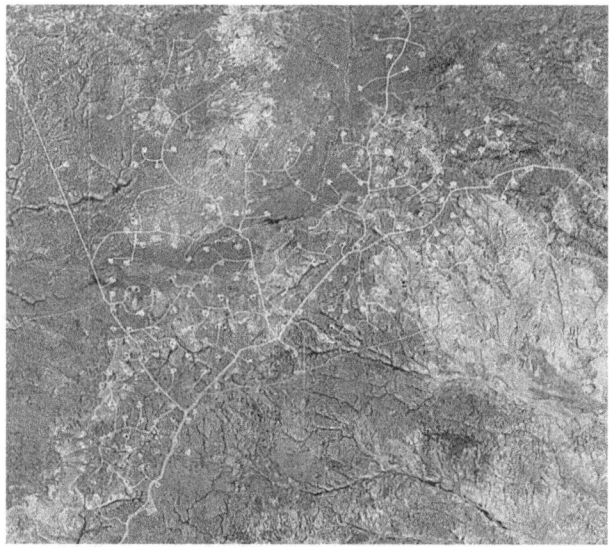

Fig. 1. View from space of the Jonah gas field.
(https://www.epa.gov/outdoor-air-quality-data/interactive-map-air-quality-monitors)

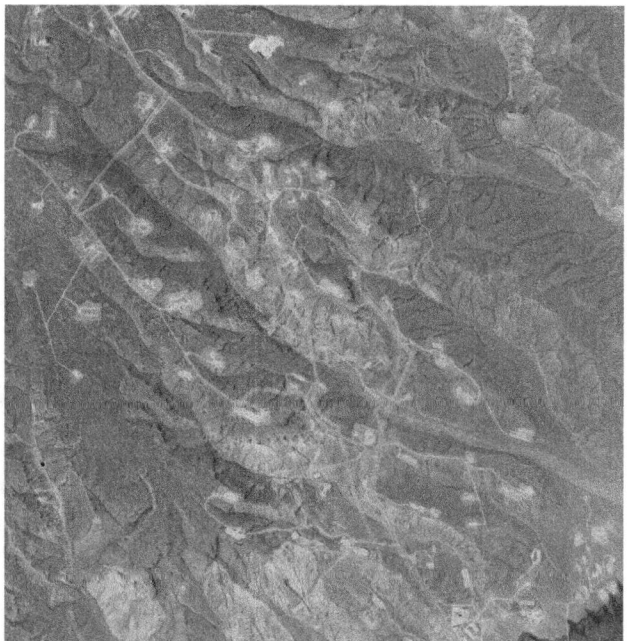

Fig. 2. View from space of the Pinedale Anticline gas field.
(https://www.epa.gov/outdoor-air-quality-data/interactive-map-air-quality-monitors)

Fig. 3. Aerial view across the Jonah gas field showing density of well pads.

Fig. 4. Normally Pressurized Lance (NPL) gas field in July 2013 before development illustrating the original state of the Anticline and Jonah regions.

Fig. 5. Typical drilling operation underway. Note the battery of storage tanks.

Fig 6. Example of heavy construction underway in the gas fields.

Fig. 7. Industrialization of what had been open prairie and Pronghorn Antelope and Sage Grouse habitat.

Fig. 8. Example of multiple dehydrator combustor stacks on a single well pad in the Anticline gas field.

Fig. 9. Typical well completion flare.

Fig. 10. Examples of well completion haze scattering light from sun hidden behind post: No haze (left), medium haze (middle), heavy haze (right).

a.

b.

Fig. 11. Haze 6 hours after a well completion flaring event (a.) and a clear view for comparison (b).

Fig. 12. Examples of drill rig smoke created by engines and burn operations.

Fig. 13. Visible (left) and infrared (right) views of a pipe fitting methane fugitive leak.

Fig. 14. Example fluids storage tank battery and flash vapor combustor.

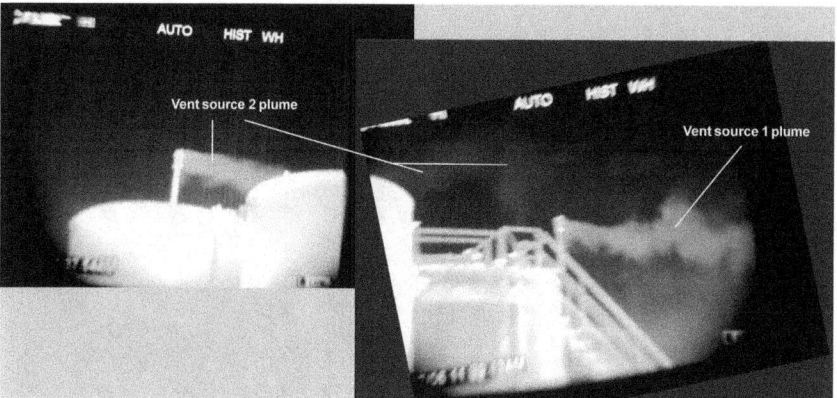

Fig. 15. Composite infrared view of tank vent sources 1 and 2 in previous figure.

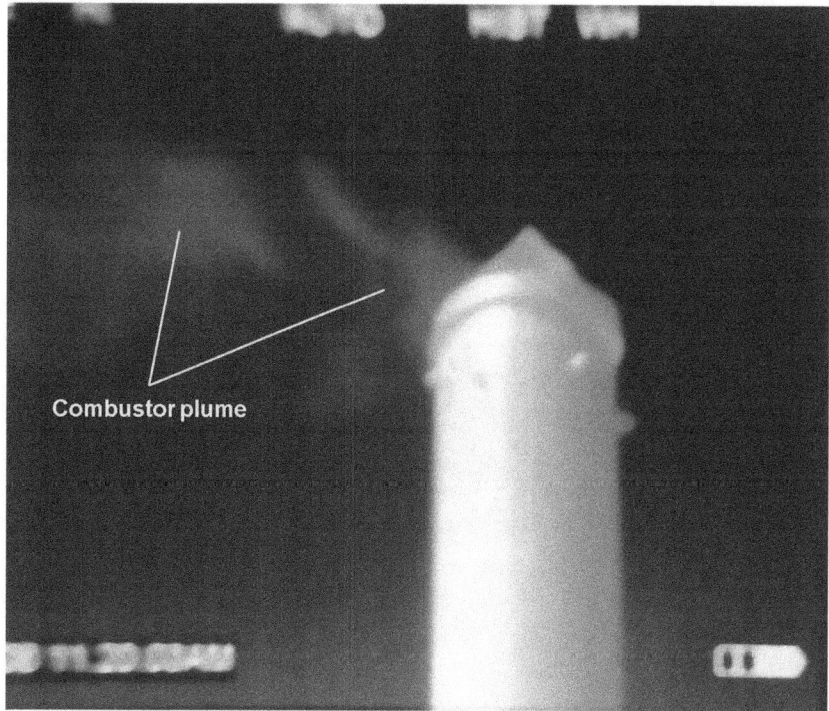

Fig. 16. Infrared view of tank flash vapor combustor plume from previous figure.

Fig. 17. Miniature optical spectrometer setup in field portable configuration and camera used to document flares.

Fig. 18. Raising ozone monitoring tower.

Fig. 19. Controlled environment cabinet containing Dasibi ozone monitor and support equipment.

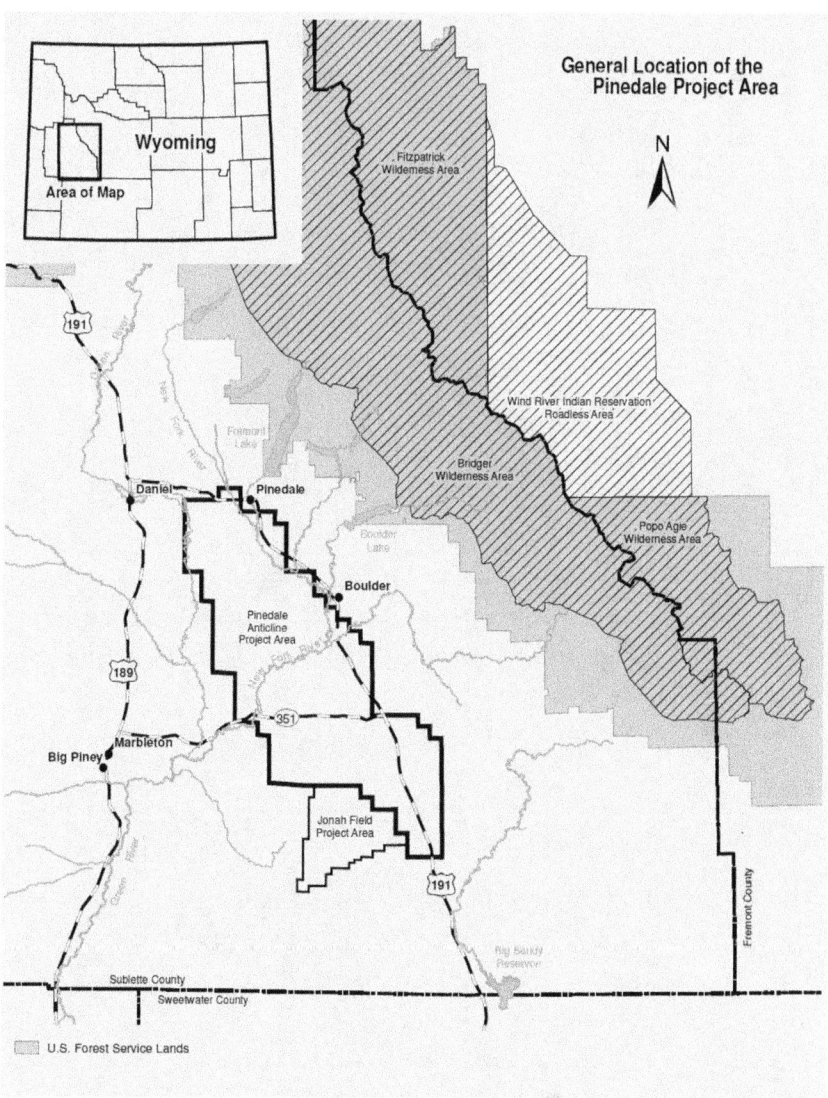

Map 1. Locations of Jonah and Anticline gas fields in Sublette County.

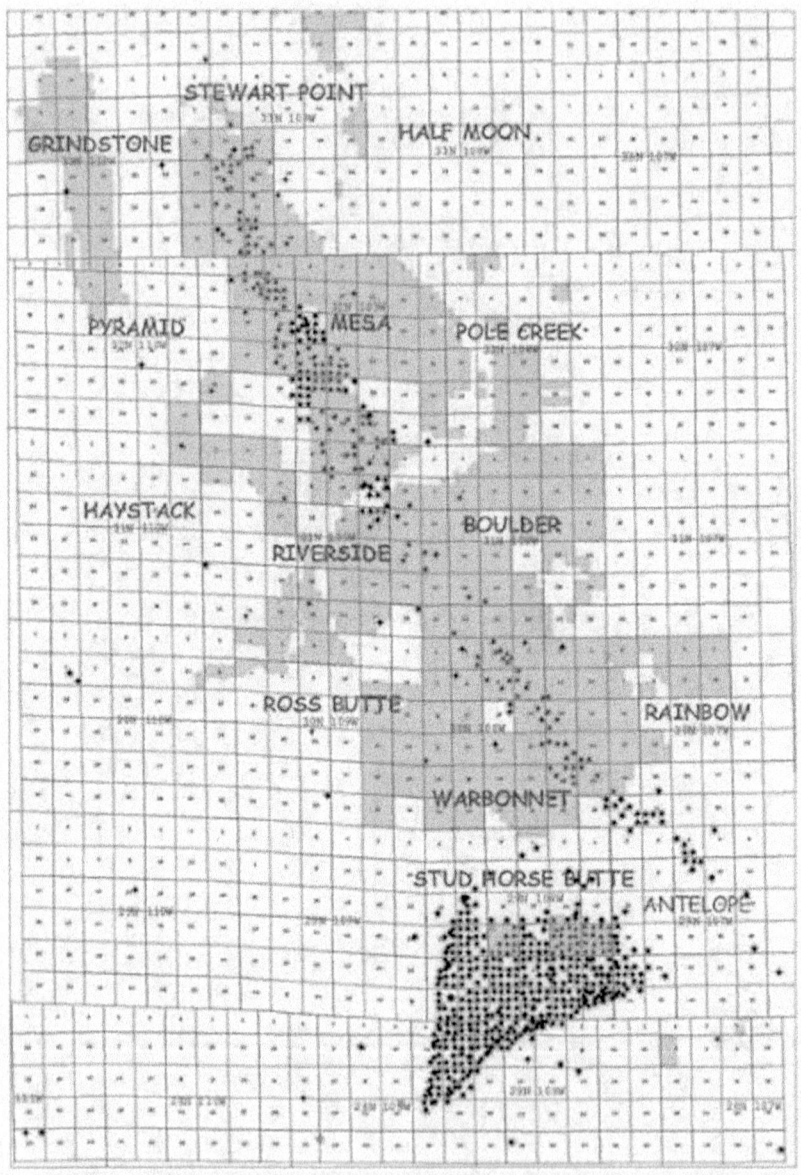

Map 2. Locations of gas development lease units within the gas fields.

By the Rasping in My Lungs

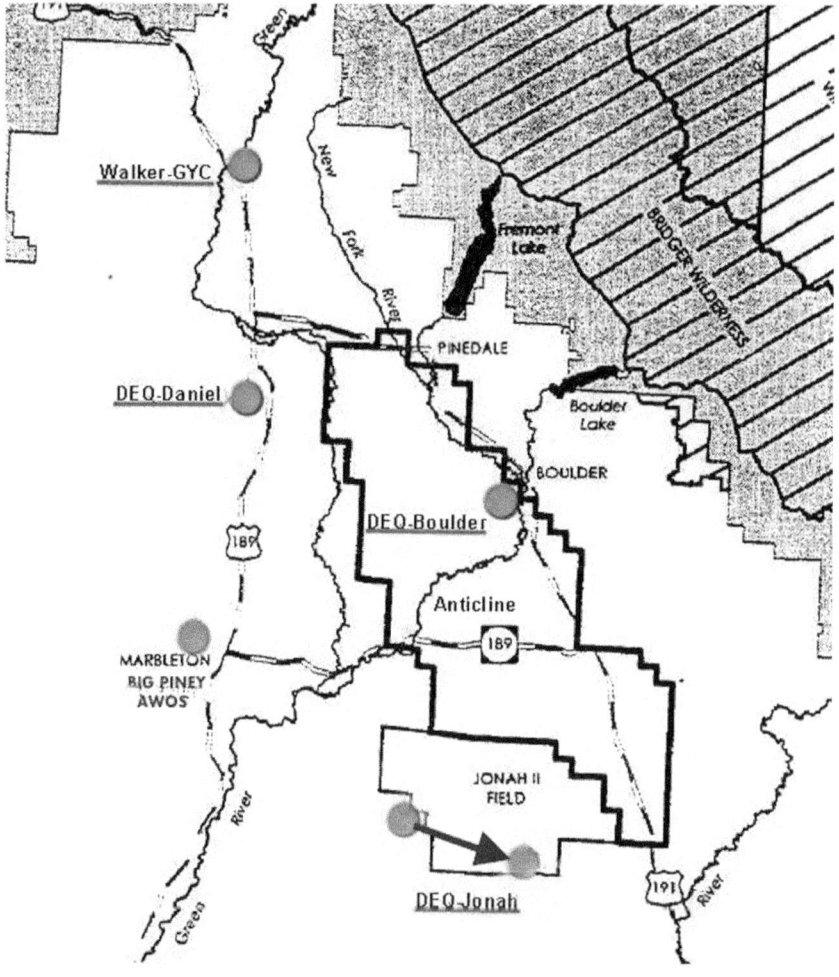

Map 3. Locations of state meteorological stations relative to the author's. Note relocated DEQ-Jonah station.

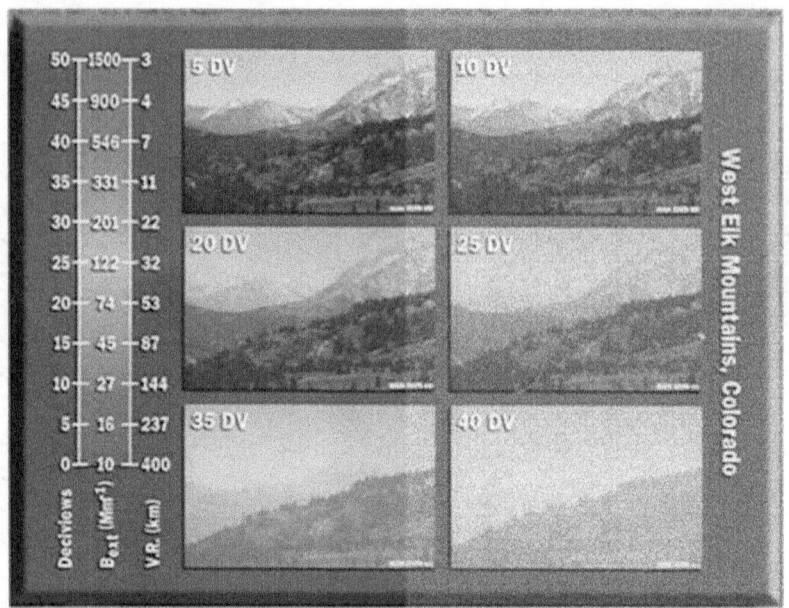

Chart 1. Comparative chart of visibility reductions due to particulate haze. (Source: USFS)

Wyo Range	2010	2009	2008	2007	2006	2005	2004	Wind River	2010	2009	2008	2007	2006	2005
Jan	39%	30%	32%	40%	31%	37%	37%	Jan	27%	21%	43%	25%	24%	27%
Feb	29%	49%	23%	27%	36%	44%	32%	Feb	31%	12%	38%	24%	27%	25%
Mar	21%	29%	21%	24%	22%	16%	22%	Mar	30%	37%	49%	31%	9%	30%
Apr	20%	24%	13%	21%	23%	27%	28%	Apr	36%	30%	58%	42%	41%	27%
May	5%	19%	20%	25%	21%	23%	21%	May	74%	39%	47%	41%	45%	41%
Jun	22%	28%	16%	26%	15%	21%	26%	Jun	40%	38%	60%	48%	50%	43%
Jul	30%	23%	21%	28%	21%	17%	25%	Jul	42%	34%	49%	46%	21%	43%
Aug	37%	26%	31%	26%	26%	27%	25%	Aug	40%	33%	40%	50%	42%	39%
Sep	30%	29%	36%	37%	22%	24%	31%	Sep	38%	37%	32%	47%	37%	36%
Oct	26%	27%	28%	31%	24%	32%	26%	Oct	30%	23%	29%	43%	33%	28%
Nov	22%	30%	24%	25%	37%	29%	29%	Nov	33%	24%	24%	45%	37%	35%
Dec	42%	31%	35%	24%	37%	29%	23%	Dec	25%	26%	26%	50%	27%	32%

Note: Values for May of 2010 as well as April and June of 2008 in the Wind River table are unusually large excursions and therefore are suspect due to possible station malfunction.

Table 1. Percent of time that winds travel toward Class I and Class II regions.

By the Rasping in My Lungs

> Fracking fluid compounds are marketed under trade names such as: Clearfrac, Fracsal-Ultra, Fracsal-G, Spectra-Frac-G, HPF, Flowzan, Clarizan, Greenbase, Prod XL-xxxx, Prod CX-xxxx, and Protectozone, to name but a few. The chemicals used are employed in a specific relationship to the geology and down-hole conditions prevalent in a particular location. Examples are:
>
> | | | |
> |---|---|---|
> | Sodium Persulfate | Sodium thiosulfate | Sodium hypochlorite |
> | Lithium hypochlorite | Potassium Chloride | Ammonium Chloride |
> | Sodium Chloride | Sodium Formate | Calcium Chloride |
> | Sodium Bromide | Potassium Formate | Calcium Bromide |
> | Cesium Formate | Zinc Bromide | |
>
> A major area of activity is in the realm of gel cross linkers which employs exotic sounding chemistry. Examples are Borated alkali metals (Sodium tetraborate decahydrate, Lithium tetraborate, Calcium borate). Other example metals used are titanium, zirconium, and chromium.

Table 2. Information the author collected from the internet in 2006 to identify emission peaks observed in spectra of well completion flaring, dehy, and combustor emissions.

Spectral Peaks Seen in Glycol Heaters

Wavelength	Possible Emitter	Wavelength	Possible Emitter
215nm-256nm	NO	627nm	?
308nm	OH radical	670nm	Li (lithium)
387-390nm	CN, CH radicals	692-694nm	?
429nm	NO_2, CH radical	716-721nm	?
467-470nm	NO_2, C_2, C_3	766, 769nm	K (potassium)
511-514nm	C_2	814nm	?
556-561nm	O^+ ion	823nm	?
588nm	Na (sodium ion)	829nm	?

Table 3. Matches between observed wavelength and possible chemical sources.

DEQ #	AP Date	Operator	Compressor Stations Facility	Ctld VOC	Ctld HAP	Ctld NOx	Ctld CO	
AP-4050	Nov05, Jul06		PetroGulf Cmprsr Sta	18.6	1.1	18.6	15	
AP-1405	4-Jun		Luman Cmprsr Sta					
			Existing in 2004	110.9	15	140.7	65.6	Note 1
			Proposed Change	143.4	18.3	175.1	87.4	
AP-3169	5-Jun		Luman Cmprsr Sta					
			Existing in 2005	143.1	20.9	175.3	87.7	
			Proposed Change	178.5	26.6	224.9	105.5	
	6-Jan		Final JIDP EIS sting Permitted in 2006			121.5		Note 2
			Projected			198.3		Note 3
AP-2448	4-Sep		Bird Canyon Cmprsr Sta					
			Existing in 2004	185	22.5	241.3	138.3	
			Proposed Change	187.2	22.8	244.3	139.6	
	6-Jan		Final JIDP EIS sting Permitted in 2006			104.3		Note 4
			Projected			177.7		
		Dec06, Dec07, Jun08 Draft, Revised,Final Ant.EIS-Projected				98.3		Note 5
AP-3170	5-Jun		Falcon Cmprsr Sta					
			Existing in 2005	142.8	21.4	183.9	92	
			Proposed Change	139.8	21.9	188.9	75.1	
	6-Jan		Final JIDP EIS sting Permitted in 2006			77.7		
		Dec06, Dec07, Jun08 Draft, Revised, Final Ant.EIS-Existing				185.3		
		Dec06, Dec07, Jun08 Draft, Revised, Final Ant.EIS-Projected				250.3		Note 6
AP-2600	5-Feb		Paradise Cmprsr Sta					
			Existing in 2005	109.5	16.6	149.3	56.1	
			Proposed Change	128.5	17.8	165.5	59.7	
Ap-3171	5-Jun		Paradise Cmprsr Sta					
			Existing in 2005	129.3	17.9	167.4	58.2	
			Proposed Change	182.4	26.4	241.8	84.9	
		Dec06, Dec07, Jun08 Draft, Revised Draft, Final Ant. EIS-Existing				161.2		
		Dec06, Dec07, Jun08 Draft, Revised Draft, Final Ant. EIS-Projected				1,232		Note 7
(no AP)	6-Jan	Final JIDP EIS	Yellow Point Cmprsr Sta					
			Existing in 2006			7.4		
			Projected			7.4		
(no AP)	6-Jan	Final JIDP EIS	Jonah Field Cmprsr Sta					
			Existing in 2006			32.4		
			Proposed Change			57.2		
(no AP)			Gobblers Knob Cmprsr Sta					
		Dec06, Dec07, Jun08 Draft ,Revise , Final Ant.EIS-Existing				125.7		
		Dec06, Dec07, Jun08 Draft, Revised, Final Ant.EIS-Projected				207.7		Note 8
AP-3484	5-Dec		Birch Creek Cmprsr Sta					
			Existing in 2005	25.8	5.2	40.5	48.9	
			Proposed Change	22.3	5	45.1	59.3	

Notes:
1. emission calc. based on 8700 hr/yr run time
2. Luman existing Nox emissions not included in any Anticline EIS documents
3. Luman projected Nox emissions not included in any Anticline EIS documents
4. Bird Canyon existing Nox emissions not included in any Anticline EIS documents
5. 98 tpy NOx by 2011 in all Anticline EIS documents
6. 250.3 tpy NOx by 2011 in all Anticline EIS documents
7. 1,232 tpy NOx by 2011 in all Anticline EIS documents
8. 207.7 tpy NOx by 2011 in all Anticline EIS documents

Table 4. Compilation of reported and projected changes to compressor station emissions resulting from capacity increases. Derived from various EIS documents.

AP No. & Unit	Well Count	WOGCC "Missing Gas"-2010 mmscf	DEQ Permit Listed 2010 Vol Burned mmscf/yr*	Count of Htrs & Cmbstrs
		Encana		
Cabrito				
AP-10149	1	4.2	28.3	7
AP-11276	18	24.3	46.0	10
AP-11400	24	9.1	36.6	9
AP-11435	17	9.7	30.5	8
Yellow Point				
AP-9476	8	5.3	31.7	7
AP-8357	13	6.9	31.7	7
Studhorse				
AP-11015	4	3.8	36.9	5
AP-11150	15	6.2	34.0	7
		Shell		
Riverside				
AP-9953	5	32.2	74.9	13
AP-10499	10	56.1	91.1	14
AP-10796	8	38.8	43.2	6
Mesa				
AP-12396	13	25.5	75.5	11
AP-12612	10	30.7	66.6	9
Warbonnet				
AP-7804	8	29	62.5	10
AP-8656	16	35.3	75.5	11
AP-9090	16	36.5	62.5	10
		Ultra		
Riverside				
AP-10023	8	20.7	176.4	31
AP-10947	17	16.5	40.1	8
Warbonnet*	9 months of 2011			
AP-12262	10	21.6	120.8	22
AP-12263	20	37.7	164.6	28
Mesa				
AP-11672	12	33.3	119.8	18
AP-11675	11	21.7	125.9	22

Table 5. Comparison statistics of gas unaccounted for versus gas burned in heaters and combustors on example well pads. The two were not the same.

AP No. & Unit	Well Count	WOGCC Missing Gas-2010 mmscf	DEQ Permit Listed 2010 Vol Burned mmscf/yr*	Count of Htrs & Cmbstrs
		Questar		
Mesa				
AP-9147	14	51.65	147.6	22
AP-9220	21	65.4	153.7	25
AP-10024	17	38.2	138.9	22
AP-11453	2	1.45	33.9	6
Stewart Point				
AP-11427	6	12.35	67.4	11
AP-11429	3	23.57	42.5	8
AP-9223	10	28.15	90.1	14
AP-9234	3	4.63	38.0	6
		BP America		
Corona				
AP-8137	3	1.21	32.2	7
AP-7145	14	15	61.5	11
Studhorse				
AP-8192	4	4.38	35.7	7
AP-9358	8	8.76	48.7	10
AP-9360	7	6.17	49.9	10
Cabrito				
AP-8193	1	0.71	7.1	4
AP-8194	3	3.29	18.2	4
AP-10847	5	5.48	49.9	10

Notes: Each operator uses a different pilot gas heating value
  Encana: 1065btu/scfm
  Shell: 1075btu/scfm
  Ultra: 1050btu/scfm and 1061btu/scfm
  Questar: 1000btu/scfm
  BP America: 1109btu/scfm
  DEQ: 1050btu/scfm

Findings: 1 - VOC cmbstrs consume only a few percent of total burned gas
  2 - Large number of heaters and BTEX cmbstrs consume most gas and create large volume of emissions vented directly to the atmosphere
  3 - BTEX cmbstrs process dehy vapors; VOC cmbstrs process tank vapors
  4 - There appears to be no relationship between WOGCC "missing gas" and gas consumed on-pad by burners and cmbstrs; Missing gas is always less than burned gas by a factor of 2 to 10.
  5 - This unaccounted-for gas and widely differing company reporting methods of combustion-caused emissions add to reasons for doubting DEQ emissions inventory.

Table 5. Continued

By the Rasping in My Lungs

Ozone Badge Results 2008
April

Day of month	6	7	8	9	10	11	12	13	14	15	16	17	18	19	20	21	22	23	24	25	26	27	28	29	30
Center of Pinedale			54	56	55			33		38		29						67					63		
East Pinedale	51		43		70			34		69		55			51	47	62		48			66		71	
Middle Butte			57		74			37		41		59					46		42		41		40		
Upper Green River Valley					48			35		58		55						86					76		
Cora		41	52		66			42		78		75			63		60		67		59		69		
Black Butte			58		28			33		66		59													
West Pinedale	6		61		56	78		80							73		66		62						
Boulder							63											60							
Cora	41		49					24			45	36						67							
New Fork Road				72		92		60			45	64	44	69		69		70	80	66		73			
5 mi S. of Pinedale																		38					76		73
New Fork Social Club											72														
Cora					83			56		74		80							88	60		99		85	
Marbleton			42																						
Big Piney					80			49		76		105			81		106		44			93			
New Fork River Valley						82		63		69		75			61			71	80		80		73		
Boulder				66		88		74		53		73			67		47	71	59		76	83	65		
Boulder									72			72			47		50	53		68		64	67		
Merna				46		52		40			56		39	38	30		44		63				46		
Daniel								34			40	50	50				55								
Big Piney (?)											55		70				51		66		69				
three locations			47		81		62																		
Big Piney Elementary																		55		54		51			48

May

Day of month	1	2	3	4	5	6	7	8	9	10	11	12	13	14	15	16	17	18	19
Center of Pinedale																			
East Pinedale		80			52		61		61										
Middle Butte																			
Upper Green River Valley																			
Cora		72																	
Black Butte																			
West Pinedale																			
Boulder																			
Cora																			
New Fork Road																			
5 mi S. of Pinedale									39										
New Fork Social Club																			
Cora			77			48		46		68			79		55		39		
Marbleton																			
Big Piney	78																		
New Fork River Valley			66				69	42		43			66		54		55		68
Boulder	47					60		43		43			55				54		
Boulder	56				58	53			67			53		51		67			
Merna				47				43			61		43		40		37		
Daniel																			
Big Piney (?)		80	85																
three locations																			
Big Piney Elementary					63														

Tables 6 & 7. Ozone badge results from CLOUD participants. Gaps in data reflects difficulty in achieving 100 percent participation 100 percent of the time. Nevertheless, many high readings near or exceeding the EPA limit of 75 ppb are apparent.

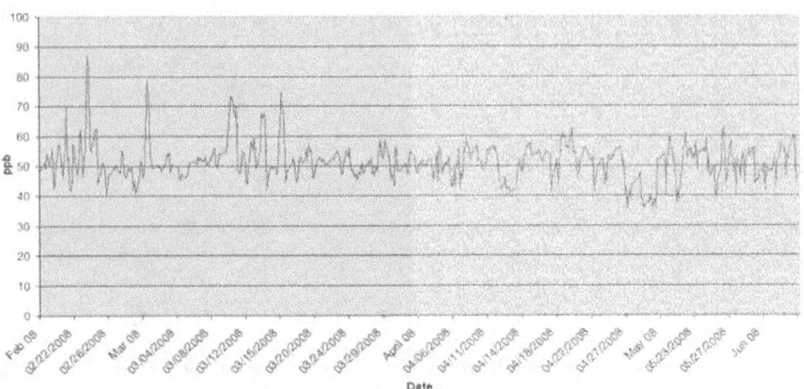

Graph 1. Results from author's ozone monitor for the period between Feb 23 and June 27, 2008. The series of five peaks register 87, 79, 73, 65, and 75 ppb four of which came close to or exceeded the new EPA maximum ozone health limit of 75 ppb.

Air Quality Statistics Report

	O3: 1-hr 2nd max	O3: 8-hr 4th max
2005	120	79
2006	90	74
2007	80	68
2008	120	101
2009	80	66
2010	80	67
2011	150	103

Table 8. Reporting compiled by EPA on Sublette County gas field ozone history.
Source: U.S. EPA Air Data http://www.epa.gov/airdata

The State

Chapter 6

Wyoming Department of Environmental Quality (WYDEQ)

Probably the darkest experience I had during the years I tried to make a difference in protecting Sublette County air quality was a result of the contest between WYDEQ and myself. From the very beginning of my interaction with the staff of that department I encountered unyielding resistance rooted in protection of turf and protection of public image. Adding to the problems of this interaction process was, except for the Director of WYDEQ John Corra, the turnover in staff. Corra did not leave and was to be my adversary for the duration. However, due to retirements, I would be dealing with three different directors of the air quality department of WYDEQ and each time I felt I was starting from scratch with the new guy.

This chapter contains examples of heated exchanges between the Air Quality Division directors and myself. Rather than summarize their content, I chose to relate them almost verbatim. My reasoning was that I want to preserve fidelity of content for the sake of those readers having a more academic research based interest in my story.

Over a period of 10 years, I transitioned from initially trying to work with them, to a posture of hammering them publically to force them to actually implement real regulatory controls over the operators. They worked assiduously to *appear* regulatory over the gas fields but actually imposed minimal controls over the operators in the early years. WYDEQ's preferred approach was to rely on the operators to "volunteer" implementation of best practices to reduce their well pad emissions. Apparently even some "gas patch" workers saw that as absurd because around 2005 I received an anonymous email from one of them asserting that he had seen enough to convince him that his employers would only do what they were *forced* to do.

The Early Days

In the beginning WYDEQ demonstrated a near loathing toward becoming involved in the growing problem of air quality impacts from the huge uptick in drilling and well completions. I was one of the few voices that publically criticized WYDEQ for failing to realize this was going on. I wrote several op-eds for area and state newspapers decrying WYDEQ's failures to realize the magnitude of the damage becoming visible. I pointed out as early as 2003 that there should have been air quality monitoring stations in place at least two years earlier to establish a baseline data reference against which to evaluate future emissions impacts from gas development.

I tried early on to engage my first director of the Air Quality Division, Dan Olson, and persuade him to address the situation developing in Sublette County. At that time one could drive south of Pinedale to where the highway topped a high hill and view a panorama that took in a large stretch of the New Fork River Valley. Above that was a ridge where no less than 15 drill rigs were in operation, belching huge plumes of black diesel smoke and burning off gas being vented from well bore holes. These smokes were being trapped along the valley and nearby foothills of the Wind River Mountain Range where they transformed into what we locals came to label the "brown cloud."

I wrote to Olson and Corra more than a few times trying to get them to believe there was a problem but to no avail. Finally, enough voices were raised that they both were compelled to respond with a few visits to our area to assure us there was really not a problem. They were confident this was a startup aberration that would level out as the drilling activities matured. I considered this to be an absurd premise and took to hammering that idea with op-eds in the local newspapers of nearby Rock Springs and Pinedale.

Initially I was relentless in my criticism of the two directors by name. I accused them of engaging in pabulum-based assurances that had no foundation in empirical science. Furthermore, I challenged them to get cracking and build some much needed air quality measurement stations around the Jonah and Anticline gas fields. This had to happen while there was still an air mass in place resembling its original clean state of a few years earlier. However, this idea was unpopular because of cost.

My public criticisms began to draw blood because a staffer to the governor with whom I had developed a strong advocacy found it necessary to pull me back. I have more to say about her in the section about the office of the governor. I will simply repeat here that she begged me to end my scathing criticisms because when she conducted staff meetings in Cheyenne that included WYDEQ they would as she put it, "lock up at the very mention of my name."

Her advice caused me to realize that WYDEQ's reaction to my editorial attacks was perhaps too negative to produce positive outcomes and that vinegar had to be replaced by honey. Accordingly, at a public forum held a few months later in Pinedale by those same two directors I stood up and acknowledged that I had been perhaps unreasonable in my criticisms. I went on to state that I was, therefore, apologizing before the local citizens sitting in the room. Corra then waxed effusive.

He declared his pleasure at so courteous a gesture, praised me for demonstrating gentlemanly behavior, and called for applause. In the coming months and years I grew to regret what I had done. I felt as future events involving WYDEQ unfolded, my gesture had been turned into a propaganda opportunity for him to continue with actions that were long on appearance and short on substance.

An example of the contest between the Air Quality Division (Dan Olson) and myself illustrates the tension that existed between us. In March of 2004, I penned a rather blunt set of comments in response to WYDEQ's call for public comment on its proposed approval of a set of permit applications from BP America to modify production equipment on its Corona 14 Pad.[1] My comments in this instance became the pattern for subsequent comments I directed at WYDEQ for the next eight years. I explained that in studying the permit details about the Corona 14 modifications, I thought I had discovered at least four fundamental flaws.

Looking back now, I realize I committed a somewhat heavy-handed act. I warned that WYDEQ could legally disregard some of my arguments on the basis of strict interpretation of statutory guidelines. However, doing so would illustrate a lack of foresight and determination to execute its mandate to protect air quality. I also asserted that failure to take a longer view could be construed as failure to properly enforce existing federal environmental air quality law. I was worried about the impact of gas facility emissions upon the adjacent Wind River Mountains Class I Air Quality Protection Zone.

I continued by asserting that the most telling factor to be extracted from any permit approval was that it would illustrate a weakness of the permitting process. This weakness was the seeming total reliance upon an honor system by the gas industry to comply with environmental law. The permit applications contained the phrases *according to the applicant, the applicant reports, applicant must notify, applicant must certify,* and *reportedly installed*. Nowhere was it spelled out what action if any, WYDEQ would execute to insure compliance or accuracy of applicant input pursuant to requirements.

The next most telling factor but of equal importance was the apparent absence of any mechanism within WYDEQ to verify input. Quantitative information submitted by industry to WYDEQ for the purpose of adding up emission volumes was entirely industry-generated. Furthermore, the applications contained words like *estimate, simulation, average, calculated, assumed, predicted,* and *projected*. In no instance had WYDEQ explained how it would validate submitted data.

For instance, heavy reliance upon 98% control of VOC emissions was stressed in the permit approval document. Missing was an explanation of how WYDEQ would validate this expectation and how regularly it would revisit facilities to maintain validation. This concern was legitimate as I describe in the chapter about the operators.

By now it should be clear to the reader that I believed the absence of WYDEQ inspectors conducting empirical measurement programs in the gas fields had been glaringly obvious from the beginning of operations. Also there appeared to be no requirement levied upon industry to conduct meaningful quantitative measurement programs of its own to validate the emission information submitted to WYDEQ. The entire process seemed to rely too heavily upon industry computer modeling which itself relied upon estimates, averages, and assumptions.

I warned that I had heard one WYDEQ staffer opine that WYDEQ was confident it understood the emissions issue based upon modeling but by so doing I worried this was confusing predictions with fact. Simulations, i.e., modeling is doubtful without empirical validation. I backed this up by citing my experience whereby I had managed to get a confession from an industry contact that indeed, little if any effort directed explicitly at validation of their models had been conducted. Additionally, this contact had confessed that this absolutely had not been done for Sublette County gas field visibility impacts on the Wind River Mountain viewshed.

I next charged that yet another failure of the permitting process could be found in the consideration of the emissions numbers themselves. The permit approval referenced federally enforceable trigger values of 250 tpy VOCs, 10 tpy individual HAPs, and declared that the applicant's sites would fall well below these limits. I insisted that this approach was flawed because it treated each site *in isolation* as if each was all that existed or would exist.

I pointed out that 3000 wells were being predicted (in fact this number was actually reached in 2012) in the Jonah and 10,000 wells projected in the Green River Basin overall. The 3000 figure and simple arithmetic applied to the applicants' declaration of expected controlled emissions of 3.0-5.5 tpy VOCs and 0.1-0.2 tpy HAPs for this one facility was scary. Total annual emissions would grow into 9000-16,500 tpy VOCs and 300-600 tpy HAPs. Similar outcomes would occur with NOx and CO.

This begged the question that WYDEQ had thus far avoided. What level of visibility degradation toward the Wind River Mountain Range would result from such a prodigious input of visible haze precursor chemical constituents? Also, what health effects upon the human and wildlife population would be created? I warned that here again there was no empirical data to draw upon.

My criticisms continued with the proposed permit conditions of approval. One item provided for one year of 98% control of VOC emissions following the date of installation of controls. After that the control could be removed as long as the previous 30-day uncontrolled, annualized VOC emission rate was "less than 30 tons per year."

I insisted this annualized accounting process would distort actual emissions assessments on a day-by-day, week-by-week, and month-by-month basis because it would mask wind transport effects across those smaller time scales. I supported this concern by pointing out that I had been analyzing monthly wind patterns for year 2003 and had developed evidence that BLM and Forest Service outsourced wind transport modeling was seriously in error for this very reason. I explain my wind pattern research in considerable detail in the science chapter of this book.

I warned that they were relying too much upon limited data that was further averaged across year-long periods for self-admitted reasons of simplifying the modeling effort. I ended by warning that even the minimum projected 3000 well figure would generate 9000 tpy of visible haze precursor VOCs into the local airshed, thereby rendering removal of the 98% control ill advised.

My comments then moved into the realm of suggestions for improvement. Recognizing that gas exploration and production were in Sublette County to stay, I made the following recommendations as first steps toward achieving better balance between environmental protection and continued but well-regulated gas production:

1. WYDEQ should quickly evaluate its capacity to validate industry information coming from industry, identify its weaknesses to do so, and formulate corrective action.
2. WYDEQ should craft an extensive field measurement program designed to place efficient air quality measurement instrumentation in strategic locations throughout the Sublette

County gas fields. Such instrumentation should by chosen to monitor VOCs, HAPs, NOx, CO, CO2, O3(ozone), benzene, and solid particulates.

3. Measurement programs should be undertaken to confirm emission volumes for the specific purpose of validating models used by industry; also, to validate models used by government contractors.

4. WYDEQ should evaluate the level of funding necessary to accomplish recommendations 1 through 3 and submit as soon as possible a budget request to the Governor and to the Legislature for approval; emphasize the need to use a portion of gas revenues to protect the region being tapped by the operators.

5. Discontinue suspension of any provisions requiring use of control devices. This will become crucial as the well count climbs toward the 3000 mark.

I concluded my comment document with a vinegar-laden set of observations that, again in retrospect was not smart. I observed that the permitting process as it currently existed and had been practiced was anachronistic. This was due to the enormous increase in development activity adjacent to the premier Wind River Mountain Range. It threatened the wellbeing of humans and wildlife in the region if allowed to continue unregulated.

I was concerned that regulation would adhere to existing law that was inadequate. There needed to be more room found in that law for WYDEQ to execute its stewardship responsibilities in a more proactive manner than had thus far been evident. In the meantime WYDEQ needed also to be recommending additional statutes that would provide necessary tools for more effective regulation of loophole areas.

I advised that a government scientist had specifically warned me that WYDEQ would never enforce environmental law with conviction because of the fear that doing so would drive the operators away. I disagreed on grounds that this was likely an unfounded fear because the Sublette gas reserves were lucrative and not a resource that industry would easily abandon.

To back this up, I cited the president of Ultra Petroleum who had declared in an oil and gas trade publication "At today's gas prices, for example, our Jonah wells pay out in around six to eight months."[2, 3, 4, 5] Given this very real financial incentive and my faith that technology can solve most physical problems, I suggested that it should be quite feasible to hammer out an arrangement between the State and operators that would develop and apply appropriate technology in a manner that balanced the interests of both parties.

Dan Olson's terse reply was three months in coming and went straight to my list of recommendations.[6] He responded to my first recommendation by noting that in his opinion I had referenced

> "... the Division's seeming absence of any mechanism to verify quantitative information submitted by the industry." He then stated: "... emissions from wells are a function of gas and condensate composition and production rates and operating parameters such as temperature and pressure."
>
> "The Division requires industry to take speciated condensate and gas analyses from the wells and submit the analyses with the permit applications. The Division checks reported gas and condensate production rates in the application with production rates as reported to the Wyoming Oil and Gas Conservation Commission. While I would not claim that the Division cannot improve on information collection and review, there is information submitted in each permit application specific to the subject well and the information submitted is reviewed and checked against other databases."

To my second recommendation he responded

> "... The Division has required McMurry Oil Company to establish an air quality monitoring site for NOx, ozone, continuous PM10, meteorological data, and a digital camera in the southern portion of the Jonah Field. This site should be set up and active by this summer. The Division has also been granted funding during the last legislative session for additional ambient monitoring throughout the

By the Rasping in My Lungs

state to monitor impacts from energy development. Part of this monitoring effort will be in the Sublette County area and the Division is in the preparatory stages of initiating the additional monitoring."

Regarding my third recommendation Olson expressed his irritation that again I had

"... referenced the Division's seeming absence to verify the industry's quantification of emissions. The information on production rates, gas/condensate composition and operating pressures and temperatures are input to emission models to determine uncontrolled or potential emissions. These emission models include HYSYS (formerly known as HYSIM), PROSIM, KFLASH, API E&P TANK and GRI-GLYCalc. These are trade names for process simulators used by various industries to design, optimize and troubleshoot industrial processes."

"They are based on rigorous thermodynamic principles, or in some cases, upon kinetic models, all well accepted by Industry, the U.S. Environmental Protection Agency (EPA), Gas Research Institute (GRI), American Petroleum Institute (API), colleges, and universities. With regard to control efficiencies, the Division relies on information from the manufacturer's demonstrating sound engineering practices to design and construct the devices for the destruction of VOCs evolving from wellhead production and storage equipment."

"For example, J.W. Williams Inc., a worldwide engineering and fabrication company states their combustion devices "meet or exceed the requirements set forth by the EPA Regulations in Appendix D 40 CFR 60 Subpart Kb, paragraph 60.113b(c)(1)(i) demonstrating that the control device will achieve the required control efficiency during maximum loading conditions and Appendix E 40 CFR 63 Subpart HH, paragraph 63.771(d)(I)(C), which specifies minimum residence time of 0.5 seconds at a minimum temperature of 760 °C for enclosed combustion devices. Similar to

> *the response to item 1, I would not claim that the Division cannot improve on emission calculation and verification, however the methodologies laid out in the permitting guidance and currently utilized by the Division in reviewing projected emissions from the wells have a sound engineering basis."*

Addressing my fourth recommendation, he repeated his earlier assurance that

> *"... the Division included an expanded monitoring program to monitor impacts from energy development in the last budgetary session. The proposed budget was approved by the Governor and the Legislature and we are preparing to proceed with the implementation of additional monitoring statewide during the coming biennium.*
>
> *"We are constantly reviewing anticipated staffing needs with respect to permitting and compliance activities and monitoring strategies considering likely future development scenarios statewide so that appropriate requests for additional resources can be submitted to the Governor and the Legislature during subsequent budget cycles. We also continuously review our permitting guidance, our permitting processes, and our compliance oversight activities to insure they are and remain effective. Contrary to your assertion that our permitting process is 'anachronistic', you will find, if you choose to do the research, that Wyoming's permitting program, in particular for oil & gaswell [sic] field operations, is the most restrictive and comprehensive process of any oil & gas producing state in the west."*

To my last recommendation, Olson replied

> *"... The Division is currently in discussions with the industry on lowering the required control levels for single well installations and to require all emission sources to be controlled at multiple well or PAD sites. This involves an addition to the current Oil and Gas Permitting Guidance. These revisions are being made due to the*

number of wells planned in the area. Subsequent to the changes, conditions contained in future permits will reflect the revised control levels. Information regarding the revision to the guidance will be the subject of an Air Quality Advisory Board (AQAB) meeting in the near future. As you have expressed concerns regarding the permitting process, you will be provide notice of the AQAB meeting."

He ended his letter with an caustic observation that

"... you have covered a long list of perceived ailments in the Division's permitting program. I cannot agree that all of the seeming shortcomings of the existing program are factual or as serious as you describe. I also cannot agree that current permitting guidance and practices is allowing unregulated growth in the area. However, I can agree with you that the development in the area will in fact continue and will require continued attention and evaluation of potential impacts."

I was not reassured by his responses. While he did invoke seemingly legitimate gas production measurement processes in terms of physics and chemistry, his department, nevertheless, was relying on outside expertise. Furthermore, his department seemed to be self-assured that its administrative oversight tools supported by industry modeling would provide necessary compliance enforcement. I still remained unconvinced that his staff was sufficiently populated and sufficiently versed in the technicalities of what they were chartered to oversee. However, I could not challenge any further.

In an attempt to influence the state legislature's funding of WYDEQ, I engaged in something of a "Hail Mary" maneuver by writing to every senator and representative. The result of my experiment in legislative lobbying produced results I had not expected. I was invited to testify before the state finance committee but that developed into a debacle I describe more fully in the section on Wyoming politicians. Suffice it here to say that during the finance committee hearing on the WYDEQ budget requests it was revealed that Dan Olson was retiring.

This was the first I had heard of it and I was stunned. By now I was beginning to think I was making progress in winning him over to my thinking. Not too long before, I had sent him several field photographs I had taken showing clear evidence of stratified layers of carbon soot belching from the exhaust stacks of the drill rig diesel engines. They resembled black undulating curtains trailing across the prairie directly from those exhaust stacks. He replied with thanks and commented that my photos served to confirm his longstanding suspicions that rig exhausts were a significant source of regional pollution.

Now he was gone. As the room emptied out, I went over to him and congratulated him but also expressed disappointment. To my surprise, he paid me a high compliment by requesting that I continue the fight to protect air quality. This and his reaction to my carbon soot photographs were the first and only signals I received from him indicating he had regarded my activities with anything but loathing.

A New Guy Enters the Fray

I eventually learned that his replacement was to be Dave Finley whom if I recall correctly had been in charge of solid waste management for WYDEQ. I had met him a year or so earlier in a public forum in Pinedale before his appointment. During a break, I introduced myself and we conversed about the general issue of regulatory activities involving oil and gas in Sublette County. He seemed forthright and even opined that there was always a need to be on guard against environmental abuse by the operators. In the following years, I found myself wondering if he forgot that bit of sage advice.

In his defense, I must point out that Finley was quite suddenly dropped into the proverbial alligator-infested swamp. Political cross currents and esoteric topics surrounding drilling operations in the Jonah and Anticline were a very distant environment from what he had dealt with in his former position. I was impressed by the speed with which he seemed to absorb his new regulatory diet. However, I also learned to dislike his equally developing skills in regulatory double speak and determination to circle his wagons around his staff even when they were demonstrably wrong. Indeed, this latter behavior would lead to some very heated exchanges between us.

The field of combat that developed between us was the realm of well pad production emissions permitting. Recall that operators were required to submit applications for well pad emissions to be analyzed by AQD staff for compliance with Best Available Control Technologies (BACT). This was a reasonable sounding requirement for production emissions but in truth, the actual details of what BACTs the operators were willing to embrace had a heavy influence upon what AQD was willing to impose.

However, as the months went by and AQD file cabinets in Cheyenne relentlessly filled with approved air quality permit applications, I noted a pattern that took me from suspicion to anger. Each permit application continued to address future predicted emissions history in complete isolation. AQD was continuing to address performance of each pad facility with regard to its compliance with maximum allowable emissions under state and federal regulations as though it were the sole emissions source. No weighting was being assigned in the larger context of the hundreds and eventually thousands of production pads facilities and the total volumetric emissions coming from them.

Eventually, AQD did pursue the construction of a total emissions volume inventory but that came way late and was itself a highly dubious product. The thread of my suspicions carried across every aspect of gas field activity I was challenging in the field, in the air quality task group, and the Pinedale Anticline Working Group. That thread was modeling.

Over the years between 2003 and 2012, I collected copies of as many AQD air quality permit applications as I was able to lay my hands on. At first, I had to make copies of the paper documents that were placed in the Sublette County Court House for public review in accordance with state public review regulations. These applications could be as small as ten pages and as large as 50 pages or more. I came to be in possession of several boxes of these things as time went on.

As portable digital office equipment improved, I substituted electronic scanning of the permits instead of copying paper. I spent around three man-hours twice per week in the courthouse with my laptop and a portable scanner copying hundreds of pages into my own air quality permit archive. I believe I can safely state that my archive is a close duplicate of that maintained by AQD in Cheyenne.

In parallel with this effort, I set about archiving every old, new, and pending well being added to the well inventory record in Cheyenne for the Jonah and Anticline gas fields. This was harder to do because I had to thumb through thousands of well statistics documents and laboriously write down the data on a yellow legal pad and then type the information into my computer

Not knowing in advance what statistics might be valuable, I captured the well name designations assigned by operators, their latitude and longitude coordinates, their start date, their state application permit number, the quarter/quarter section they resided in, their section and township location, and a statistic called "footage." Later on as I discovered the difference in permit numbering between the Wyoming Oil and Gas Conservation Commission (WOGCC) I added columns in which I logged the AQD permit number and the WOGCC API number.

Thus, over time I built up a comprehensive archive of statistics that allowed cross-referencing of wells characterized by AQD methods and those characterized with the WOGCC's completely different methodology. I suspect in that regard, my archive is unique. All of this was intended to provide me with information I could use to challenge WYDEQ. Because the information was theirs, I believed they would be hard pressed to refute it. However, WYDEQ directors instead adopted a defensive strategy that employed marginalization and refusal to recognize my work.

Over the years I was collecting and analyzing AQD's permit applications from operators, their format became a moving target. Initially the format contained several potentially useful sections. Within each section, details of operating conditions were summarized and characterized as to methods, impacts, and compliance with regulatory requirements.

These application forms started with information detailing the applicant's company name, the type of operation on a pad being discussed such as a drilling operation or modification of an existing equipment setup like a condensate production facility. The latitude and longitude coordinates, facility name and its startup date were also listed. The facility name originates from lease unit names of which there were many colorful sounding examples. The Jonah field is made up of units with names like Cabrito, Corona, Coronado, Hacienda, Studhorse, and Yellow point. The Anticline field is made up of unit names like Antelope, Boulder, Green River, Mesa, Rainbow, Riverside, and Warbonnet.

The main business of the application is given next. This is where the operator explains his desire to, for instance, modify a gas/condensate production facility. The modification might involve the replacement of dehydration equipment. More specificity is then given such as "… the current Desi-Dri dehydration unit with one 4.0 million cubic feet per day (MMCFD tri-ethylene (TEG) dehydration unit with a Kimray Model AO 15PY glycol pump, 0.750 million Btu per hour (MMBtu/hr) reboiler heater, reboiler overhead condenser and smokeless combustion device."[7]

The permit history of the site is then enumerated. Details included original permit number and date of issuance, the original requirement for control of VOCs and HAPs being produced, and a complete list of equipment items operating on the site. Typical item lists can look like this:

- one (1) three-phase separator w/0.500 million Btu per hour
- (MMBtu/hr) heater
- one (1) 500-barrel (bbl) condensate tank
- one (1) 500-bbl produced water tank
- one (1) Graco pneumatic pump (heat trace)
- one (1) 20-foot smokeless combustion chamber (combustion device)

The proposed new equipment list then follows:

- one (1) 4.0 million cubic feet per day (MMCFD) tri-ethylene glycol (TEG) dehydration unit w/Kimray Model 4015PY glycol pump, 0.750 MMBtu/hr reboiler heater and reboiler overhead condenser
- one (1) 28-foot smokeless combustion chamber (combustion device) w/continual pilot flame recording/monitoring device

Next comes a diagram of the site process flow sheet followed by additional details about specific devices. In typical application examples the estimating process for reboiler emissions venting is described. The key word here is "estimating." The basic purpose of this document is to provide quantified data on the volume of emissions to be expected during operation of the facility in question. At this point in the document it is explained that:

> "Potential uncontrolled VOC and HAP emissions are estimated using GRI-GLYCalc V4.0 software based on the maximum glycol circulation rate of 0.67 gallons per minute (gpm) for the Kimray Model 4015 glycol pump, reported equipment operating parameters, the extended hydrocarbon composition of wet gas from the well and the current daily gas production rate."
>
> "Controlled VOC and HAP emissions (... Process Flow Diagram) were estimated in the same fashion except that a condenser and combustion unit were added to the overhead still vent. The condenser is proposed to operate at a temperature of 120^0 F and 12 psig. NOx and CO emissions are based on AP-42 EF for flares and the calculated volume of incinerated vapors."

Here is where I routinely challenged AQD. I complained that passages in the permit analyses like "potential," "estimated using GRI-GLYCalc," "VOC and HAP emissions ... were estimated" all constituted assertions of dubious validity until the numerical models being used had been validated for local altitude and atmospheric conditions.

There were several of these models used by operators to supply WYDEQ with estimate data. GRI-GLYCalc is used to calculate VOCs and HAPs from re-boilers. EPA_TANKS is applied to calculating HAP and VOC emissions released from produced water in storage tanks due to the reduction in pressure between the borehole and the open atmosphere (called flash tank emissions). HYSYS is used to compute flash tank emissions of VOCs and HAPs. ProMax is used to estimate gas volumes released from combustors, and EPA AP-42 can be applied to estimating CO emission factors from combustor flares.

For a long time I expected someone in AQD or the operators to tutor me about these computer tools if for no other reason than to shut me up. I was expecting them to bury me in reasons why these models were in fact reliable but that never happened. In fact, it took a while before the subject was brought up either one-on-one, or in public. Thus over time, I began to wonder if I was on to something they preferred remain in the realm of their own closely held esoterica.

Finally in 2008, Finley felt compelled to respond but his explanation was couched in general assurances that themselves invoked industry assurances. When I challenged Finley's reliance upon modeling of emissions from a proposed EnCana Oil and Gas Corp. water disposal facility he replied as follows:[8]

> "The American Petroleum Institute (API) E&P TANKS V2.0 program you referred to in your letter is an estimation tool used by regulatory and industry persons to estimate volumes and components of vapors flashed from pressurized hydrocarbon liquids as they are subjected to atmospheric pressure and is based upon known physical properties of hydrocarbon gasses and Peng-Robertson equations of state."
>
> "This program and similar others is widely accepted are used [sic] to estimate vapor stream volumes and compositions when no other method for doing so is available. As API E&P TANKS V2.0 is no longer supported by its developer the Division recently invested in ProMax® software, a tool used by business, educational and regulatory groups and institutions to design processes and predict process streams. This software will replace the Division's current tool, API E&P TANK V2.0. You may learn more about the program at http://www.bre.com/. Although we will be using a new software tool to estimate emissions from flashing and other O &G production sources, you should note that both software programs are fundamentally based on known thermodynamic and physical properties gas/liquid mixtures."

R. Perry Walker

In reply to my closing comment in which I expressed concern that his staff may have been not as well trained as necessary to understand the technical details of the permits they were routinely approving, Finley stated the following:[9]

> *"In response to the closing summary of your comment letter, let me stress that while I honestly respect your concern and active participation in the protection and preservation of Wyoming's Air Quality resources I, also respect and continue to be impressed by the competence and professionalism of my staff and their ongoing efforts to maintain and improve the NSR (New Source Review) Permitting program for O&G activities."*

This comment became a routine component to his replies to me. In retrospect, I probably was something of the pot calling the pan black here because I too lacked detailed understanding of the foundations of these software models. Nevertheless, my experience over many years of monitoring modelers' struggles to make their models conform to reality conditioned me to be very skeptical of their assurances that their virtual reality was in fact actual reality.

Another class of new emissions sources many of us began to press WYDEQ on was waste water (called produced water) disposal facilities that were sources of hydrocarbon emissions. We wanted to know why and how WYDEQ was justifying a surge in construction of these facilities. It looked like each operator was determined to build independent facilities and we questioned why they couldn't be joint operations. This activity could no longer be ignored by WYDEQ so Finely wrote to assure me that:

> *"Wyoming's New Source Review (NSR) permitting program for oil and gas (O&G) production Facilities has been positively developing over the last decade, primarily in response to concentrated development in the Jonah-Pinedale area. The program is recognized by both the regulating and regulated communities nationwide as a very effective program and serves as a model for other State's programs."*
>
> *"One such improvement to our NSR program for O&G activities is our foray into the permitting of produced water handling*

facilities, previously unaddressed ... As we gained knowledge indicating regulated air pollutants are associated with these types of facilities, the Division initiated action by soliciting information from operators that could be used to qualify and quantify emissions. Operators were informed to expect eventual permitting of these facilities and asked to actively participate in the gathering of valid, scientific information to be used for effective permitting."

"In response to unsatisfactory participation by industry, the Division is moving forward with this matter by proposing permits requiring the scientific gathering of information. The Division believes this permitted gathering of scientific information is a reasonable step forward in qualifying and quantifying potential emissions. With this step, the Division continues to demonstrate why the NSR permitting program for O&G production facilities is recognized as an effectual and innovative program."

WYDEQ was fond of claiming to be in the forefront of regulatory initiatives but that view was not the same for people in the county. Even if true, they were always behind the operators and playing catch-up. That was our main source of dissatisfaction with WYDEQ. Too little was always happening too late.

Returning to my discussion of the WYDEQ permit application format, next in the application text came brief statements of emissions estimates from pneumatic pumps, heaters, and truck loading operations for removal of condensate and produced water. Here again, generalizations were invoked in the form of "... constant values based on historical data reflective of average equipment emissions," and reliance upon EPA's general guidance called AP-42 that my EPA field contact advised me should never be applied as other than a very general guideline.

The next sections in the application addressed emissions volumes in the context of national and state regulatory requirements. The subsection on prevention of significant deterioration (PSD) *always* stated that emissions from the facility were less than the major source levels defined and allowed in Wyoming air quality regulations. I saw this statement literally thousands of times and demanded that AQD address the aggregate buildup of these emissions from all well sites and did so more heatedly as the years passed.

I attempted a challenge of some of these provisions in 2008 by arguing against a proposed action by another operator regarding completion/recompletion activities at wellheads. This activity involved finalizing the tie-in of new wells to gathering lines or processing old wells to reestablish useful flow rates of gas to those gathering lines. Venting and flaring were involved which introduced raw and combusted hydrocarbon products into the atmosphere. This period saw the operators engaging in more and more recompletion work as the original wells in the Jonah and Anticline began to drop off in their production rates.

The WYDEQ approval document for this activity began as others did with the statement that no public comment had been received within the allotted 30-day comment period. I found this odd because WYDEQ's reply to me explicitly noted the date my comment was received and that it met the 30-day criterion, so for sure, they had received at least one comment from *me*. This was not a one-off occurrence either. Nevertheless, WYDEQ in its reply to the operator advised that due to the lack of comments, and on the basis of information provided by the operator, approval was granted.[10] There were conditions, however.

Some of the conditions were administrative in nature such as a WYDEQ inspector having the right of no notice inspection of the facility. All paperwork would be submitted to the WYDEQ point of contact in Casper, Wyoming. The operators were to notify WYDEQ 15 days in advance of intent to perform a completion operation. Within 40 days of first production, operators were to report total volumes of gas and liquids processed in the completion operation.

Monitoring by the operator was to be directed toward assessment of the composition of gas recovered to establish marketability or non-marketability of the gas. Also, the operator was to monitor pressure parameters associated with the well and the surface recompletion equipment. Records were to be kept of all gas and liquids volumes processed, as well as volumes in tons per year of VOC, HAP, NOx, and CO emissions resulting from these operations using WYDEQ spreadsheets.

Other conditions sounded more serious. Operators would adhere to Best Management Practices to eliminate *to the extent practicable*, emissions of

VOCs and HAPs due to flaring and venting of recovered fluids. I never saw any clarification of "to the extent practical" and came to realize this was essentially an operator's call pursuant to WYDEQ's preference for operator voluntary compliance.

Other serious conditions involved flaring of gas. Operators were warned they could not invoke excuses to flare on the basis of (1) lack of pipeline connection to gathering systems, (2) inadequate water disposal facilities, or (3) absence of or inadequate inventory of flowback equipment and operating personnel. Also, opacity of visible emissions from flaring gases and liquids could not exceed 20 percent pursuant to 40 CFR Part 60, Appendix A, Method 9.

Despite the seemingly serious tone to the permit approval conditions, it remained apparent that the whole exercise was one of numbers gathering and paperwork collection. Wells and other source facilities were being created at a rate of hundreds per year but WYDEQ still had only a handful of staff to review them. It was humanly impossible for them to carefully scrutinize every permit application without going numb. Under such circumstances it was inevitable that regulatory oversight would become an exercise in record keeping that replaced comprehensive field inspection and enforcement.

Not surprisingly, WYDEQ disagreed. In a follow-up reply to my challenge which contained duplicate paragraphs to those cited earlier about the modeling codes, the following added assertions were offered:[11]

> *"Around 2003, concentrated oil and gas drilling activities occurring so close to a populated area like Pinedale spawned public comment regarding the highly visible emissions associated with well completions and required action by the Division. Until this time, air emissions from flaring and venting episodes occurring during well completions were not considered under the New Source Review (NSR) permitting process since they did not originate from a stationary source.*[12]
>
> *"The Division worked together with the largest petroleum industry trade association and its members in Wyoming to identify and implement technology to reduce or eliminate well flaring and venting emissions which was then incorporated in to NSR permits.*

> *Permits were issued to each company conducting well completion activities in the Jonah-Pinedale development area, requiring the use of Best Management Practices to reduce regulated air emissions including visible emissions associated with these activities by at least 90 percent. It is the Division's view these permits have been extremely effective in reducing the visible emissions which had been generating public comment."*

In a verbose way, WYDEQ was denying the presence of the "brown cloud" problem many of us were complaining about. Implicit in these explanations invoking data collection and filing was the unstated premise that the mere act of collecting all this information and filing it was itself curative of the problem of the brown cloud. I hit WYDEQ hard over this absurdity in public forums but without effect.

Furthermore, I was always annoyed by the fact that in the next section on conditions for operators, emissions control requirements (such that they were) did not become binding for a period of four months after startup of the facility. That was 120 days they could dump emissions into the atmosphere with freedom and there were always scores of such new facilities operating at any one time.

Next came discussions of items like condensate storage tanks and pressure vessels. There was a standard stipulation that emissions control devices be installed. Emission controls devices were required to reduce the mass content of VOCs and HAPs in the vapors vented to the device by at least 98 percent by weight for at least one year following the date of installation of the control device.

After that time the control could be removed provided it could be demonstrated that the previous 30-day, uncontrolled, *annualized* VOC emission rate was less than 20 tons per year. That was later reduced to 15 tpy. Control device removal required the operator to notify WYDEQ of the removal date and certify the uncontrolled emission rate, including method of calculation, within thirty days of removal.

What stood out to me was the complete reliance of AQD upon the operator to retain the control device in its operation until that operator deemed it

By the Rasping in My Lungs

unnecessary. Furthermore, AQD was relying upon the operator to do his own certification of the uncontrolled emission rate and to assure WYDEQ that the method of calculation as well as the actual emission rate all fell within compliance. This was one of my big wake-up calls that WYDEQ lacked the staff or expertise to do the compliance field inspection work itself. The foxes were in charge of the henhouse.

Lastly in regard to these tank emissions, I challenged WYDEQ that any removal of the control device allowing up to 20 tons per year of emissions was wrong. My worry was that 20 tpy multiplied by thousands of dehy units would create a local air quality disaster. Also, I was unhappy that the certification process was one of calculation rather than actual stack emissions measurement.

These concerns were repeated in the following sections that addressed additional guidance for production emissions. Much discussion was inserted regarding smokeless combustion emissions control devices. Here again the same latitude was granted regarding one-year operation, removal of the control device, and 20 tpy emissions limit. There was an added kicker, however.

Emissions from these individual facilities were not to exceed 100 tpy or more of any regulated air pollutant, 10 tpy or more of any individual Hazardous Air Pollutant or 25 tpy or more of any combination of Hazardous Air Pollutants. I felt that these emissions volumes multiplied by thousands of dehys would essentially set up the local population to experience atmospheric visibility that would become opaque. Furthermore, the vague phrase "or more" increased my concern.

The document ended with a table summarizing emissions to be expected from the facility. Both controlled and uncontrolled emissions of criteria pollutants were tabulated for the supposed purpose of demonstrating how controls would reduce production emissions. Emission volumes of VOCs, NOx, and CO were presented for the key equipment like condensate tanks, dehy units, re-boilers, and pneumatic devices. Also tabulated were HAPs that included benzene, toluene and xylenes. The total predicted volumes for controlled and uncontrolled emissions of those pollutants were given at the bottom of the table.

This table was where WYDEQ thwarted my plans. My original intent had been to tabulate VOC, NOx, CO, benzene, and toluene, emissions with an eye toward trends showing whether WYDEQ was in fact controlling and limiting emissions in a meaningful way. Rather suddenly however in about 2009, the table format changed with the result that information I had begun to accumulate in sufficient quantity to reveal patterns suddenly disappeared and was replaced with a new category of data called "Offset Requirements." This reflected an emissions cap and trade scheme that EnCana proposed and WYDEQ eagerly accepted. I have more to say about this scheme later on.

Another format change that I particularly chafed over was the removal of a detail that was part of the operator well production analysis in the earliest years of permitting. This detail was an impressive mathematical formula developed by the operator for each individual well that allowed calculation of gas output over the life of the well. It was an exponential expression that offered someone like myself the chance to look into the future of that well's production behavior and its emissions behavior too. An example of what I describe is the following equation for Warbonnet 8-9D-30-108:

$$-1143.9 \ln(x) + 3640.3$$

I had hoped that with all these wells equations in hand and fed into a computer program I wanted to craft, I could project any number of years into the future and calculate a total emissions inventory from all wells.

I was most unhappy when this piece of information suddenly disappeared from the application format, thereby, ending my plan. However, I began to realize I had been naive about keeping up with the increase in well population. As drilling hit full tilt, I found myself struggling to just keep up with the lists of new wells entering the WYDEQ and WOGCC inventory at the rate of 350 per year. I would never have time to develop the modeling tool I was dreaming of.

Emissions Accounting - A Shell Game

Starting in 2005, WYDEQ invested its entire regulatory defense in the concept of an emissions inventory. As I have stated several times, it is important to remember that operators' emissions numbers are not field data from actual instrumented measurements but are instead derived from computer calculation estimates of on-pad production processes. They are in fact modeled values.

I had been assured by industry and BLM personnel that such inventory calculations are to be trusted because, as I explained earlier, they relied upon industry standard models like GRI-GLYCalc for dehydrator heater VOCs and HAPs, EPA TANKS or PROMAX for storage tank vapor VOCs and HAPs, and EPA AP-42 for combustion flare gases and leaked gases.[13] These assurances seemed dubious to me until actual empirical validation had been obtained for inefficiencies inherent in combustion at altitudes of 7000 to 9000 feet above sea level where the Jonah and Anticline fields are situated.

In conversations with EPA personnel from the Region 8 headquarters about this issue, I pointed out that AP-42 does not contain any references to adjustments in modeling methodologies relating to altitude. Only temperature variations are addressed. The Region 8 folks agreed on this point and another official repeated that AP-42 is a "very general" methodology that should be applied with caution.

Admittedly, actual measurements of all emissions sources in two gas fields where there are thousands of such sources would be a significant task. I suggested that the needed measure of credibility could be achieved by collecting actual measurements from a statistically representative sampling of all source types under a realistic spectrum of operating conditions. These measurements could then be compared against the modeled data for the express purpose of evaluating the accuracy of each model being used.

This approach seemed justified by a BLM comment in the Anticline 2006 draft environmental impact statement. BLM admitted that due to concerns that the air quality monitoring network in the region was insufficient for quantifying Anticline development impacts, modeling was used to estimate air quality impacts for the previous year 2005.[14]

It was common to see comments in later environmental impact statement documents that alluded to "... modeled actual project emissions ..."[15,16] derived from the dubiously established 2005 baseline. There were other references to "modeled reductions,"[17] and "modeled future emissions."[18] How, I asked in my public comment reply, was it possible for emissions values to be "actual" when they were in fact being estimated and modeled?

Furthermore, how reliable were the values when there were no error bars attached and no field measurement corroboration of the model output? BLM in its 2005 Draft Jonah Infill Environmental Impact Statement added reason for doubt by noting that recent NOx emissions in the Anticline Project Area were already greater than had been projected in the earlier 2004 environmental impact statement. Compounding this mistake, BLM stated that further effort to quantify development impacts to air quality would be "too time consuming and costly."[19] I thus argued that WYDEQ/BLM "actual" emissions had been virtual reality from the beginning.

By the Rasping in My Lungs - Ozone

The winter of 2005 proved to be a watershed period for WYDEQ, the operators, and concerned citizens in the region. That was when the first signs of a new experience began evolving in the region because of gas development. That experience was ozone pollution.

Elevated levels of ozone began to be detected in the winters of 2005 and 2006 but not in 2007. Then in 2008, winter levels were surprising. The issue became impossible to dismiss when in February of 2008, ozone concentrations were monitored at a level of 122 parts per billion (ppb), well above the EPA recommended 8-hour exposure level of 75 ppb.[20] In that same winter, WYDEQ on five separate days issued warnings to residents about potential harmful ozone levels. This was apparently the first time such a winter warning had ever been issued anywhere in the United States.

To its credit, although for ulterior motives, WYDEQ initiated an aggressive campaign of study to understand the ozone issue. Up to this point, air pollution researchers had assumed that ozone was an exclusively urban summer phenomenon. Ultimately it was concluded that Sublette County hosted key winter conditions that when present simultaneously, create high levels of ozone. These conditions are bright sun, a calm air mass especially favorable for a temperature inversion, continuous ground snow cover, and ample nitrous oxides and volatile organic compounds from vehicle engines, gas field operations, and other sources.

Experts concluded that intense solar ultraviolet radiation that drives ozone creating chemical reactions, streams down through the lower atmosphere performing its effect and then reflects upward from snow covering the surface of the region. In so doing, the UV energy passes through the lower atmosphere's pollutants twice and thereby doubles its effect.[21]

Winter ozone events in the region from 2008 were not encouraging regarding the effectiveness of mitigation actions approved by WYDEQ and initiated by the operators in that year. EPA data supported this in a compilation of reported values that exceeded the EPA standard before and after WYDEQ's imposed new requirements.[22]

Media reports noted that in 2011, Sublette County had experienced ozone levels that on several occasions exceeded that of any U.S. city during that year. According to EPA, its own data showed 13 days in February and March with levels above its standard of 75 ppb. One 8-hour reading was 124 ppb.[23] The chronology was as follows:

- In March of 2009, the governor, acting under pressure by concerned Sublette County citizens, recommended to EPA that the area be declared in ozone non-attainment.[24] An area is in non-attainment if the air quality is considered worse than the National Ambient Air Quality Standards set by the EPA under the Clean Air Act (CAA).

- March 2011 was the worst month for ozone in the county since 2008.[25]

- In the same month, high ozone levels caused operators to implement "voluntary" measures that postponed non-essential field operations in an attempt to hold down emissions. WYDEQ met with operators to determine more that could be done to curb emissions during times of WYDEQ-issued ozone health-threat advisories.[26]

- In 2011, ozone advisories were issued for the last day of February and nine days in March.[27]

- In October 2011 a local citizens' group calling itself "CURED" notified EPA in writing that it intended to sue the agency "unless the Upper Green River Basin is formally designated as a non-attainment area."[28]

Ultimately, Sublette County was nominated for designation as a non-attainment area. Adding to the health issue, ozone was a main ingredient contributing to the growing local problem of visible haze that had become obvious in the area.

Tales of the Dark Side

The most vexing aspect of all that has been discussed thus far is that it very possibly could have been avoided had WYDEQ and BLM not been so inflexible. Gas burned by on-pad process heaters, emissions from those activities, and ozone all could have been greatly reduced or even eliminated in some instances. The solution to which I refer is a state-of the-art gas dehydration process developed by a small firm in Farmington New Mexico.

This technology had been field tested in Colorado by the EPA, which issued a very positive report on the test results.[29] I have earlier noted that the technology called the Quantum Leap Dehydrator, effectively reduced emissions of NOX, VOCs, HAPs and CO_2 and a few other species from hundreds of tons per year to a few tens of pounds per year.

EnCana showed interest and initiated field testing of the technology but then I made one of the biggest mistakes of my life. I invited WYDEQ to observe our project. It wasn't long before a WYDEQ staffer challenged the technology as being no better than WWII vintage technology currently in use. Understandably, EnCana was not eager to invest large sums in a technology that WYDEQ was refusing to acknowledge as being effective.

Thus ended a promising technology that very likely would have relieved Sublette County of considerable emissions volumes by as much as 1000 percent. A much more detailed treatment of this affair is provided in the chapter about the operators.

By the Rasping in My Lungs

• • • • •

It should be noted that over the years, WYDEQ has expanded its number of monitoring stations to form a credible network. However, it continued to seemingly collect data for the stated purpose of "understanding" the phenomenology at work rather than applying the network toward enforcement of mitigation actions. Instead, WYDEQ tried to jaw bone the issue.

Under its umbrella, yet another committee called a task force was created by the governor in 2012. It consisted of town, county and state government personnel, the Bureau of Land Management, the U.S. Forest Service, and citizens.[30] Not surprisingly, the number of positions for citizens was much smaller than those from the government groups. Furthermore, on the various committees and task forces created by the state, including this one, there were no medical or public health officials included. Therefore health related issues were rarely discussed.

The ozone problem has continued to persist as was reported in 2014 by that air quality citizens' task group. Their report declared that state regulators had made progress in limiting ozone levels due to *new* gas development operations but had yet to implement measures to deal with pollution from *existing* facilities.[31] This was discouraging because existing well facilities numbered in the thousands already.

• • • • •

An affair I rank as one of the most ingenious examples of bureaucratic inventiveness in the annals of doublespeak occurred in 2008. EnCana submitted a permit for a large group of wells and proposed that emissions of NOx be offset by its reduction of VOCs in the form of emissions credits. EnCana felt that VOC reductions it had applied to existing sources should become a tradable commodity applied to new wells. Finley praised this approach in a public forum in Pinedale and as I listened I could hardly believe my ears. When he finished, I raised my hand and went into attack mode.

255

I challenged his authority to unilaterally initiate such a fundamental policy change. I argued that he had circumvented all the state guidance stipulating how rulemaking was to be undertaken. I ended by declaring that I doubted he had the legal latitude to implement what was a cap and trade scheme nor was he justified in implementing an idea completely crafted and handed to him by an operator. I could clearly see that he was staggered and after a bit of tap dancing, he said he would have to get back to me.

A few months later in another forum Finley took up the cap and trade subject. His solution was truly innovative. He declared that upon further review back in Cheyenne it had been determined that EnCana's credits approach had been incorporated as a part of his division's "interim policy" and as such, was exempt from public challenge.

I was so impressed by his out-of-the-blue explanation I couldn't help being envious of this display of skill at escape and evasion. Ordinarily I would have been volcanic but I instantly recognized I had been beaten in a way I could never reverse. Later I actually told Finley this was one of the most brilliant maneuvers I had ever seen.

Curiously, this bit of "interim policy" remained in force as a permanent fixture for years thereafter.

• • • • •

As I grew ever more frustrated over what I perceived to be WYDEQ's refusal to look more directly at emission sources I came to the conclusion that perhaps this was due to an absence of expertise in the Division. In a public forum meeting held by WYDEQ in Pinedale, I had urged Finley to undertake my statistical analysis project idea involving selection of a properly sized representative sample-set of the increasing population of emission sources in the Jonah and Anticline. This sample-set would then be subjected to statistical analysis and species characterization. Armed with that information, extrapolation to the entire population of emissions sources would hopefully then yield an accurate emission inventory.

Finley seemed a bit taken aback by this and after a short pause, admitted he had no staff qualified to undertake such a project. Being a believer that universities are a great source of technical support of all kinds, I concluded that the University of Wyoming math department might be just what we needed.

About a week later I placed a personal call to the Chair of the Math Department and was courteously received even though I had no special reputation that he might recognize. I explained the project I had suggested to Finley and he responded immediately that he had just the right guy on staff to do such a thing. I inquired if he would agree to my advising Finley of our conversation and be willing to directly provide more detail regarding the nature of a support contract and a cost estimate. He replied that this would not be a problem.

A few days after that I called Finley and recounted my conversation with the Math Department head. I told him the man was awaiting his call to explore the idea further and I commented that I hoped he was not offended that I was pursuing this track. He assured me he was not but sadly, to my knowledge Finley never followed through.

Dance of the Regulators

On June 20, 2007, Finley issued a memorandum to all oil and gas operators and also made it available to the public for comment. This action had become necessary because of the steadily increasing rate of gas and oil development across the state and especially in Sublette County. The memo explained that in June 2006, a meeting of WYDEQ had been convened for the purpose of reviewing proposed revisions of the Division's permitting guidance contained in Chapter 6, Section 2 regarding production facilities permitting. The resulting guidance proposals were submitted to industry for comment that prompted a second review meeting by WYDEQ in August 2006.[32]

Those comments resulted in modification of the June proposals. The memo presented the list of originally proposed guidance changes as well as the modified version resulting from industry input. Of note was extra attention toward the Jonah and Anticline gas fields. I elected to submit comments because I found the language of some of the provisions to be difficult to make sense of and also because I was concerned that industry was being given too much opportunity to influence how it was to be regulated.

The memo began by addressing glycol dehydration units on production pads. The initial proposal would require installation of 98% control for all dehys with projected HAP emissions greater than or equal to 5 tpy and would eliminate options for operators to limit emissions by simply reducing glycol pump rates or glycol pump model change-out. The revised proposal language remained the same but with the added stipulation that in the Jonah, the 98% rule would be effective upon first production. Also, further explanation was included which appeared to be designed to assuage operator worries rooted in concerns that the guidance conflicted with other guidance under EPA standards known as Maximum Achievable Control Technology or "MACT."

The added explanation asserted that WYDEQ's assessment methodology would always produce higher emissions estimates than actual since they were based upon maximum "... glycol circulation rate of the pump in service and a gas throughput based on the projected decline of the well." However, the MACT methodology was "based upon emission estimates using actual glycol circulation rate[s] of the pump and actual gas throughput. As a result, most dehys subject to the [EPA] area source MACT would be subject to the control requirements in the [proposed new] guidance."

In my comments I asked about the proposal that 98% control be applied to units whose emissions would be greater than or equal to 5 tpy. If this meant that 98% of the 5 tpy (equal to 4.9 tpy) would be eliminated, then the proposal was necessary and appropriate. I further stated that given the rapid rate of increase in the population of wells and dehy units, the measure was long overdue and might in fact be eclipsed by further increases in dehy population, necessitating even tighter restrictions.

I complained that WYDEQ worried too much about economic impacts of regulatory actions on the operators. Furthermore, I reminded Finley that there was a solution available that I had been advocating for over two years that would eliminate all fugitive, tank, and combustor emissions. It was the Quantum Leap Dehydrator. This was as close to a silver bullet solution as one could hope to find. I reminded him that one of his staffers in his WYDEQ Casper office had singlehandedly blocked that initiative and I was blunt.

I asserted that to my great irritation, it had appeared to me that the staffer had single handedly interfered with any objective evaluation of the unit solely on the basis of her laser focus upon her flawed analysis of the unit's VOC, and NOx performance. Test data was in hand which dismissed her NOx argument and the VOC argument was eliminated by the unit's diversion of all emissions into the market stream for profit.

Seemingly in need of some kind of defense to save face, she took that last point of VOC removal and distorted it into an odd business observation. She deemed the potential elimination of VOCs by sending them to market in the gas stream as "*… not our concern what operators do regarding their choices of economic trade-offs.*" Thus, on one hand, WYDEQ seemed to worry that over-regulation would economically burden the operators and yet on the other hand, claimed indifference to operator business models even though this particular situation presented a win-win solution.

The next big source of emissions, flash gases, was then addressed. Remember, these are the emissions that exit from liquid condensates coming out of the well as a result of reduced atmospheric pressure compared to the pressures present down-hole. These condensates are routed into storage tanks for subsequent transport elsewhere. The initial new guidance proposed 98% controls upon first date of production and to remain in place for one year. After a year, the controls could be removed once emissions dropped below 15 tpy. However, the new guidance tightened this to read total emissions from all wells on a single pad must fall below 15 tpy before removal of controls.[33]

I warned of the possibility that even if consolidation of well count under single large capacity tankage took place (pads), those tank batteries would produce higher total quantities of emissions. Absence of language quantifying the number of wells per tank battery appeared to leave open the barn door for rates of growth that would overwhelm any projected improvements. I had objected in April 2004 to WYDEQ's refusal to consider these fields in the aggregate rather than as individual sources independent from one another but it appeared they still intended to proceed in that manner.

The other flash emissions source, that of gases exiting produced water coming out of the bore hole and routed into storage tanks was next addressed. While no controls were proposed on a statewide level, here in the Jonah, the exception would be made that vent lines would be tied into vent lines from condensate tanks. The combined gases from both tanks types would be routed for disposal, usually by burning in combustors.

This initial proposal was later dropped. WYDEQ noted that one operator, BP America, had reported emissions of 0.75 tpy in the Jonah which the agency was unable to refute due to absence of data. Thus, although as was stated clearly here by Finley that he was unconvinced that *"water tanks were not a significant source of VOC emissions,"* the proposal was removed pending further review.[34] It seemed that the operators had won this one.

The good news was that while Finely put the operators on notice that he would continue to study the issue, in the meantime operators were urged to discontinue their use of open roof tanks. I was quite happy that he had addressed this practice. In my field research, I had climbed on top of a few of these tanks and was stunned at the strong aromatic benzene odor billowing out of water.

It seemed clear to me that thousands of cubic feet of these gases were venting into the atmosphere through the course wire mesh safety cover over the opening. Adding this volume to that of the many other tanks of this type was literally creating a sweet smell for hundreds of yards away from these pads. In fact, one set of BP America tanks I visited had a sign posted requiring safety masks before getting on those tanks due to the poisonous breathing hazard.

In my formal comments on this topic, I complimented Finley on his thinking by noting that this was the correct attitude to take. I pointed out that my own study of production startup permit applications revealed that such tanks were assessed by the applicants to be producing 45-55% of all VOC emissions. I noted also that he had opined that operators should "… *take a proactive approach and discontinue using open roof tanks.*" I complained that this statement was far too lenient. I urged him to dictate an end to the use of these tanks and retrofit existing batteries with seals and emission capture apparatus.

The next source of emissions of concern to be addressed was that of pneumatic pumps on the pads.[35] I was keen on this one as well because I had observed the leaks in progress with an infrared camera which I advised him of and which I discuss thoroughly in the science chapter. The initial proposal was to eliminate these emissions outright but the modified version softened this to suggest less stringent equipment replacement options with total elimination being encouraged. In the end, the proposed requirements for "concentrated development areas," i.e., the Jonah and Anticline were not incorporated into the revised guidance at all. Instead, WYDEQ agreed to consult with operators for purposes of identifying areas to be incorporated in future revisions. The operators appeared to have won again.

Regardless, I went on to compliment him for this implied attempt to regulate methane leakage. It revealed that he was willing to go beyond EPA statutory guidance by regulating a non-criteria, unregulated specie. I urged him to carry this forward to other species such as CO2 and get ahead of what I thought would be the inevitable requirement to comply with the Kyoto Protocol. I was far too optimistic about this unfortunately.

The memo contained an interesting discussion about the give and take that apparently ensued between Finley's division and the operators concerning 95% versus 98% control of VOC and HAP emissions from glycol dehy units. At an administrative level the change to a higher percentage would remove confusion about permit conditions needed to account for efficiency behavior of combustors used to control both types of flash gas emissions. Also, it was argued that such changes would afford credit to operators for actual emissions control efficiencies versus assumed efficiencies.

However, operators apparently tried to argue that the 98% stipulation might not be achievable especially when gas production through a unit started to decline. Finley stood his ground by citing results from reference method emissions testing on dehy units. Those results showed 98% efficiencies were in fact being achieved and that gas flow to the dehy units had no bearing on control efficiencies.[36] I wrote in my comments that this response was excellent. Even *if* the law of diminishing returns began to kick in, operators would not improve their operations unless they were stressed to do so.

The last substantive section of the memorandum addressed what was implied to be desires from operators to eventually remove dehy emission controls. Finley stated that such removal was not being considered. He noted that dehy emissions were dependent upon glycol circulation rates and the hydrocarbon composition of the wet gas from the borehole. That combination did not noticeably alter the emissions rate as wet gas throughput declined. Absent any guarantee that possible and actual emissions would decline with a drop in gas production, he had no interest in considering control removal.[37]

I commented that his observation disputing a linear relationship between production and dehy emissions was excellent. In my conversations with operators, they had long insisted there *was* a linear relationship but again, they invoked modeling that lacked validation for our altitude/climate circumstance. What I often got from them was a "trust me" approach. My desire of WYDEQ was that it "trust but verify."

A final issue that the memo commented on had to do with green completion. This was a technique that required operators to bring in a new well by capturing the fracking fluids and vented gases rather than burn them off in the open air at the surface. Apparently a commenter had requested this technique be required in the Jonah but Finley declined. It was stated that WYDEQ would not consider the idea at that time.[38]

I suggested that his decline to apply green completion in the Jonah seemed to indicate a reluctance or at least preference by WYDEQ to follow a lenient track for the time being. If correct, this simply delayed the inevitable. The very phrase "concentrated development areas" carried an implication of a need for immediate and forceful regulation. If operator costs were the concern, it would seem logical to conclude that corrective actions now would be cheaper than in the future.

Somewhat to my surprise, Finley responded to my comments document via email not long after I submitted them.[39] He started out by agreeing that his new guidance contained language that could be difficult to understand if one was unfamiliar with oil and gas production as he had been the previous year when he was new at the job. He explained that this guidance document was the fifth revision since 1997 and the first revision since 2004 of the more stringent guidance applicable to the Jonah-Pinedale Development Area (JPDA).

He went on to explain that prior to 2004, operations in JPDA were subject to statewide Best Available Control Technology (BACT) requirements. He recited differences in the June 2007 version of draft guidance as compared to initial proposals for changes that were made when WYDEQ started its revision process in June 2006. Thus, comparison of the changes in the memorandum under "Initial Proposal" and "Current Proposal," revealed only those changes from the June 2006 proposal to the current June 2007 proposal.

His key points of interest for me were:

> (1) Options to limit hazardous air pollutant emissions from dehys by setting limited pump rates were being eliminated. Under previous guidance, operators could reduce the calculated HAP emissions from dehys by limiting the glycol recirculation rate, and if the reduced HAP level was below the control threshold, no emission control was required. As a result, there were many applications where HAP rate was limited to 6.9 tpy that was below the (old) 7 tpy limit thus skirting the need for controls. Under the revised guidance, if calculated HAP emissions were to be 5 tpy, controls must be installed, lowering HAP emissions to 0.1 tpy (a 98% reduction); formerly a pump rate could just be "dialed down" to 4.9 tpy and no control would be required.
>
> (2) The control threshold for HAPs in the JPDA for single wells was being eliminated so all dehys would now require controls. The time allowed for control installation in the JPDA under old guidance allowed between 40 and 120 days to control dehy emissions but the revised guidance would require controls from date of first production. Furthermore, control efficiency requirements were increased from 95% to 98%.
>
> (3) In the JPDA, tank emissions would be controlled sooner and remain longer than under old guidance.
>
> (4) The proposal to require control of VOC's from produced water tanks was indeed deferred until additional data was in hand. The current guidance at the time did not require controls on VOCs from produced water tanks.
>
> (5) Controls on emissions from pneumatic pumps in JPDA would be required for the first time.

(6) WYDEQ had proposed to extend more stringent JPDA emission control requirements to "other concentrated development areas" (Pinedale Anticline for example), but would not be included in the June 2007 guidance document revisions. Therefore, statewide control requirements would govern. This meant that there were controls on oil and gas development in "other concentrated development areas," but not JPDA-type controls.

At this point Finley circled his wagons around the staffer I had accused of interfering with the Quantum Leap Dehydrator project. He proceeded by claiming:

> *"Your email criticizes Cynthia Madison for 'interfering' with your efforts to get companies to install Quantum Leap Dehydrators. We have absolutely no objection to a company's use of a specific technology, so long as it meets BACT requirements. It is tiresome to me that we continue to provide you with our objective comparison of emissions from conventional dehys and Quantum Leap units, which shows no appreciable difference in emissions to the air, yet you continue to lob criticisms at my staff for interference with your efforts to convince industry to adopt this new technology. We have no quarrel with you—if you can convince industry that a Quantum Leap dehydrator is better than a conventional dehy, have at it."*

My reaction to this was to return fire and I did so with the following blunt rebuttal:

> *"You are NOT providing 'objective' comparisons. In fact, I have received little on the topic from anyone in WYDEQ. What I have received from you (aka, Cynthia), to all indications, is EPA statutory scripture-based objections to the performance of the QLD which focuses on ONE regulated specie (I got this from Cynthia last October) and that IS WRONG. The fact is, according to recent test*

data the QLD pump engine emits 1/3rd NOx per horsepower hour that I see attributed to other engine types.[40] Secondly, it eliminates ALL methane fugitive leaks. Third, that the QLD is 'no better' in VOC's is ridiculous. They are captured and sent to the market stream so they are eliminated to a major extent, if not totally. And by the way, elimination of the VOC combustor ALSO eliminates ALL of the combustion gases exiting the combustor stack into the atmosphere as a by-product of the destruction of those VOC's. My growing anger and frustration with WYDEQ/Cynthia is your seeming refusal to evaluate the TOTAL picture."

"Furthermore, you couch the issue in the context of MY convincing industry to use the technology. This comment dodges reality. My efforts are ultimately irrelevant due to the fundamental fact that EnCana and all operators are unwilling to adopt a technology UNLESS WYDEQ blesses it first."

"This is fact, Dave. I have been told personally that EnCana field managers were very frustrated by Cynthia's resistance because they believe the unit does all that has been advertised. However, their superiors would not continue with the original plan to purchase 50 units in April 2006 until WYDEQ/Cynthia is convinced. That has a simple meaning: for bureaucratic reasons WYDEQ has indefensibly delayed deployment of an advanced BACT by a full year....so far."

"So understand this....you can get as tired of my 'lobbing' as you can stand. I am equally tired of trying to advance the state of the art up here and having to do so by fighting not only my industry enemies, but also my regulatory 'friends.'"

This became the character of the relationship between me and WYDEQ and Dave Finley from that time forward.

• • • • •

An area of regulatory concern that ultimately assumed more immediacy than visibility was that of ozone in the region. As explained in previous pages, this problem began to assume urgency by 2006 and escalated to such a level in the minds of Sublette citizens that the Governor was compelled to get involved. His involvement translated into pressure on his director of WYDEQ and Dave Finley to address the issue and put to rest the growing alarm being expressed within Sublette County.

The seriousness of the problem was revealed in a study done by NOAA and published January 18, 2009, in the journal *Nature Geosciences*. Looking back, NOAA concluded that due to air conditions in the area of Sublette County and intensive natural gas production, incidents of high concentrations of ozone in winter had occurred rivaling those experienced in cities in summer months. An EPA official even warned that the number of incidents in which the Jonah/Pinedale area exceeded the EPA standards of 75 parts per billion over an eight-hour period was likely to prompt the agency to extend existing ozone regulations to the area. Furthermore, he warned that if the exceedances continued, a worse case situation could develop whereby production in the gas fields would have to be shut down during such events.[41]

This situation had been a recognized anathema four years earlier but not expected to actually develop into such a looming threat. In February 2006, ozone spikes occurred which EPA warned WYDEQ in writing were *not* due to weather patterns.[42] In fact, similar spikes had taken place in February 2005.

Dave Finley publically stated that his division was taking the issue seriously and was engaged in talks with operators. He noted too that two months earlier, his division had initiated a revision of its oil and gas permitting requirements to effect tighter controls over ozone precursor emissions. This was referencing the very same memorandum of proposed revisions I had commented on and described at length earlier.

Faced with the scary prospect of seeing state gas production revenues at risk if they did not act, the directors of WYDEQ and WYDEQ scrambled. They initiated a field measurement program and so called voluntary restraints by the operators that would buy time until a real understanding and solution to the problem could be developed. In addition to these actions, WYDEQ tried a delaying tactic against EPA involving regulatory maneuver.

Air quality rule making regulations require that a health-threatening situation be identified and monitored for three consecutive years. In that period of time records must be maintained that chronicle exceedances. These exceedances must themselves meet certain statistical relevance criteria to be judged useable for purposes of justifying regulatory action. If such exceedances fail to be detected in any one year of the three, the clock must be restarted and the monitoring process restarted too. In 2007, WYDEQ appeared to take a shot at restarting its clock.

Through my government sources I learned in mid-2010 that WYDEQ was on the verge of filing a petition with EPA-Region 8 in Denver requesting that a particularly surprising exceedance event be classified as a natural anomaly. This would throw into question the influence of gas well emissions on ozone creation and also allow WYDEQ to restart it monitoring clock.

WYDEQ's petition was a 54-page document containing copious charts, tables, and graphs all aimed at persuading EPA that the event in question was a natural aberration. Specifically, WYDEQ was maintaining that two stratospheric intrusions during the period of May 23-26, 2007, caused three periods where ozone exceedances occurred at the South Pass, Wyoming monitor located at the southeastern tip of the Wind River Mountain Range in Fremont County, Wyoming.[43] Such intrusions involve ozone-rich air from the stratosphere invading the lower troposphere where our weather primarily occurs. The mechanisms and conditions for this can be complex but not abnormal.

WYDEQ maintained that On May 21, 2007, a stratospheric intrusion associated with an upper level atmospheric disturbance allowed ozone-rich stratospheric air to enter the troposphere over northeastern Oregon. That storm system and ozone-rich parcel moved east over Idaho and arrived over western Wyoming. Subsequently, monitored 1-hour average ozone values at South Pass started to increase. This happened a second time the following day.

Two days later, a second upper atmospheric disturbance and attendant stratospheric intrusion located over southwestern Canada moved over western Wyoming. The same monitor registered a third period of elevated 1-hour average ozone levels which continued into the following day. WYDEQ then went on to assert:[44]

> Quality Assurance/Quality Control checks of the South Pass ozone monitor during this period confirm that the monitor was running properly. Calibration results and an independent audit conducted are consistent with 40 CFR Part 58, Appendix A, Section 3.2 and the Quality Assurance Project Plan for the South Pass monitoring project.
>
> With the preceding points in mind, the WYDEQ/WYDEQ submits May 23-26, 2007 South Pass ozone exceedances as a case for EPA's concurrence regarding the events as being exceptional events as outlined by the final Treatment of Data Influenced by Exceptional Events Rule. The WYDEQ-AQD presents supporting evidence which clearly shows that the exceptional events passed the four required tests A-D under 40 CFR 50.14 (3)(iii). Specifically:
>
> *(A)* The event satisfies the criteria set forth in 40 CFR 50.1(j);
> *(B)* There is a clear causal relationship between the measurement under consideration;
> *(C)* The event is associated with a measured concentration in excess of normal historical fluctuations, including back ground; and
> *(D)* There would have been no exceedance or violation but for the event.

The monitor at South Pass was located there to monitor long-range transport of pollutants and a serious potential source of those pollutants was the Jonah-Pinedale Development Area. This connection immediately raised my suspicions. By now I mistrusted WYDEQ greatly. I wondered if this was a gambit designed to relieve pressure that was forcing the division to be more proactive about ozone. I feared that were WYDEQ to be successful, the Pinedale area citizenry would be set back three years in its fight to harness the growing ozone threat.

Unfortunately for me, this subject was even further removed from my comfort zone than had been visibility issues. I knew next to nothing about the science of ozone and I realized I needed help to initially ascertain if the WYDEQ argument was credible, and second, to explain to EPA why not if it was indeed not credible. I thus turned to an academic on staff at the University of Wyoming, Derek Montague.

I had become acquainted with Derek a few years earlier as a result of my contacts in the University of Wyoming Haub School of the Environment and Natural Resources. He was an atmospheric science researcher and teaching professor at UW. An additional plus was that he was from Britain. I have a fondness for these guys because of my work with another British scientist in my Air Force days as well as yet another who came on the scene at UW a bit later. I find them marvelously competent and rigorously logical. Even better, they were willing to talk to me in a manner that even my limited intellect could keep up with.

Thus, I contacted Derek and asked if he might be able to assist me in my quest. He replied that he too was weak in his knowledge of stratospheric ozone intrusion dynamics but had doubts about WYDEQ invoking it. He considered the hypothesis interesting but had to be persuaded of its relevance to the Upper Green River Valley region. That said, he passed a copy of the WYDEQ petition to a more knowledgeable colleague with the request that he reply directly to me as his time would allow. What I received was wonderful and it became the basis for the challenge I sent to EPA in December 2010 arguing that the WYDEQ petition should be rejected.

I began by stating that WYDEQ's case for an explanation of high ozone events based solely on stratospheric intrusions was far from established. While there was indeed evidence supporting such a conclusion, the framework for the arguments had not been properly crafted. It seems that only evidence supporting WYDEQ contentions had been presented (i.e., "cherry-picked") which was behavior I thought characteristic of WYDEQ. I then listed the points of contention as given to me by my UW coach.

Previous examples of ground level monitors recording high O3 levels possibly resulting from stratospheric intrusions were usually indicative of high altitude mountain top sites. Examples could be found in Switzerland and were associated with other characterizing observations. Even when all of these indicators are considered, it's still quite difficult to unambiguously identify stratospheric intrusion events, and for the Wyoming cases, the evidence presented was not very extensive.

It was also relevant to know about existing precedents for non-mountain-top site high-O3 observations attributed to stratospheric intrusions. Could it be established that the South Pass cases were unique occurrences? If so, it was incumbent upon WYDEQ to be more scrupulous with the evidence when presenting their proposal, so that it better supported a commanding argument.

Another point was that the events presented were described as exceptional. This seemed unproven, i.e., how exceptional were they really? WYDEQ's meteorological analyses were confined to a very short time period (9 days). A much longer (at least a year) period needed to be reviewed. It was really necessary for one to look at a number of synoptic events over a longer time period, some with even stronger intrusions, to see how South Pass ozone is driven. Thus there may be many cases of stratospheric air being found in the troposphere, with no associated high O3. As of the time of this petition, this matter was poorly understood.

A third point was that the WYDEQ ozone data exhibited a diurnal pattern, with ozone peaking in late afternoon which happened to be after the time of observed peaking at many other stations. Since stratospheric intrusions events are not diurnally modulated, this strongly suggested that there must have been a component of tropospheric origin.

In summary, the WYDEQ case was unproven, largely because of inadequate supporting evidence and because the events' exceptionality was not well established.

I then warned EPA to regard the petition cautiously. This was not the only time that Wyoming officials had interceded with dubious environmental protection process to shield the energy industry from regulatory inconvenience. If there was any exceptionality to be considered perhaps it was that since 2006, WYDEQ appeared to have petitioned EPA no less than 15 times for relief under the "Exceptional Events Rule."

I recommended that Region 8 impress upon WYDEQ that it should expend its resources more toward the task of protecting air quality from impacts resulting from industrial development and less on seeking ways to avoid execution of its responsibilities. I ended by stating my suspicion based upon a seven-year-long science contest with WYDEQ that it looked like it was seeking to throw out the South Pass 2007 events solely to reset the clock. The likely purpose was to avoid crossing the threshold of ozone non-attainment thereby shielding the energy industry.

A few months later I received an email reply from EPA. It was terse and basically scolded me for not having submitted my comments through proper channels. This referred to the procedure of routing my letter through WYDEQ, so that they could have an opportunity to rebut for EPA consumption. That was exactly why I went over WYDEQ's head. I did not want them to get the opportunity to marginalize my comments before EPA could even see them.

Following this wrist slap, the EPA author delivered the ultimate irony. It had rejected the WYDEQ petition not because of flawed science but because WYDEQ had missed the deadline for submission. There was a specific period of time after a supposed event when documentation had to reach EPA and WYDEQ had missed it by something like a week. Thus, the petition was dead. WYDEQ had itself become the victim of bureaucratic process.

Repel All Invaders

Another dark episode involved the BLM's Jonah Inter Agency Office or JIO. This group was formed as part of requirements to mitigate environmental damage done as a result of gas development in its namesake field. The Final EIS for the Jonah Infill (JIDP) contained details on how the JIO would function.

The basic purpose was to implement adaptive management. It's charter stipulated that it was to execute plans, monitoring, and other activities needed to advance land management initiatives, reclamation actions, and mitigation in and adjacent to the JIDP in accordance with the Record of Decision (ROD) for the JIDP. In addition, the JIO was to provide oversight of funds available for reclamation monitoring and mitigation (offsite and onsite). Of relevance to my interests, one of its charter tasks was to ensure compliance with Wyoming WYDEQ Air Quality rules and regulations.[45]

The operators provided funding for the JIO and its projects. Chief among those had been EnCana for six years. Estimated annual cost was put at

$600,000 to be applied toward office facilities, computers and furnishings, and transportation. Funds from the operators would also be applied toward "wildlife habitat improvement, resource monitoring and/or other mitigation."[46] Eventually, the JIO issued a twice per year public call for project proposals to be considered for funding.

Charter members of the JIO were the Wyoming Department of Agriculture (WDA), The Wyoming Game and Fish Department (WGFD), the Wyoming Department of Environmental Quality, and the Bureau of Land Management. Their powers included approval of all disbursements of funds contributed by the operators for the purpose of wildlife habitat improvement, resource monitoring and/or other mitigation. A key point of which I would run afoul was that these charter members had sole authority to select mitigation projects they judged worthy and also, they could submit projects of their own for consideration.[47]

Around this time, WYDEQ had assigned a permanent staff member to the Pinedale area for the purposes of performing field compliance inspections. I soon made her acquaintance as a result of my visits to the Jonah and Anticline fields as well as our mutual attendance at meetings of the Air Quality Task Group. At one of these meetings I described to her my plan to set up an EPA compliant ozone monitoring station on my property north of Pinedale.

In the course of telling her how I planned to do this and the costs I anticipated to do so, she suggested that I should submit a proposal to the JIO for the purpose of obtaining funding assistance. She further offered that she was confident in her capacity as the county WYDEQ field representative to successfully advocate my proposal and get it approved. It turned out that she promised more than she could deliver.

The eight-page application document was typical of the government. The applicant was required to supply statements describing the project, its objectives and benefits to wildlife, potential gains offered by the project, timelines and completion date, and potential for future project expansion. As for administrative details, explanations were required listing any other project participants and contributions, monitoring and reporting protocols, and sources of funding and amounts.

I explained that I wanted to fund an independent ozone monitor station on my ten-acre home sight 21 miles north of Pinedale. My project would involve continuous measurement of ambient ozone utilizing an EPA approved equivalent method ozone monitor (Dasibi Model 1003-PC). The monitor would be located and operated in compliance with 40CFR Part 58 "Ambient Air Quality Surveillance" guidelines.

I stated my intention to monitor ozone presence in the vicinity of a locally known landmark called the "Rim." I explained the Rim was a narrow wind transport pathway for winds traveling from the gas fields toward the Hoback and Jackson Hole Class I and II airsheds. My proposed project would also be situated to detect ozone being transported from sources located northwest of the Jonah and Anticline gas fields. My ultimate objective was to contribute to WYDEQ's ozone measurement data- base from a location not currently being monitored.

As for participants and contributors, I advised that the environmental group Greater Yellowstone Coalition (GYC) contributed $1000 to my effort and the Wilderness Society contributed $2000. Actually, the $1000 sum had come from a local part-time resident and former PepsiCo CEO, Donald Kendall who contributed to my project through the GYC. I ended by noting that in my position as the principal investigator, I would be providing the monitor device, the monitor site, all consumables and all construction materials.

Under the sections on reporting and potential value of the research I explained that resulting data would be collected along with meteorological data and documented per EPA protocols. This data would be made available to anyone desiring it. It would be available in raw form and in quality controlled form in keeping with EPA QC protocols.

The budget breakdown was to include $450 for a surplus ozone monitor device, $4,900 for an ozone calibration source, and some pocket change for construction and consumable supplies. The total price tag was put at $7500. Having filled in all the applicable blanks, I delivered the document to the JIO office in March 2007 and thereby launched a contest with the WYDEQ charter member of the JIO I had not expected.

Rather than evaluate my proposal locally, the JIO forwarded it to none other than Dave Finley. Five months later in early August 2007 I was provided a copy of Finley's reply.[48] He recommended non-approval and justified his opinion on the grounds that his division had previously sought JIO funding for a monitoring program of its own. Apparently he had been rejected because he then opined that JIO programs were intended to address mitigation rather than monitoring. In short, if he couldn't get funding, neither should I. He followed that short declaration with an extensive explanation of the ozone monitoring station he claimed was being planned by WYDEQ for the Wyoming Range on the other side of the Jonah and Anticline fields.

I felt I had been shut out by a state official who had just signaled a conflict of interest by seeking funding from an organization of which he was a charter member and perversely denying that same avenue to me. I immediately resolved to not take this lying down. On September 4, 2007, I shot a lengthy letter of rebuttal to the JIO chairman Mike Steiwig challenging my rejection and demanding reconsideration.[49]

My demand was based upon the grounds that the proposal was routed for review in a questionable manner and on the grounds that the reasons given for rejection were highly subjective rather than objective and empirical.

I also criticized the time element that accompanied the proposal review. The deadline for submission was early March 2007 but I did not receive the notice of disapproval until mid-August. I argued there could be no excuse for such a protracted review period given the single line of authority through which the proposal was routed, i.e., from the office of the JIO to the office of Air Quality within the Dept. of Environmental Quality in Cheyenne. Furthermore, the proposal was straightforward and contained no complexities; so the long delay could not be attributed to difficulty of evaluation. Finally, that the proposal was routed directly to WYDEQ-AQD was itself a questionable action.

I reminded the chairman of a telephone conversation sometime between May and July in which he commented to me to the effect that there was no one in JIO qualified to evaluate the proposal so it was forwarded to WYDEQ-AQD. I asserted that the more correct path should have been directly to the BLM air resource scientist Susan Caplan on staff at the state BLM headquarters in Cheyenne. After her, it might be appropriate to forward on to WYDEQ-AQD if there were remaining issues to be discussed.

I charged that BLM's routine deferral to the state because of the latter's regulatory authority was not sufficient reason to bypass its own in-house expertise. I reminded him that BLM itself had maintained that it had statutory responsibility to participate in the protection of the atmosphere, a point tacitly acknowledged by the fact that it had an air resource scientist on staff in Cheyenne.

I moved on to a second line of argument relating directly to the explanation for recommended disapproval by Dave Finley. I called up Finley's admission of having applied for and been denied funding for his own project and stating that the mitigation grant program was *"intended—for air quality purposes—to fund mitigation projects and not mitigation projects."* I warned this strongly suggested that Mr. Finley's objection was rooted more in the failure of his department to obtain funding for some unnamed previous project. He had, thus, resolved to obstruct any other entity from gaining acceptance of an air quality proposal. This seemed to violate standards of both actual conflict of interest as well as *appearance* of conflict of interest.

My third line of argument addressed the final paragraph in Finley's disapproval recommendation. I noted that he recited a list of actions WYDEQ-AQD planned to execute through a process of contractual arrangements. These included establishing a monitor station in the Wyoming Range. I pointed out that WYDEQ had been promising this to the Air Quality Task Group for two years.

I then charged that Finley had failed to grasp the value of leveraging qualified citizen participation in any program designed to define the behavior and characteristics of the local airshed as well as support the other listed activities. I reminded the chairman that while awaiting word on the decision about my proposal, I had already constructed my station to meet EPA requirements for form and citing. Also, some months earlier Finley had provided me with a copy of quality control protocols developed by his department that I intended to implement.

While I admitted that my data would probably always lack some aspect of quality in WYDEQ circles, it should not have been difficult to recognize that the data could be of sufficient utility to serve as an additional and badly needed source of baseline information in the event that gas leases upwind from my location were actually developed. In fact, a project <u>was</u> in the works that became known as the Eagle Prospect Project.

Furthermore, he cited WYDEQ intentions to contract temporary ozone monitoring that would lead to better ozone modeling capabilities in the Upper

Green. I suggested that Finley had an opportunity here to acquire additional airshed information at no expense to his own departmental budget.

I closed with observations about some implications conveyed by Finley's tone. His recitation could be interpreted as declaring that: "Air quality monitoring is the purview of WYDEQ-AQD and non-contracted outsiders shall be deemed unwelcome." His assertion that "… *my home*…" may not in his eyes constitute "… *a needed location*" for "*supplemental ozone monitors*" was a red herring. I maintained there could never be too much monitoring especially given the slowness with which WYDEQ-AQD had established its boots on the ground here. I felt Finley was still playing catch-up and I maintained that I offered an additional capability that he should be able to embrace as a serendipitous source of additional data.

I ended by pointing out that the JIO's web site explicitly stated that *JIO manages* "… *Jonah Field on- and off-site monitoring and mitigation*" contrary to Finley's assertion that monitoring was not included in the JIO charter. Thus, I requested that JIO reinitiate evaluation of my proposal and that it be routed for evaluation to BLM's air quality scientist in the state office. Should there be any issues warranting coordination with WYDEQ-AQD, my proposal should go to the director of WYDEQ, John Corra. I learned not many weeks later this Corra suggestion was a bad idea.

Corra called me to arrange a meeting but he did not state the purpose. We met in Jackson, where as it happened, my wife and I were on personal business at the same time he was in town for official meetings of his own. He arranged to meet us in a local restaurant and in the course of our two-hour conversation declared that the JIO was still a work in progress. He also stated categorically that WYDEQ policy was that no monitoring outside of explicit WYDEQ control would be supported. He admitted that he knew of few people more technically qualified than myself but that would not alter his policy.

I maintained my calm and replied that I could understand his not being able to accept my work for regulatory purposes but at least use it to guide WYDEQ monitoring station placement.

Weeks later in a public forum I again challenged Corra about this attitude. He responded that "The JIO is still a work in progress." I eventually learned through private sources in BLM that my challenge had unnerved Corra and

Finley because no one had ever done so and they had no procedures in place with which to react. However, they apparently weren't staggered for long because that oversight was soon corrected and procedures crafted.

Trojan Horse

WYDEQ tried to assure citizens of its concern for their wellbeing while facilitating the larger goal of drilling and production. It exhibited a proactive public image by engaging in outreach forums but in the end, most citizens felt they had been merely offered doses of pabulum. When challenged in these forums about its process and procedures, WYDEQ danced and when that failed, it invoked inventive interpretation of its guiding statutes.

The role of WYDEQ as a steward of air quality must be rated as inconsistent at best. It was slow to get involved in the beginning of gas development in the Jonah and Anticline when it should have immediately realized that a heavy monitoring presence was going to become unavoidable.

For ten years WYDEQ wandered back and forth between slow implementation of protective measures and almost desperate reliance upon the operators for voluntary actions to address the problems. It exhibited petulant resistance when it perceived that BLM and USFS were straying into its regulatory turf. On occasion it exhibited a "not initiated by us" syndrome and killed promising technology solutions through bureaucratic stonewalling.

To this day, WYDEQ chants self-praise by declaring it to be the breaker of new ground in air quality protection that other states clamor to emulate. If true, many of us shake our heads and question the fate of other regions experiencing what we have gone through.

The saga of gas development in Sublette County is a case study in ineffective oversight and excessively close relationships between regulatory agencies and those they are supposed to regulate.

Chapter 7

The State Apparatchiki

The County Commissioners

This chapter begins at the bottom of the government pyramid with the Sublette County Commissioners. These guys (they were all guys) were as rigidly supportive of the gas industry as you could find. Their sole focus was on the economic benefits bestowed. It wasn't until the local population began to express outrage over the growing ozone problem that they finally realized they had better show some interest in the human factors.

Periodically, they engage in an update of the county's comprehensive management plan. This supposedly affords the public some opportunity to voice its desires about what it wants to see prioritized. I crafted a comprehensive set of written recommendations pertaining to the growing footprint of gas production, all the while realizing I probably was wasting my time for two reasons. First, I doubted they would agree and second, most of the activity was taking place on federal lands beyond their administrative reach. Nevertheless I took my shot.

I first addressed the growth of industrial support facilities in the field. I pointed out that one need only drive the Lumen Road in the vicinity of the

EnCana field operations center and warehouse/shop yard to see a large and growing collection of industrial buildings that provided office and shop spaces for company personnel.

Additionally, literally every well pad included a tank battery to store produced water and condensate gas as well as a dehydration shack and in many cases there were four or more such dehydrator shacks. Finally, there were large parking yards for heavy equipment and big-rig trucks. I wanted the Commissioners to incentivize the operators in Jonah and the Anticline to consolidate their office and shop facilities in just a few locations.

Next I urged action on air quality concerns. I wanted the Commissioners to create incentives encouraging operators to engage in proactive efforts at seeking out and implementing improved methods of emission reduction for both drilling operations and wellhead production facilities. Although air quality regulation is in fact the chief purview of WYODEQ, it is, nevertheless, a fact that the Department can be influenced to take action as a result of urgings expressed by the citizenry and certainly the County Commissioners.

A common argument offered by industry and its supporters was that they were not to blame for *all* visibility degradation that was taking place. This was true. However, empirical evidence accumulated showing that industry was a growing significant contributor to the total problem. It seemed logical to counter argue that there was no reason to allow industry the freedom to exacerbate a clearly increasing visibility problem, and indeed it would be good business for industry to undertake efforts addressing those emissions.

It was appropriate that operators be persuaded to address their production emissions in light of the many assessments by industry that 10,000 to 15,000 wells were a distinct possibility in Sublette County. Each tank battery and dehydrator combination emitted, by the operators' own calculations, VOCs and HAPs in the range of 4 to 12 tons per year.[1] Simple arithmetic showed the potential to be tens of thousands of tons of these pollutants per year even though wells would be in various stages of initial production and production decline. Any efforts by the County Commissioner to address this long-term issue would lay the foundation for air quality protection on behalf of literally three future generations of Sublette citizens.

I ended by asking the Commissioners to press for a more comprehensive monitoring system that would place monitoring stations up wind, down the centerline of the Anticline and Jonah, and downwind along the foothills of the Wind River Range. Actually, the monitor stations along the centerline were already in place so the concept I advocated had a head start.[2]

In November 2008, issues surrounding the public's growing concern over ozone prompted me to make another try at persuading the Commission to take action. This time I wrote directly to the chairman and advised him of events I felt warranted his awareness.

I told him how in 2004 I had been pressing EPA and WYDEQ, Dan Olsen in particular, to act upon my findings of the presence of sodium, lithium, and potassium from fracking fluids in well completion smoke columns. I also recounted how following the infamous February 2004 pollution event by Anschutz across the river from Pinedale, the Wyoming Outdoor Council and I double teamed WYDEQ with a request for rulemaking to control well completion flaring. For that reason and not my spectroscopy fieldwork, WYDEQ issued a rule mandating green completions. By my count, flaring completions were reduced from 12-14 per month to around 4 per month because of that rule.

I also told him the story of my effort to introduce new dehydrator technology in the Jonah. I described what took place between 2005 and 2007 when I brokered a meeting between EnCana and a small company in Farmington N.M. that had invented a new gas dehydrator technology. I explained that this dehydrator had been tested by EPA and judged to reduce emissions of NOx, CO, and HAPs by a cumulative 1000% compared to current dehydrators. In 2006 EnCana field-tested two units that ran almost flawlessly into 2007 but WYDEQ interference stopped the effort.

I related these tales to the chairman for the purpose of showing him I had background to support my opinions. Of equal importance, I wanted him to realize that I was trying to assist the operators in my own way to advance the process of getting gas out of the ground while also protecting local air quality. Frustratingly, all of these attempts at persuasion were not even acknowledged.

The closest event I can point to as an indicator of brief Commissioners' interest in seeking my input came in November of 2008 when one of them actually called me at home to request my help. They had recently received a technical report of findings and recommendations from Sonoma Technologies, Inc. titled "Southwest Wyoming Air Monitoring Network Assessment."[3] It had been commissioned by WYDEQ. The individual who called expressed some frustration over the difficulty of making practical sense of what had been written in the report and he wanted my assistance in reducing it to straight-forward meaning.

I responded to him by citing each key observation and recommendation contained in the report followed by my evaluation of each point. This approach resulted in my addressing 32 specific issues cited in the report. My major focus was centered on comments about adequacy of wind measurements because they were in strong agreement with my own wind study findings.[4] On that topic I commented as follows:[5]

- *The maps depicting areas served by monitoring stations with respect to wind monitoring correctly illustrate the incompleteness of that coverage. This is supportive of my long-standing insistence that modeling of visibility impacts on the Class I/II regions has been inadequate regarding accurate wind behavior input to CALPUFF.*

- *My own research supports the "key finding" for the surface network that "seasonal and diurnal differences and dramatic shifts were observed in the wind patterns at most sites. This indicates a complex system." Also, this constitutes a validation of my longstanding criticism of the BLM practice of providing antiquated met data that is averaged over as much as three years for use in CALPUFF.*

- *I strongly agree with the surface network recommendation of adding a station between Jonah and Rock Springs and I urge higher priority than is stated for doing the same thing in the western Upper Green River Valley. This area is very likely a conveyor to that Class I area and perhaps more pertinent to this assessment exercise, toward the Jackson Hole population center.*

- *The statement that "some sites are greatly influenced by local conditions and are not representative of a larger area (important to modeling) or of boundary conditions (i.e., the Hoback Canyon re-*

gion)" is in total agreement with my own research. Furthermore, this again validates my previous criticism of the BLM use of old met data. Another point is the BLM data is wrongly assumed to be indicative of the entire region between the two mountain ranges down to Rock Springs.

- *The meteorological upper air network assessment section cites a special upper air radar wind profile done in Feb-Mar 2007. This period is far too brief to be of rigorous utility. My own conversation with a Riverton meteorologist knowledgeable in atmospheric modeling cautioned that applying upper air behavior measured at Riverton is inapplicable for our area below 12,000 feet. Encouragingly, a recommendation [by Sonoma Technologies] urges at least a year-long measurement period.*

- *The recommendation to establish 1 to 3 additional surface met sites to cope with "...winds [that] are highly variable in the region..." is absolutely sound.*

I also expressed my support for several recommendations pertaining to strengthening the existing monitoring system:

- *Recommendation that the Jonah monitor be considered for re-adjustment of its objective to become a monitor of source precursor emissions supports my own recommendation [to WYDEQ]. I stated: DEQ has recently commented that the Jonah monitor must now be relocated because it no longer represents ambient conditions...the air mass within a gas field...is [now]...ambient for the conditions created by the hundreds of well facilities present. Monitoring these emissions in situ is essential to learning specifics about how that regime differs from the regime outside the boundaries of the gas fields.*

- *The statements that the existing monitoring network does not support future emission assessment, reconciliation, and modeling studies and that there are parameters needing added support are sound; corrective action should be a high priority.*

- *The statement that diesel particulate matter, in particular, black carbon should be considered for monitoring attention is sound. This agrees with a conversation I had with a climatologist researcher at Lawrence Livermore Labs in 2004 expressing concern for long term impacts from that specie.*
- *It is correct to add ozone, NOx, and meteorology measurement capability to Rock Springs and Pinedale to fill gaps in monitoring.*
- *The recommendation to add 1 to 3 surface met stations to the network and to add UV sensors at another site are sound; I would only recommend UV sensors at all sites in the Jonah and Anticline vicinity instead of just one as is recommended; also, hourly met measurements added to air quality sites is sound and needed.*
- *I strongly agree with the assessment that sites of strong wind variability represent possible gaps in the surface network "...between Jonah and Rock Springs, West side of the Upper Green."*
- *The recommendation not to eliminate any surface met site due to acknowledged surface wind field complexity is excellent.*
- *The statement that the existing monitoring network does not support adequate characterization of particulate matter is sound. Proposed solutions are worth implementing.*

I ended my reply to the Commissioner by explaining my overall impressions of the findings and recommendations in the Sonoma report. Generally, all statements and recommendations were well founded. The study was performed at substantial cost to the state and represented an objective third party evaluation that WYDEQ felt was required. It, therefore, behooved WYDEQ to apply the recommendations to the maximum extent. This would be necessary to accomplishing effective pro-active corrections to our air quality dilemma. It was also needed to earn the trust of Sublette County citizens who remained skeptical of Cheyenne's dedication to correcting the air quality situation here.

For good measure, I sent an information copy of my reply to WYDEQ-AQD Director Dave Finley in hopes he would be less inclined to dismiss the findings and unjustly assure the Commissioners.

• • • • •

Despite cautions expressed by myself and others and my best efforts to help, the Commissioners adamantly defended the gas industry and assailed what they saw as environmentalists opposed to any development. In fact, anyone whom they deemed to be in opposition was condemned as "environmentalist." They were determined to see, speak, and hear no evil about the operators. They were effusive over mineral tax revenues coming from gas production and gas development bringing jobs. However, they said little about the invasion of migrant field workers I called petro braceros who strained community schools and law enforcement. Nor did they show any initial concern over growing air quality problems.

Eventually, community uprising over ozone increases and ozone health warnings forced them to step up to that growing concern. Realizing they had to appear proactive, the Commissioners launched a project to award a contract for the purpose of evaluating the presence of harmful pollutants in our local air mass.

From the beginning the effort was suspect. Little could be learned about its gestation outside their inner sanctum. I learned of it from a person on the inside who expressed concern that at least one of the Commissioners might be trying to rig the process. The reason for this concern was an obstinate senior Commissioner in particular who was heard to assert that his goal for the study was to prove that any pollutants were wafting in from Salt Lake City, far to the southwest.

I obtained the draft request for proposal (RFP) document and took it on myself to raise concerns about certain stipulations therein. I say "took it on myself" because weeks earlier I had offered my assistance to the chairman of the Commission to help craft the RFP but received no acknowledgement of my offer. Consequently, I volunteered my points of concern.

I noted the stipulation calling for once-per-week sampling and use of non-reference method ozone monitors. I suggested that once per week seemed too little given our wind and terrain characteristics. Also, I questioned the use of non-reference ozone monitors because WYDEQ had been traditionally unyielding on that point. They had blocked my own ozone monitor program on grounds of it being no-reference, which was in fact untrue.

The RFP invoked what had become an infamous Garfield County Colorado air toxics history and maintained that once-per-week sampling was sufficient to avoid a similar outcome.[6,7] However, I challenged on the grounds that Garfield County meteorology had not been demonstrated to mimic our own. In particular I sighted our environment of wind direction variability and speed behavior.

I next challenged the notion of county boundary-line sampling to measure transport into the county. I cautioned that monitors would have to be placed in valid down-wind locations from assumed external sources. I offered my years of wind data derived from federal meteorological sites in the area as a starting point.

There was a reference to budget based sampling on a less frequent basis than once per week at up to five sites for purposes of gauging transport into the county. I challenged this again on the basis that wind pattern behavior and sufficiency of sample frequency should dictate monitoring architecture.

In mid-2008, a formal contract request for proposals was issued. The scope of work contained four elements that differed in small ways from the first draft.[8] First, the contractor was to design a sampling program sufficient to characterize exposure of the general population to air toxics and ozone and to evaluate if air contaminants were being transported into the county.

Target pollutants were to be HAPs listed as COPCs in up to ten sites once per week.[9] Formaldehyde was to be measured with the same frequency and spatial distribution. HAPs and formaldehyde at four county boundary sites were to be sampled weekly to assess any transport into the county. Finally, ozone monitoring was called out for five locations in populated areas of the county.

Other elements addressed selection of sampling sites acceptable to WYDEQ, crafting a quality control system that would insure data compliance with state criteria, and stipulation of "routine operations." This latter element described use of county staff as operators, and monitor sample placement/retrieval training provided by the contractor.

In April 2011, a public forum was convened in Pinedale by the Commissioners to review the results of the study. Many residents expressed skepticism over the fundamental conclusion indicating there were no significant risks of adverse health effects resulting from nearby gas development activities. One citizen worried aloud that the meager population being exposed was acceptable collateral damage in the eyes of the industry and the Wyoming Department of Health.

That comment was the result of the statement from a representative from the Wyoming State Health Department in attendance. He had just explained that the group of residents' suffering adverse health effects they believed were caused by gas development was a sample-set still statistically too small from which to distil significant scientific conclusions.[10] This supported my own realizations in later years that the statistics game was a ready-made hiding place for officials and industry.

Medical conclusions are by necessity derived from statistical studies that require sifting through a variety of variables. Furthermore, it is a near impossibility to trace a carcinogenic molecule back to the exact rig or well from which it came. This provides the escape route that industry then invokes when it ritually declares, "there is no evidence linking (this) to (that)." These caveats also satisfied the Commissioners' needs.

The senior Commissioner boasted that the Commission had spent $1 million on air quality monitoring and this health assessment project. Also, the Commission had (magnanimously) released the report to the public and had afforded this chance to submit written questions beforehand to be addressed at the forum. Left unsaid was that a requirement for written questions circumvented unexpected questions that could prove embarrassing.

Many of these citizens ultimately concluded that the study had been a waste of the $1 million, in part because the study contract statement of work had been crafted behind closed doors. Others felt that although the study addressed 50 toxic air species, it had been set up to spend too much effort looking for the wrong species, thereby missing the more relevant issue of growing ozone presence. Another Commissioner addressed this point by observing that the county paid for a study of as many pollutants as it could afford.[11]

Figuring into the awareness of those in the audience, there had been 13 days of elevated ozone in the previous month alone.[12] To acknowledge awareness of citizen unhappiness, the Commissioners' assured listeners that $500,000 in added monitoring was being considered. Lastly, State officials declared they would return for another public forum in about three months with results from the current winter's ozone monitoring findings.

The same senior Commissioner expressed his conviction that more study would be a waste of money. Instead, he insisted that concentrated attention had to be focused on emissions reduction. This was not unreasonable and it served two purposes. First, the Commissioners had come away looking proactive. An equally valuable second accomplishment was that they had now punted the ball back into WYDEQ's court. By invoking reductions, the Commissioners were signaling they had done all they intended and now it was up to the State to fix the problems.

Ultimately, a comment voiced in 2004 in a Pinedale paper has proven to be exact prophecy twelve years later. In a letter to the editor, the writer warned "When the oil prices plummet, and eventually they will, the oil companies will be gone quicker than fire through a gas line, and all of us will be left picking up the pieces of our destroyed local economy — not to mention the plunder and pillage of our public lands."[13]

Ditto for air quality.

My Backchannel to the Governor

My first contact with the governor's office came in the form of an informal conversation while standing in the snow outside the meeting hall in Daniel, Wyoming. This happened in 2003 and my conversation was with Mary Flanderka, then a mid-level policy analyst for Governor Freudenthal. Her task was to address issues of land use, forestry, cultural resources and public access.[14]

She had been sent to the Pinedale area by the governor to obtain feedback from the local population concerning our growing worries over environmental impacts from gas development. I managed to assert temporary dominance in a direct way by latching onto her while she was outside and away from the main crowd.

I started by giving a fast description of my background in science and transitioned to a statement of what I believed was underway in the form of air quality degradation due to the recent drilling onslaught in the Jonah and Anticline. To my surprise, she listened attentively. When I finished my monologue, she asked a few salient questions seeking clarification of some of my points and actually expressed interest in what I had told her. We parted with the assurance that she would convey what she had heard to Cheyenne and would be in touch in the future. I could not foresee that this was the beginning of a long period of excellent collaboration between the governor and myself through her.[15]

Our professional relationship proved a continual surprise to me because I had fully expected to be dismissed as soon as I had driven away. In fact, Mary was soon to be promoted to the position of Director of the Governor's Office of Planning Coordination. This would in turn afford me a direct pipeline to the governor that resulted in he and I being able to converse on a first-name basis. I would describe the period of this, his first term in office as my golden years. Unfortunately, his second term evolved into a less successful period due in part to the operators cranking up their unrelenting pressure to drill year-round.

My first opportunity to sit down across a table from the governor came in November of 2005. I learned that during a scheduled visit to Rock Springs, he had agreed to set aside a one-hour period in a meeting room at a motel to listen to input from Pinedale citizens. The individuals who participated were there by invitation and I was one. As I recall, something like eight to ten of us made the two-hour drive and assembled in the conference room.

Soon after we sat down, the governor strode in accompanied by a hulking Highway Patrol bodyguard with no neck and who hung back near the door during most of the proceeding. Governor Freudenthal thanked us for coming, opened his large notebook and invited us to individually offer comments. He started by going around the table from my left. This suited me just fine because I wanted to hear the tone of what the others had to say so that I could craft my own input accordingly.

The comments were generally similar in that worries were voiced about impacts on schools from the kids of gas field workers. Impacts on the economy and housing were noted as well but overall, concerns over air quality impacts were center stage. The most memorable input came from Pinedale's gritty family medical practitioner Doctor Tom Johnson whom I cited in the first chapter of this book. He briefly described what he was seeing in the way of respiratory ailments and their apparent increase coincident with the heavy drilling that was underway around Pinedale. He ended by bluntly declaring that something had to be done about the growing brown cloud that was hanging over the county.

Freudenthal scribbled some notes but then skeptically commented that the brown cloud had likely been around for some time before the gas guys had come on the scene. This was the clue I needed to form my own comments and my turn came soon thereafter.

I very briefly outlined my background in the Air Force as a scientist doing work that had introduced me to atmospheric dynamics. I reinforced Doc Johnson's viewpoint about the brown cloud by noting that indeed it had become an increasing presence in the past year of heavy drilling. I had engaged in a personal project of photographing the sky and was chronicling regular occurrences of strong light scattering from a thick particulate haze suspended in the lower atmosphere. Under dawn and dusk sun angles, this haze became visible as a reddish brown cloud. It was real and it was a recent addition.

My turn at speaking was near the end of the allotted hour so quite soon the governor closed his notebook, stood up, and thanked us all for coming. After he left the room, we all milled around a bit and speculated on the effect we hoped had resulted, then we made the two-hour drive back home. My next opportunity to speak personally with the governor wouldn't happen again until 2006.

One of my perceptions coming out of this meeting was that in response to requests to the governor by two lawyers with the Greater Yellowstone Coalition and the Wyoming Outdoor Council the governor was not encouraging. They had urged that drilling in the county be slowed so the impacts could be better assessed and mitigated. However, the governor opined that there was no statutory or regulatory mechanism by which the state could impose such restrictions. Thus, the growing desire of the residents in this regard would likely go unfulfilled and I worried that they might simply give up.

During the next few years, my interactions with Mary Flanderka increased and improved. In the period between 2005 and 2007, I had to make several trips to Laramie on personal business and I made a point to dedicate at least one day for travel to Cheyenne and the offices of WYDEQ. My mission was to comb through its file cabinets holding hundreds of drilling permit applications and copy well information. While there, I always took time to visit Mary.

These visits afforded me the opportunity to brief her on the developments I had witnessed around Pinedale, particularly with regard to the BLM and the AQTG. In return, I was able to learn from her the current attitudes and actions at work within the governor's circles as they related to the air quality problem in Sublette County. In fact, I found myself reporting to her in memo form what would transpire in BLM hosted open house meetings it held to initiate each latest round of drilling project proposals being pressed by the operators.

A particularly contentious example happened in November 2005. I reported to her the following: On November 8, I attended yet another public meeting hosted by BLM–Pinedale the purpose of which was to provide information to the public regarding the Shell-Ultra-Anschutz winter drilling proposal. The first surprise was that the meeting room was packed with Pinedale residents (perhaps 50 or 60) and very few if any industry persons. The crowd was even standing in the entrance.

The meeting was led by a woman who stated she was a spokesperson for the operators. She gave a lengthy presentation complete with a slick slide show that touched on all of the key points of contention, i.e., habitat fragmentation, socio-economic impact, air quality impact, and surface disturbance. There was really little information that was new, if any.

When she finished, a young new BLM employee, Mat Anderson stepped up to the podium and gave a somewhat broken account of the history of EIS events beginning with the first Jonah development. He commented that growth has been beyond what was first anticipated and thus a need for a new EIS was recognized. He concluded with the advice that questions would next be accepted but stipulated that those questions were to be directed to operators stationed around the room where poster displays were set up. This was the first BLM blunder.

Several citizens spoke up and demanded to know why they were being forced to present questions in such a compartmented manner. The concern here was that the "divide and conquer" principle was being applied by BLM.

Anderson insisted this was the format to be followed, period. By then the crowd began to get very vocal in protest. A few began to shout that we should convene our own meeting to compile questions for BLM en masse. Anderson then committed a second blunder by declaring that "this is a BLM meeting …" implying that it would proceed according to plan, period. At that point a couple of citizens confronted Anderson and angrily pointed out that BLM worked for the public and not the other way around. Several other of the audience then surrounded Anderson and a few other BLM staff and angrily protested.

By then, the rest of the crowd had formed into small groups and engaged in heated observations about how they felt they were being denied a hearing. Several voiced a need for someone to take charge and organize a response and others suggested a need for a petition to protest our being denied a voice in the proceedings.

At that moment, I don't know why, unless it was a recognition that someone had to speak up, I stepped forward and got the crowd's attention. I declared that I had a practice of writing a critique of these meetings and sending them to a contact I had on the governor's staff. I advised that if they were as dismayed as I regarding the way the meeting had been conducted, I would be willing to include their petition in my critique to my contact.

At this point the BLM office director, Prill Mecham, apparently sensed disaster in the making; so she had one of her staff quiet everyone down so she could talk. She then lectured the crowd by stating that we apparently failed to understand what the meeting was intended to be. She informed us that this was an "open house" which meant that it was NOT "a hearing." She explained that as such, comments of a brief nature, which simply outlined individual concerns were all that was being solicited.

Forms were handed out on which we were to write our comments and return to BLM staff. They would be summarized, written up, and serve as the agenda for another meeting to be held in about six months when they would supposedly be addressed in detail via more questions from the public. However, no one was interested in such an antiseptic approach. Thus, after her comments, a petition was circulated and I attached it to the memo that I sent to Flanderka.

Afterward, in sideline discussions, several folks came to me to express their intense anger over the rate of development that was being thrust upon us. They expressed feelings that this must be stopped and BLM made to be more open and responsive. Sadly though subsequent BLM decisions in following years continued to support the operators' drilling agenda and the idea of limits never gained serious traction.

In August 2005, I again hammered out a letter to Mary detailing the latest developments involving the BLM and the operators. I had been motivated by a story in the *Pinedale Roundup* telling of a recent decision by the Wyoming Oil and Gas Conservation Commission approving a denser 10-acre spacing for down-hole drilling separation on the Anticline.

I pointed out this would nearly quadruple the well density over what was originally proposed. Consequently, Questar was asserting that more wells meant it would be in operation for one-third less time as a result of winter drilling due to this increase in well density. When Questar, presented its original request for winter drilling, it argued that its time on site would be halved and its impacts on the environment brought to an end accordingly. I insisted that Governor Freudenthal should be deeply suspicious because he had supported the first proposal on the basis of Questar's original presentations. I summarized by stating that Sublette County was now presented with the specter of accelerated death by 12,000 cuts.

During an earlier BLM/Questar winter proposal open house, I questioned what was to prevent the other operators from rushing to the door with similar requests and receiving rubber stamp approval.

Prill Mecham assured us each application would be viewed on the basis of individual circumstances but I knew that there would be no legal basis to deny to one what had been given another. It was logical to assume that Shell, Ultra, Anschutz and no doubt BP America were making their applications. The aggregate potential from all operators would be 10,000 to 12,000 wells.

A local BLM geologist commented in a manner that cast doubt on the likelihood of such a large well population but BLM had a strong track record of marginalizing the facts to avoid any need to alter course. Also, BLM had no credibility by then and I warned that any reassuring mutterings by its officials should be taken as nothing more than noise.

I then expressed my own frustration over this newest event on the grounds that I had invested a lot of hard work developing a dialogue with the inventors of the Quantum Leap Dehydrator and with EnCana who would field test it. I explained that I had worked hard at maneuvering EnCana to perform the testing which was being scheduled to take place during the coming winter. Even given the fact that the dehydrator reduced the volumes of NOx, VOCs, and BTEXs to the realm of less than ten pounds per year versus the 7-15 tons per year common to dehydrators then in use, such a multiplier reported by the *Pinedale Roundup* could render the QLD a mere footnote in Sublette County air pollution history to come.

I next transitioned to more pointed comments regarding BLM's general behavior. I began with the Air Quality Task Group and my original motivation to participate. My specific purpose had been to measure the seriousness demonstrated by the BLM in soliciting stakeholder input. I was skeptical then and I could now declare with authority that my skepticism had been well rewarded.

I reminded Mary that in the beginning, Prill Mecham had sung the praises of the PAWG concept and assured us that BLM would not interfere nor could it interfere in the workings of the PAWG. This was blatantly reversed in the last PAWG meeting when she received the proposed agenda from the PAWG chairperson and promptly turned it back in a completely reworked form. She had deleted some topical discussions she/BLM felt were inappropriate due to the "pre-decisional" nature of the subject matter. I pointed out in my memo to Mary that this definition of out-of-bounds for an issue because it had yet to be decided upon was absurd. What good would result from discussing an issue if it had already been decided upon?

I next recited to Mary the story of Prill"s approach to adaptive management that I repeat as a reminder here. During the Questar winter drilling proposal open house, Prill had been questioned sharply as to the likely impact on the deer and sage grouse on the Anticline. She waxed eloquent about the adaptive management process and assured all that when BLM detected negative impacts, it had the authority to suspend operator activities pending a solution.

However, in the most recent PAWG meeting, I pressed her to invoke adaptive management regarding the sage grouse in light of definitive research findings that serious impacts were well underway. She evasively declared that the adaptive management process was an uncertain mechanism and the criteria for invoking it were not well defined.

I went on to explain that BLM had been frustratingly perverse in the realm of statute interpretation. I cited BLM's about-face regarding reimbursement for travel by AQTG members and the AQTG event when the EPA representative argued futilely that BLM needed to have a management level person in attendance with us to provide management decision advice. His justification had been that we were unavoidably being drawn into issues of air monitoring that transcended individual gas field borders.

I further opined that BLM had become adept at "liturgical recitation of statutory scripture." To a man and woman, they would invoke the existence of statutory authority to intervene in environmental matters but NEVER actually initiate meaningful actions that could be justified from those statutes. When challenged, they made excuses of the type that I illustrated with Prill. In one case of land reclamation they cited vagueness in the guidance as a reason for holding back and in another they worried about making life difficult for small operators with less operating revenue.

I then moved on to a story I was told by one of my EnCana contacts. There was an EnCana field manager on site who had a reputation for thinking out of the box on environmental matters. He went to BLM and presented a request to try an arctic tundra technique of conducting operations on top of a network of cushioning pads covering the ground. He believed this would reduce disturbance of the ground surface.

However, BLM rebuffed the proposal on the grounds that an EIS would be required and due to low staffing the EIS for the Jonah Infill Project would have to be put on hold. Of course EnCana would not risk that so they canceled the idea. Curiously, a few months later the *Pinedale Roundup* carried a call for public comment on the same proposal thus suggesting that something had changed BLM's mind.

In spite of this, or more correctly, because of this kind of behavior, I characterized BLM-Pinedale and BLM-State as *the* fundamental obstacle to environmental change and progress in the Jonah and the Anticline. As a result, we were hearing only vague assurances that energy development was being done with "environmental responsibility."

I concluded my report to Mary with observations about Wyoming's bureaucracies. I opined that there were too many players who seemed to operate in isolation from each other. Examples were WOGCC, WYDEQ, and the State Legislature to name a few and I cited the appearance of lack of coordination in the exercise of interdisciplinary decision making. For instance, the WOGCC had issued approval for operations here that would quadruple air quality impacts because their charter did not include environmental impact criteria.

I speculated that his would have a knock-on effect for WYDEQ's Dan Olsen because his life would be complicated by resultant non-attainment status unless he doubted that 12,000 more wells would develop into such a situation. This became prophecy because non-attainment did in fact become reality under Olsen's successor.

I closed with a confession of a growing sense of defeat and resignation that Wyoming was losing its traditional identity. I blamed this on State Legislature complacency and addiction to mineral revenues, a fragmented approach to controlling the State's future, dismissive attitudes of the feds, and huge political momentum on the part of the BTU prospectors.

My successful interaction with Mary resulted in a rare one-on-one meeting with Governor Freudenthal. I had accumulated over a year of field observations of well completion flares and dehydrator emissions using my optical spectroscopy approach. I had shown this work to her when I visited her office and it seemed to impress her and others in the office. During one such visit, I was chatting with her when the governor's Chief of Staff and some others wandered in. She introduced me and asked me to describe what I had been doing.

Always eager to explain my methods and findings, I launched into a fast paced monologue while being careful to explain the science when I feared I might be sounding too esoteric for their non-science level of understanding. As I was nearing the end of my explanation, I noticed the gentleman seeming to have a blank look on his face and I feared I had lost him. Suddenly he became animated and said: "The Governor needs to hear this." I was stunned.

Soon they all departed, leaving Mary and me alone. She was thrilled by my performance. I was oblivious though because I was in an arena that was quite alien to me so she proceeded to enlighten me. She explained the direct line to the governor the man represented and assure me what I had just described would be passed on to the governor. This event turned into a meeting with Freudenthal I describe next.

Walking a Tightrope

The governor visited Pinedale in September 2005 if my memory is correct. He flew into Pinedale on the state business jet but had first done a fly-over to inspect the Wyoming Range where additional drilling was being proposed.[16] Mary had notified earlier that the governor had agreed to speak with me directly and she had arranged for this meeting to take place in the waiting room of the little Pinedale airport. I would be given a half-hour to speak my piece and then he would head on into town for his series of meetings with town officials.

The day came and the governor's plane taxied almost to the door of the terminal building. I walked out to greet him and commented, "Boy, nice wheels, Governor." He smiled, shook my hand and we settled down on the sofa in the waiting room. At this moment, I realized with some surprise that the meeting was going to be private. The entrance doors were shut and a highway patrol officer stood guard at each.

The only people in the room were myself, the governor, Mary, a few others of his staff and my wife whom I had invited for the chance to see him up close. I did this because her father had been very active in Cheyenne and personally knew a previous governor; so these circles were familiar to her. I am glad I did because after the meeting, she informed me of details I had missed.

Realizing I had no time to waste, I launched into my monologue about my optical spectroscopy and the fact that I had been routinely detecting the presence of potassium, lithium, and sodium in the smoke plumes from well completion flares. I also outlined the presence of several hydrocarbon combustion gases and particulates I was detecting in the exhaust gases from well pad dehydrators.

I explained my suspicion that fracking fluids were the source of the completion flare alkali metal emissions. I also warned that thousands of dehy units emitting so much combustion products that were precursors to visible haze and ozone portended pending problems for the citizens of Sublette County. I showed him my field notebook with its spectrograms and photographs of the flare fires from which the spectrograms were taken. Also, I showed him the spectrograms from dehy burners and photos of those burners in action.

When I finished, I was amazed to realize we had been sitting there for 45 minutes and not a single interruption from outside had intervened. I had been granted uninterrupted access to the Governor and the officers at the doors had guaranteed it. After we parted company, my wife asked: "Did you see what the governor did?" I replied that I had not. She then explained that during the entire time I was talking, he had been madly writing in his legal-size notebook and by her reckoning, filled up two and a half pages with his notes. I had indeed captured his attention.

• • • • •

During his two terms in office from 2002 through 2010, Governor Freudenthal was generally a consistent vocal critic of both the BLM and the U.S. Forest Service over their respective approaches to regulating gas drilling in Sublette County. However, as he transitioned to his second term, he joined with the operators in ways that bedeviled us in Sublette County who saw industry gaining ground at our expense.

He often weighed in with public statements of concern over environmental impacts from drilling and as the ozone problem around Pinedale ballooned, he became even more vocal. It should be said however; his concern about ozone seemed facilitated by growing outrage of local citizens demanding that his director of WYDEQ get involved with correcting the problem.

In June of 2004 he complained to State BLM Director Bennett that the Bureau was revising its 1980 Resource Management Plan in a manner that would transform its oil and gas management strategies in our region in a way that favored industry. The Governor wanted BLM to suspend leasing in the Pinedale area until the Plan had been completed on the grounds that some parcels might be deemed inappropriate for development. BLM disagreed and issued a perfunctory reply that offering the parcels in question did not compromise the RMP revision.[17]

Then in August 2004, the governor made a decision that seemed a turnaround. On August 2, 2004, his office released a statement endorsing the Questar Corporation's proposal for winter drilling on the Anticline. He told Prill Mecham, the Pinedale BLM field manager that the proposal was a good model.

He based this assessment on what he called caveats that addressed impacts to air, water, and wildlife, as well as other impact categories. Those caveats included the operator's promise to employ flare-less well completions, and a liquids gathering system.[18] The flare-less part satisfied me because of what I had observed and measured in the enormous smoke plumes that accompanied completion flaring. The gathering system also pleased me because that would eliminate the batteries of storage tanks for produced water and condensate gas located on every pad. This would eliminate most of the emissions of flash gas from the water tanks and fumes from the condensate tanks.

In the spring of 2005, he took a welcome action to those of us outside the drilling industry. State Director of BLM Bob Bennett issued an edict that threatened to shorten the period for public comment about oil and gas leasing projects and to require any protest documents challenging a proposed lease bear an original signature. No fax or email documents would be accepted. His explanation for this was that receipt of a lease protest with as little as a day remaining before the sale deadline was delaying the agency in its review process. Therefore, the public comment period would be shortened by 15 days.[19]

Governor Freudenthal weighed in with a letter to Bennett. He noted his understanding of the difficulties faced by BLM in that the agency had a legitimate need for ample time to analyze such protests. However, cutting into the time allowed for the public to participate carried the appearance of the BLM placing higher priority on its own needs to the detriment of the needs of the public.

Furthermore, Bennett's demand for an original signature on protest documents would have the effect of further shortening the public's response period by the extra days needed for delivery of hardcopy mail. Freudenthal urged Bennett to apply a 60-day review period or institute some kind of preliminary public notification naming the parcels to be included in the sale. The normal practice was for notices of lease sales to be issued 45 days before the sale, sometimes earlier. Critics and the governor could see what 15 days taken off that period would do to public input.

It was reported that the State BLM had acknowledged receipt of Freudental's letter but would offer no comment.[20] Apparently, BLM was no more interested in attending to his concerns than it had demonstrated toward the general public.

On the flipside in October 2006, the governor took an action that appeared inexplicable. In his capacity as the Chairman of the Interstate Oil and Gas Compact Commission (IOGCC) during 2005-06, he officiated over a ceremony in Austin, Texas, which bestowed an outstanding environmental stewardship award on EnCana Oil and Gas (USA).[21] This was touted as "the commission's highest honor for exemplary efforts by the oil and natural gas industry in environmental stewardship and challenge organizations, companies and individuals nationwide to demonstrate innovation, dedication and passion for the environment."[22]

The story went on to portray EnCana as "... demonstrating (in the Jonah gas field) that production could be accomplished in the area with minimal environmental impact." Examples cited were flare-less well completion, centralized fracking operations, natural gas-fired drilling rigs, road dust control, off-site mitigation, surface reclamation, and low impact development.[23] These proceedings were insulting to those of us who had been trying to pressure the governor and the operators to address these problems.

What we saw was a cynical move by the governor to inflate to philanthropic status that list of environmental protection measures which in truth EnCana and the other operators had been forced by growing public pressure to implement. Indeed, his director of WYDEQ was constantly reacting to that pressure by urging operators to adopt specified "voluntary" measures, and implied through unspoken signals that public-forced regulatory solutions could become unavoidable otherwise. Thus, in the main, operator voluntarism was in fact coerced.

Alternatively, in May 2008 Governor Freudenthal again criticized the BLM for the way it addressed oil and gas drilling in its revised RMP. His 5,000-word letter to the Pinedale office accused BLM of failing to consult cooperating agencies and Wyoming State Government. He warned that the draft revised plan failed to go far enough in addressing development effects on air quality, water quality, socio-economic issues, and winter range and migration routes for wildlife.[24]

As usual, BLM responded with assurances that all the governor's concerns would be addressed. Some refinements to the draft RMP had been done to address community impacts and BLM was certain that it was creating a plan that worked for the BLM as well as for cooperating agencies. One had to wonder what "cooperating agencies" it was referring to in light of the governor's accusation that none had been consulted.

Again in May 2008, the governor criticized the Forest Service for, "allowing an energy company undue influence in a process that would allow oil and gas leases in the Wyoming Range." Earlier in April he had blasted Bridger-Teton National Forest managers for including Stanley Energy, Inc. in meetings about the environmental impact statement that would determine if 44,700 acres of contested leases should be reissued. Furthermore, Stanley Energy had approved the contractor that would write the EIS and was going to pay for the analysis.[25]

Although Bridger-Teton officials subsequently admitted to a questionable relationship with Stanley Energy, they, nevertheless, said they would proceed with the document despite the governor's opinion that a new analysis was needed.

R. Perry Walker

• • • • •

In the winter of 2011, the contest between the citizenry, the BLM, and regulators reached a major juncture. Since 2004, the three operators on the Anticline (Shell, Questar, and Ultra) had been relentlessly pushing for authorization from BLM to drill year round. They argued that they could drill out the Anticline a decade sooner and in so doing lessen environmental impacts across the board. This was met with great skepticism because the wintering deer herd on the Anticline had already suffered a documented severe decline in population and because of the advent of regional ozone alerts.

In February 2011, BLM revealed a mitigation plan that basically tinkered with Anticline vegetation mitigation methods rather than alter drilling schedules and methods. Furthermore, the three operators committed to creating a $36 million mitigation fund to be targeted against monitoring and mitigation of impacts directly related to the PAPA SEIS Project. These funds were to be applied in the areas of air quality monitoring, wildlife, livestock, vegetation, reclamation research, analysis, monitoring, and mitigation. What stood out was the proviso that funding contributions would be based upon the pace of development.[26,27]

The fund was titled *"Pinedale Anticline Operators Mitigation and Monitoring Fund"* in Appendix 11 of the Pinedale Anticline Final SEIS, Alternative D. It would be made up of $4.2 million already donated by the operators and an estimated annual contribution based on the proposed rate of $1.8 million per year. The resulting total anticipated contribution was placed at approximately $36 million.[28]

Opponents immediately warned of loopholes I had also discovered in the PAPA SEIS appendix that described the proposal. The language contained elasticity that would afford an escape clause to the operators. I saw the pace of development proviso as a veiled threat actually signaling a warning that if the proposal were to be rejected and the operators not allowed to drill at the accelerated rate, they would exert no further effort to address citizen concerns beyond a level required of them by existing law.

This increased drilling rate would create 3700 new wells above the 700 originally authorized by BLM and do so through the next decade. EPA ozone standards had already been violated several times in previous winters so the prospect of so many new wells adding their ozone precursor emissions to the emissions inventory was not a welcome concept to us.

BLM's Caleb Hiner seemed unmoved, however. He insisted publically that industry had committed to maintaining ozone levels below what would initiate an air quality violation but in his next breath admitted that BLM had not specified exact measures to be applied in achieving that goal. Local environmental activist Linda Baker stated our skepticism succinctly: "We are over being burned by this language play."[29]

It was with some disappointment that in September 2008, Governor Freudenthal publically approved the Anticline drilling plan. He was now in support of the federal plan managing oil and gas development on the Anticline that would allow all of those new wells. His spokesperson told the Jackson Hole News & Guide the new rules contained in the approved BLM resource management plan (RMP) were a big improvement over the preliminary rules he had criticized. In particular, significant gains in air emissions were cited as reasons for the governor's turnaround. I and others worried that the gains would prove illusionary.

• • • • •

I began to suspect the governor's approach to the Anticline drilling proposal and the Wyoming Range drilling proposal were signaling a shift in tactics. I became increasingly suspicious that he had concluded it would be necessary to sacrifice the Anticline in trade for protecting the Wyoming Range. I was never able to confirm that scenario, however; in part because my conduit, Mary Flanderka had moved on to another job in state government thus depriving me of inner sanctum clues. As you will see shortly, my loss of her presence would result in my badly damaging myself.

In the wake of this decision by BLM, Governor Freudenthal made another visit to Pinedale to address citizen unhappiness. He agreed to hold a side meeting with several local folks of which I was one. I explained my worries and objections about the proposed rate of drilling. To my satisfaction, a local rancher well known to him and Sublette County citizens who was sitting to my right spoke up and advised the governor, "You should listen to Perry because he knows what he is talking about."

Then I blew it. Instead of keeping my mouth closed and allowing that vote of confidence to back me up, I blurted out that I considered the $36 million buy-off from the operators as little more than blackmail. Governor Freudenthal clouded over instantly and shot back: "That is the best deal I felt I was able to get." In that instant I crashed and burned because this was the first signal that the fund had been *his* invention or at least his notion of best achievable victory. After the session ended, I was still reeling from my self-imposed injury and furious that no media source, staffer, or any other entity had revealed that this fund had been the governor's brainchild.

That was the last time I was offered any opportunity to talk to him one-on-one.

The State Legislature

I experienced another facet of political resistance in early 2006. I had come to the realization that one of WYDEQ's problems was that it had insufficient staff to deal with the onslaught of permit reviewing and regulatory formulation and field enforcement. In an effort to add some constructive input about this issue, I began contacting the legislators in Cheyenne by writing to every single senator and representative.

I cited the need for regulatory oversight of an industry that was building a huge presence in Sublette County and eventually all of Wyoming. I pointed out that there would be an urgent need for sufficient staff in the offices of WYDEQ as well as in the field to cope with a work load that likely could increase several fold over the next few years. Accordingly, I explained, it was necessary to allocate considerable funding to the department so that the needed personnel could be hired and trained.

I pointed out that I was not opposed to gas development. Rather, I was urging that this resource be brought out of the ground in ways that would be environmentally smart. To achieve this, I explained that the operators needed to be held accountable in a way that minimized damage to air quality but would allow them to produce gas. I was requesting that they work to "plus up" the WYDEQ budget so that it could acquire necessary personnel and tools to accomplish its chartered purpose of protecting environmental quality.

I didn't expect any particularly specific outcome from this effort other than perhaps a useful increase in the budget for WYDEQ. To my surprise, however, I received a written invitation from a senator whom I found on the legislative web site to be listed as the chairman of the State Joint Appropriations Committee. He invited me to attend the next scheduled budget hearings the following month to provide testimony. I was stunned and immediately committed to making the trip.

That trip would prove to be no small undertaking because it involved a 9-hour drive in the dead of winter across a really risky stretch of I-80 between Rock Springs and Laramie. The session was to be convened in January and I would have to drive from Daniel to Cheyenne. All of this would be at my expense and at that time, I had little money to spare outside of securing food, fuel, and shelter. I was able to lessen the expense of lodging by calling upon my privilege as a retired military officer to stay at the transient officers' quarters at F.E. Warren AFB on the edge of Cheyenne.

The next morning after arriving in town, I set out for the state capitol with little idea of how to navigate my way to committee chambers. I had never been in my own State Capitol Building nor had I had any occasion to rub elbows with elected lawmakers; so I was definitely a stranger in a strange land. Nevertheless, I arrived at the building in question and found direction signs that one by one pointed me in a random walk to where I needed to be. I arrived at the meeting place well before the session was convened, so I began to search for the senator who had invited me. This involved stopping different individuals to ask where I might find my benefactor and eventually I connected.

I introduced myself to the senator and he responded with comforting courtesy. He then suggested that I follow him so he could introduce me to the chairman of the committee. This was my first signal that things would not go well. My host was listed on the state's web site as the chairman of the committee but now I learned that had not been the case for at least two months. We wound our way through the milling crowd of committee members until we located the committee vice chairman. I was introduced as a citizen who had traveled all the way from the west side of the state to provide testimony in regard to funding for WYDEQ.

The reply stunned me. The vice chairman flippantly stated, "Everyone wants to talk so we have decided that no one will be allowed to talk." With that he walked off and my guy turned to me saying he was sorry but there was nothing he could do. In that instant I realized I had come on a fool's errand. These imperial legislators evidently had no regard or respect for those of us who populated the back-country because we lacked usefulness for their agendas.

The chairman repeated this sentiment in his remarks that opened the meeting. When he was reminded that there were visitors who had traveled far to attend, he simply declined to allow us to speak and thanked us for attending.[30] I was annoyed by this imperious attitude but took comfort in the knowledge that the four energy lobbyists next to me would be silenced as well. With that, I could only sit in one of the few chairs available for observers and watch the proceedings as they played out.

The reason I elected to hang around and watch the committee in action was that the Director of WYDEQ presented his request for added funding and answered questions from the committee. That process was itself amazingly imperial in my eyes; perhaps because this was my first experience with parliamentary rules of order. That said, what I saw unfold seemed extreme in the name of protocol.

The Chairman began by cautioning that only the Director would speak and then only when spoken to. Furthermore, all committee members were to direct their questions to the Chairman and the Director would address his answers to the Chairman. Thus, for the next 45 minutes I listened to committee members ask their questions with the preamble: "Mr. Chairman ..." followed by the Director's reply beginning with "Mr. Chairman ..." I had an uncomfortable feeling that the Chairman was enjoying a sense of royalty under the guise of parliamentary protocol.

This exercise in royal court theater mercifully came to an end. However, it was noted that Dan Olson the current director of the air quality division of WYDEQ would be retiring in 30 days. I was disheartened over this because I had spent two years in conflict with him but had begun to make real headway in convincing him that there were air quality problems in Sublette County. Now I would have to start over with a new guy.

My experiment in government participation thus ended, but one more penalty awaited me. To economize on my trip I had to depart Cheyenne late that afternoon for the 9-hour trip back home. I also wanted to beat a snowstorm that was moving in and I intended to be on the other side of the state in Rock Springs before it settled in. I was unsuccessful.

Some 40 miles west of Rock Springs I got into heavy weather and I-80 was an ice rink all the rest of the way. Driving speed was no better than 30 miles per hour and I had to watch for big-rig trucks that were going nowhere but trying hard. They were stalled in place with their drive wheels turning but getting no traction at all. Eventually I came out the other side of that mess and headed north toward home on much better roads. My 9-hour trip turned into 13 hours and I was glad to get there alive. Aside from that useful success, I had achieved nothing at all.

• • • • •

In 2009, the Wyoming Legislature demonstrated is devotion to the energy industry when it approved a resolution requesting Congress to preserve the exemption of fracking from protections contained in the Safe Drinking Water Act. The justification was that the state would incur costs of $50 billion over 25 years.

Not surprisingly, the Legislature's resolution mimicked one crafted by the Interstate Oil and Gas Compact Commission which was the same special interest group I noted Governor Freudenthal chaired in 2005-06 a few pages earlier. That resolution was supposedly based on a survey allegedly finding no known cases of ground water contaminated by fracking.[31] This carried the echo of the statistical escape clause I described in the section on the County Commissioners, i.e., "There is no evidence to link (this) with (that)."

The Wyoming legislature weighed in yet again with its bizarre behavior in 2011. In that session, State Senators, Bebout, Cooper, and Jennings along with two State Representatives, Lockhart and Stubson demonstrated their disdain for the EPA. They jointly sponsored a resolution titled "Resolution – Environmental Protection Agency Regulations." This resolution requested that Congress limit air quality regulation by the EPA.

Their proclamation began by listing ten "whereas" points of dissatisfaction:[32]

- It cited numerous air quality regulations in the making or already in effect that were likely to impact the economy, jobs, and business competitiveness.
- It characterized EPA's regulatory activities as a "train wreck" because of overlapping requirements and their possible impact on the economy.
- It accused the EPA of attempting to institute a cap and trade policy through regulation. It claimed that EPA "over-regulation" was driving jobs and industry overseas.
- It worried that no studies had been performed to ascertain the pre-supposed conclusion that the cumulative effect of regulations on jobs and competitiveness was severe.
- It objected that no study had been done to ascertain the benefits to global climate resulting from greenhouse gas regulation.
- It protested that EPA should be required to declare its intended specific regulatory proposals and assess total costs of those actions.
- It offered lip service to the idea of improving the quality of the nation's air but only so long as the costs were understood and did not interfere with economic recovery, promotion of a stable business environment, and creation of jobs.

- It opined that public health and welfare would suffer in the absence of job creation and economic improvement because employed workers can better care for themselves. Also environmental improvement was only possible in a society that generates wealth.

Then came four resolutions. First, the Wyoming Legislature should call on Congress to prohibit and if need be, defund EPA to prohibit it from regulating greenhouse gases. Second, bar EPA from issuing further air quality regulations for two years except in the case of an environmental emergency. Third, the federal government should conduct a cost-benefit study on the cumulative effects of intended air quality regulations to ascertain the impact on the economy, jobs, and competitiveness. The fourth resolution merely directed delivery of this masterful exercise in twisted logic to the U.S. President, President of the U.S. Senate, and Speaker of the House of Representatives.

This was an excellent validation of a premise well stated in a book about the dysfunction that has been growing in Washington D.C. It declares such people as those who authored the resolution to have "… become insurgent outliers in American politics." They are "… ideologically extreme, scornful of compromise, un-moved by conventional understanding of facts, evidence, and science and they are dismissive of the legitimacy of [their] political opposition."[33]

The growing presence of visible haze was further contributing to the State's discomfort over federal intervention. In 2014 the State Legislature decided the solution was to sue the U.S. EPA over the issue. The Agency disapproved of its plan to regulate nitrogen oxides but the State countered that the EPA plan didn't "… noticeably improve air visibility compared to Wyoming's plan …" and "… will cost far more to Wyoming businesses and ratepayers."[34]

In still another don't-bother-me-with-facts action, the Wyoming State legislature demonstrated its determination to protect the state's fossil-fuel economy at all costs. As a result of its legislative session held in 2014, Wyoming became the first state to block new national science guidance that was part of a set of K-12 standards developed by national science education groups. Chief opponents stated that in their view, "[The standards] handle global warming as settled science ..." and "... teaching global warming as a fact would wreck Wyoming's economy, as the state is the nation's largest energy exporter. The standards are very prejudiced ... against fossil-fuel development".[35]

Not to be left behind, one of Wyoming's previous politicians inserted himself into the fray. Former U.S. Senator Malcolm Wallop wrote in a December 18, 2005 op-ed to the Casper Star arguing that damage from Hurricane Katrina to the Gulf off-shore production rigs necessitated that Wyoming step up to fill the void.[36] This logic was also at work with the Bush Administration that put pressure on the BLM to get the Anticline EIS approved and out of the way of drilling.

However, important facts were left unstated. My own investigation of the U.S. Energy Information Agency web site revealed the following: As of November 16, 2005, national natural gas stored reserves were 6 percent above the 5-year average. Lost production in the Gulf amounted to only 13 percent of national totals. At the time of the hurricanes, the national normal stored winter reserve of natural gas was already in place. National pipeline capacity was the real weaknesses because the ability to transport new supplies was insufficient.[37]

Wallop criticized what he characterized as all-or-nothing range wars over public lands, arguing that state and federal agency protective oversight of such lands was working but his assertion was flawed. The state's primary environment protector WYDEQ was hopelessly entangled in Faustian deals with the operators who were setting the agenda. On the federal side, the chief steward of federal lands, the BLM, was busy finessing environmental impact assessments to circumvent any serious limitations on the operators' way of doing business.

It is worthwhile to review the events I have described in this and previous chapters because serious conflicts evolved. In 2005 the EPA ordered an 80 percent reduction in NOx emissions on the Anticline. In 2008 WYDEQ implemented its offset policy allowing increases in NOx to be offset by reductions in VOC emissions. In 2014, EPA disapproved of the State's approach to controlling NOx so the State sued.

Thus it seems the state and federal regulators were often at odds with each other. The State was determined to exert self- determination under its all-encompassing notion of state's rights. Not surprisingly, the operators found that very convenient for their own needs. The feds on the other hand sought to walk a fine line that didn't step on the State's toes but still exerted primacy.

A head-on collision in front of the flagpole was inevitable.

The WOGCC

The Wyoming Oil and Gas Conservation Commission (WOGCC) might seem an odd agency having a connection to the subject of air quality impacts from gas development in the Jonah and Anticline but in fact it became quite relevant. The connection is rooted in the fact that groundwater is not the only potential casualty from fracking. As I keep stating, air quality can be heavily impacted if fracking is accompanied by well completion flaring.

By 2007 I had done all I could to make regulators and county citizens aware of the presence of alkali metals in completion flaring smoke plumes as well as emissions of a host of combustion products that helped create ozone. I next took on the task of quantifying the volumes of emissions using the State's own data. My motivation was rooted in WYDEQ's insistence that it was maintaining a credible emissions inventory.

The WOGCC maintains a prodigious data collection on produced volumes of gas, oil, and wastewater from all active wells in the state. Initially, I had intended to use this data in conjunction with operators' permit applications to WYDEQ containing production decline curves. As I explained in an earlier chapter, I had developed a notion of creating a dynamic computer model of all wells in operation in order to look into the future in a way that would quantify total emissions volumes by time.

Two elements thwarted this extravagant idea. First, the sheer volume of data was more than one person could ever hope to process. Furthermore, I soon learned the data changed with every statistics update issued by the agency. Second, I learned very quickly that WOGCC was as evasive and adept at the art of doublespeak as were WYDEQ and BLM. I will elaborate shortly.

My efforts to collect the data mentioned above produced a particular surprise I had not expected. My data collection assistant necessarily had to continually review WOGCC production tables for every month of every year to add newest posted numbers. In the process of assembling those thousands of production numbers from the WOGCC website, she discovered a curious trend. Around 2012, she found the numbers she had previously posted were being changed into much higher values. This change was being post-dated all the way back to 2009 and the implication posed serious questions.

For instance, recall from Chapter 6 that WYDEQ claimed in a letter to me that it compared operator reported gas and oil condensate production rates against rates reported by the WOGCC.

My leading concern here was the accuracy of modeled emissions inventories maintained by WYDEQ. The Shell environmental representative I mention in the chapter on the AQTG told me that emissions volumes from a well pad were a linear result of the rate of gas production.

Thus, I was now doubtful that the emissions models had been correct for yet another reason and they may have been reporting emissions levels far under what was actually taking place. Frustratingly, my helper's discovery came too late to hold the operators or WYDEQ accountable.

When I mentioned this discovery to some of my government contacts their reaction was one I had overlooked. Had the State gone back and applied mineral tax assessments to those post-dated increases in natural gas output? I could only shrug and assert that this was out of my realm of expertise.

When one visits the agency's web site the charter of WOGCC is stated at the bottom:

> *"The Wyoming Oil and Gas Conservation Commission is committed to regulating oil and gas activities in a manner that ensures responsible development and management of Wyoming's oil and gas resources and provides appropriate environmental stewardship for Wyoming citizens."*

The reference to environmental stewardship is almost laughable in the context of the years I was engaged in my air quality challenges. In that period, WOGCC made it clear that it's only task was to advance oil and gas production. Environmentalism was not a part of its charter. It had far reaching powers and yet was never mentioned by BLM or WYDEQ. Here was a good example of agencies operating independent of each other even though they posed collateral problems.

A key aspect of its authority was the independent power to issue downhole well spacing determinations. Simplistically stated, it can authorize denser well spacing if the operator can make the case that more wells are needed to drain an area. A modern WOGCC slide presentation covers the topic in some detail and describes a very complex framework that governs how spacing is determined.[38] That framework looks like derivatives of the clash over well surface density that was underway when the Anticline, Jonah and Eagle Prospect projects were in play between 2003 and 2012 but that probably can never be proven.

In any case, I was concerned that WOGCC was going to be a big part of the problem of growing air quality damage taking place in Sublette County. Specifically, I wanted to know more about the chemistry contained in fracking fluids that were being flared into the atmosphere. By 2007, fracking was a national hot button issue in the context of water resource protection but air quality was not on the radar. I wanted to change that in Sublette County.

Increasingly, fracking chemicals stockpiling in the county was becoming worrisome to the locals. Eventually, the operators conceded that some information about toxicity of fracking additives probably should be available to first responders. However, that information was to be available only through the county disaster response coordinator who was in turn tightly constrained.

I discovered this when at the urging of a tenacious environmental activist, Carmel Kail, we tried to access the information. We were summarily informed that we would not be allowed access because aside from not being first responders, the information was classified as not eligible for public dissemination by none other than the Department of Homeland Security! Oh and by the way, even if we had qualified for access, we could only *read* the toxicity data sheets but could not copy or otherwise write any notes.

Resistance to any rule mandating public declaration of fracking fluid chemicals was not the exclusive domain of the operators. WOGCC regulators were reported to be in a "heated crossfire" as they too resisted public release of fracking information. Their primary concern was maintaining primacy over regulation of fracking. The supervisor was quoted as stating that "We (WOGCC) need to keep others at bay and we want to keep primacy on this."[39]

Eventually, unhappiness over the cover being given to these chemicals rose to such a level that the WOGCC was forced to require that operators provide information about what they would be using in their fracking operations. This requirement took the form of a rule announced in June 2010 that had been introduced as a motion for approval by Governor Freudenthal.[40] The rule was described as helping protect people, ground water, and surface water (air quality was not mentioned).

But not so fast ...

The actual reporting requirement had considerable rattle space. By this I mean that the rule did not require actual chemical species to be named. Instead, the trade names of the additives were sufficient. It is possible to search out some of those trade names on the internet and find more detailed information about their chemistry but it can be a hard slog. I did this in 2006 when after six months of search-and-compile efforts, I managed to assemble a list of 14 chemical constituents, and functionalities.[41] The same results can be had today but in some ways, the process can be even more frustrating.

Beginning in 2010 after the passage of the requirement that operators divulge their fracking chemistry, and continuing to the present day, I discovered a variety of roadblocks impeding my desires to identify the substances being lofted into the atmosphere. To be fair, that lofting pretty much ended with a 2004 mandate by WYDEQ requiring an end to well completion flaring and adoption of green completion but this mandate applied to the Jonah and Anticline only. I worried about the other gas fields where flaring would still be used.

Exploration of the WOGCC website in search of fracking data is instructive. I learned over the years that when the federal or state governments boast about the availability of information on their sites, they omit the fact that the information is often buried so deep and under such obscure titles as to be undiscoverable. The WOGCC site was no exception.

First, if the well is pre-2010, no information will be available. Second, the scan quality of some fracking reports is so poor they are borderline legible. Third, the information is further buried under report cover-sheet obscurities. It takes additional trial and error to learn that fracking information may be included if the "Other" box is checked rather than the "Fracture Treat/Enhance" box. Fourth, many fracking fluid additives are protected by operators' requests for secrecy under the umbrella of trade secrets.

When I attempted to test the new rule in 2010, I immediately discovered that information detailing fracking chemistry was hidden under the obscure title "sundries." I had to learn this through a grinding process of opening each relevant-sounding title until I stumbled onto the right one. The next level was and is fairly straightforward. One of five well identifiers must be entered into a text field that leads to a list of reports, the titles of which give no clue where fracking details are included. Thus, another grind must be initiated to discover the right one. However, the information may or may not be there for discovery. Now and then a report can be found that is in fact useful and even lists chemical compounds contained in the additive as well as the function of the additives.[42]

Returning to the issue of trade secret protection, the WOGCC includes on its home page a link to "Trade Secret Settlements." That takes you to trade secrets approved before 2015 and another link to trade secrets approved after 2014. The trade secrets being addressed relate to proprietary additives used in fracking fluids by the various fracking service providers. There are 24 such providers listed prior to 2015 and 21 after 2014. Clicking on each one opens up a page listing all the letters that provider sent to WOGCC requesting exemption from the rule requiring public listing of their fracking chemicals.

I focused on Halliburton Corporation and Schlumberger because they were the dominant players in the Jonah and Anticline. Before 2015, Halliburton filed 25 requests with WOGCC for exemptions and Schlumberger filed 3 such requests. Post 2014 saw Halliburton file 34 requests and Schlumberger filed no requests. Examination of those request letters reveals that both companies were well versed in Wyoming proprietary secrets statutes and they both quote them liberally. Both settled into submission of boilerplate letters where only the names of the additives needing protection were changed and scant details for those additives were provided.

For its part, WOGCC also settled into responding with boilerplate summary replies declaring almost ritualistically that the applicant had proven the need for secrecy and granting the requested protection. In the case of

Halliburton, prior to 2015, its 25 requests resulted in exemption of 40 fracking additives. Schlumberger's 3 requests resulted in protection for 4 of its additives. Post 2014, Halliburton filed 34 requests to protect 34 additives several of which had been included in requests prior to 2015. Schlumberger filed no requests.[43]

A few characteristics stand out. A search of the content of the letters requesting trade secret classification showed differences. Before the 2010 ruling went into effect, the requests contained little or even no information at all about the function or chemical family makeup of an additive. After the ruling, the function of the additive was stated and general chemistry was named although many redactions were invoked.

To be clear here, the WOGCC website is intended to make available information about fracking to the public that is cleared for release by WOGCC. Thus, by definition additives granted public disclosure immunity cannot be found there. Consequently, in those gas fields where completion flaring is still practiced, the resulting smoke certainly is loaded with those additive chemicals that cannot be easily determined by a concerned public, if at all.

When I pressed the point with EPA and WYDEQ about the chemical signatures I had identified in my spectroscopic field studies, they were a bit taken aback. This was not something they were accustomed to hearing from a mere citizen. However, they quickly recovered and argued that the heat of combustion in the flaring fireball was certainly breaking the chemical compound molecules apart and transforming them into subsequent new molecular combinations.

I agreed that was reasonable to assume but I asked, "What kind of new combinations?" Were these less health-threatening? They confessed that was an unknown but assured that the mixing dynamics taking place in the turbulent plume above the fireball would dilute the flaring products to non-dangerous levels. Dilution was the solution.

In closing this discussion of WOGCC and the operators' quest for trade secrecy, it should be pointed out that the additives these fracking service providers seek to protect are in use in various locations in Wyoming. Consequently, one of the side benefits to providers is that WOGCC protection of the additives also makes it very difficult if not impossible for citizens around the Jonah and Anticline to learn which of them are in actual use in those particular fields.

I have to admit the scientist in me causes admiration for the inventors who exhibited great genius in their pursuit of the fracking process. The chemistry being applied in the technique has all the appearances of being impressively complex, highly esoteric, and uncannily effective. The minds that conjured such a bizarre solution to release otherwise un-mineable gas must be congratulated for sheer force of ingenuity.

Sadly though, they demonstrate equal tenacity in denying the less complex and increasingly dangerous collateral effects of water contamination, air pollution, and seismic disturbance.

The State's Washington Delegation

Any discussion of Wyoming's Representative and two Senators to Washington D.C. can easily be brief. Their attitude toward environmental concerns expressed by their constituents can generally be summed up as staunchly pro-industry and near dismissive toward problems brought before them in public meetings and media. When pressed about environmental worries, they routinely sing praises of industry for their progress toward operations that are clean for the air and water.

In early 2017 an affair involving BLM played out under the Trump Administration. BLM had initiated a new planning rule called the Resource Management Planning Rule 2.0. This rule supposedly would have embraced more public input but my experience with BLM made me skeptical. However, I will never know because Wyoming denial kicked in.

Remaining true to its long established patterns of conservative political behavior, in 2016 Wyoming elected Dick Cheney's daughter to become its newest Representative to Washington D.C. One of her first actions was to craft legislation to overturn BLM's Planning Rule 2.0 and President Trump signed it with great fanfare in March 2017.[44] One of those witnessing the signing was the very same senior County Commissioner I talked about in the first section of this chapter.

She and our County Commissioners decried this rule as another Washington D.C. based power grab that placed decision making over local issues in the hands of faceless bureaucrats in the federal government. Also, Cheney's action conformed to the longstanding western mantra that state's rights of self-determination must be paramount. These beliefs have long been a founding principle driving opposition to environmental regulations promulgated by EPA and other federal agencies.

Not surprisingly, one of the champions of repeal of Planning Rule 2.0 was the natural gas industry. It opposed the rule on the grounds that it would make planning more difficult and the process more time-consuming. Likely more important to the operators was the fact that BLM estimated the rule could impact the oil and gas industry by imposing annual costs to industry of approximately $100 million.[45]

Despite all of this, there was one sterling example of synergy between the people and our senators.

In 2006, Senator Craig Thomas drafted legislation to protect the Wyoming Range from drilling. He required convincing but eventually a critical mass of public protest had to be acknowledged. The variety of voices coming into play was profound. Individuals, groups of organized individuals, and environmental groups, both small and large, state-based and national, all came together to form that critical mass. Sporting clubs and religious groups, labor unions, and politicians at the city, county and state level also contributed to the pressure.

Senator Thomas's effort was especially exemplary because 2006 was his last year of life due to his long battle with leukemia. His strategy was to place pristine portions of the Wyoming Range off-limits to future gas and oil leasing while not threatening existing production sites. His Wyoming Range Legacy Act spent 18 months in Congress but it was eventually passed in March 2009.

The Act permanently protects 1.2 million acres from future oil and gas development and that protection had a collateral benefit of causing BLM to rescind 23 leases on the eastern slope of the Wyoming Range. Public news releases cited the beneficiaries of this Act as being big game, wild game birds, and trout as well as their habitat because of an end to threatening physical disruption.[46,47] My personal relief included the removal of what I had argued so frequently to Bridger-Teton managers was the threat of unknowable health effects on those species resulting from concentration of toxic emissions that would gather across that landscape.

This affair was a rare and satisfying outcome of a years long battle that throughout its life and up until the very end looked absolutely futile.

The Others

Chapter 8

The Operators

The Jonah and Anticline gas fields hosted several operators seeking to cash in. There were some minor guys and some major guys. The minor ones included Anschutz Exploration Corp, Yates Petroleum Corp., and EOG Resources Inc. The majors were EnCana, Rocky Mountain Shell, Questar, Ultra, and British Petroleum. Of these larger operators, I most frequently interacted with the first three and developed a working relationship with their respective local environmental representatives. These guys were all field engineers tapped by their employers to serve extra duty of handling local issues of environmental concern from the State, BLM, and most worrisome to them I am sure were unhappy citizens.

Ultra

Perhaps the most dubious affair involving this operator had to do with its maneuver to sidestep any potential complications to its drilling schedule resulting from NOx restrictions. Around the start of the big push to develop the Anticline field, Ultra apparently had premonitions of this issue becoming a showstopper.

To local residents aware of this bit of history, it appeared that the company had looked around the region for other big NOx emissions sources and noticed a power plant south of Kemmerer Wyoming. That plant evidently looked to Ultra like a good emissions source to turn to its advantage. It therefore invested $2.5 million into a project to clean up the plant's stack emissions.[1]

It characterized this action as being a voluntary move on its part. The bargain was struck in 1998 with PacifiCorp who owned the plant and it was claimed that 1000 tpy of NOx were removed from the atmosphere as a result.

However, the project was not philanthropic in intent. Instead, Ultra seemed to have cynically calculated that removing the power plant's emissions from the regional air mass would free up PSD space to be consumed by its own gas development activities. In essence, as Terry Svalberg observed about such behavior a few years later, the operators and BLM were treating the PSD limit as a consumable instead of a not-to-exceed air protection buffer.[2]

An interesting footnote to this episode came to me a few years later. I was discussing the event with Lloyd Dorsey my contact with the Greater Yellowstone Coalition in an effort to learn what he knew about it. I was surprised to hear him not only confirm but tell of his personal experience. He had been in a meeting of enviros and corporate types one of whom happened to be a recently appointed CEO of Ultra.

The subject of the Naughton power plant project came up and the CEO commented that in retrospect the project should not have been undertaken. He felt the cost had been too great. The cost he was citing however was not monetary. He was referring to heated reactions the company received from other operators. Apparently, they saw this as a bad precedent and privately panned Ultra for doing it. Actually, their motivation was more predatory.

Lloyd referred me to a person whom he thought had been closely involved in episode so I contacted her. I was not disappointed. Details she provided were fascinating.[3] It turns out that Ultra interest in the Naughton project began with an earlier CEO than the one Lloyd mentioned.

In the period of 1996-1997, Ultra's CEO at that time was a geophysicist who had personal knowledge of the Pinedale area and its gas production potential. He united the financial backers of Ultra which was then just a penny stock on the Toronto Exchange, and bought up leases comprising most of the Anticline. They were eager to get started with drilling but they discovered the BLM requirement for an environmental impact statement. They knew nothing about how to accomplish an EIS but their lawyers advised it would be a several-year process.

This did not sit well with the financial backers because Ultra had only about two years of financial reserves to sustain the company. This activity was taking place in the period of the upcoming election contest between Al Gore and George W. Bush so a lot of scrambling took place. The company did not want its EIS work becoming a victim of a possible Gore victory so it hired outside talent to mentor it through the EIS process. One of the revelations this produced was a BLM expectation that impacts to wildlife and impacts to air quality would rank as major issues requiring EIS attention.

Two other players, McMurry Oil and Amoco wanted to hire a reliable industry friendly consulting firm to tackle the EIS but Ultra's consultant advised Ultra to recommend to BLM that it employ a firm having experience with wildlife and clean air issues. Eventually, BLM took this advice and hired a firm out of Denver, Colorado. The same Ultra consultant informed the company that the process should take about two years to complete which would be about a year and a half less than the norm for Wyoming. The normally longer period was the result of interventions by environmentalists, prolonged comment periods, and BLM's hesitant response process to any controversy.

The possible two-year deadline loomed large for Ultra because it did not have the financial reserves to go beyond that. Ultra's consultant was confident the process would succeed under the two-year scenario if key actions were taken. These included insuring that BLM, USFS, and the State of Wyoming could understand and have confidence in the integrity of data collection at every step of the EIS preparation effort. Furthermore, the environmental community had to be involved every step of the way to minimize possibilities of appeals at the end of the effort.

In the first year, air quality baseline data was developed and fed into models. Ultra was predicting as many as 3000 wells to be drilled but that looked problematic because air quality was going to take a big hit. Because modeling of baseline air quality had to include unregulated NOx emissions from the grandfathered Naughton Power Plant, it became clear that 3000 wells would breach Wyoming's air quality limits.

That would necessitate unwelcome restrictions on annual drilling rates, seasonal restrictions to avoid winter air inversions, or very costly mitigation measures. An attractive alternative was lowering current baseline emissions levels in the region's airshed. Put another way, they wanted to lower the accounting starting point. That would keep 3000 wells from breaching Clean Air Act limits.

Ultra's consultant had connections with PacifiCorp lobbyists; so she reached out to test their interest in a joint effort to retrofit the Naughton plant with low NOx combustors. Timing was right because PacifiCorp was facing its own pressures to clean up its act so they struck a deal to perform the retrofit with a 50/50 cost sharing agreement. This did the trick because subsequent modeling by BLM and EPA showed the 3000 wells would not now trigger a non-compliance situation. BLM also hedged with mention of a threshold that if exceeded would require the above referenced costly mitigation requirement to kick in.

A key factor figured into this entire set of maneuvers. Ultra owned about 78 percent of the Anticline, McMurry Oil owned around 11 percent, and Amoco, Yates Petroleum, and Questar owned most of the remainder. Amoco and McMurry opposed the project from its beginning but because they were not party to any costs involved, Ultra's CEO inked the deal on his own.

Those two companies knew that Ultra was cash-strapped so they dragged their feet by withholding data from BLM, reconstructing data, and engaging in a number of other stalling tactics. Their goal was to delay the EIS process until Ultra went broke. In that event, they would swoop in and acquire Ultra's leases for pennies on the dollar and develop the Anticline themselves.

As time passed, Ultra did indeed face financial problems. This resulted in its CEO being replaced by the man Lloyd had talked to in the meeting I mentioned at the beginning of this saga. This guy was more the money-raising technician who succeeded in raising required investment capital that made getting the drilling program underway not so crucial to corporate survival. Nevertheless, the Naughton deal facilitated completion of the EIS in almost two years exactly. Yates and Amoco appealed the EIS but immediately lost in the courts.

On another occasion, Ultra publically signaled the value it placed upon its Anticline field operation. At the time, slant drilling was a hot topic as a potential method of mitigating environmental problems. However, operators were resistant on the grounds that slant drilling added $1.25 million to an operation. Conversely, Ultra noted at the time that new Jonah wells were achieving profitability in only six to eight months.

A different face was on display in its 2006 annual stockholders' meeting. A shareholder proposal had been submitted recommending an annual climate change report from the Board of Directors. The goal was to be an assessment of the company's response to growing worldwide concern over $CO2$ and greenhouse gas emissions. That stockholder wanted to know how the company was reducing its own emissions and cited numerous information sources that were sounding the alarm over the threat of climate change resulting from these emissions.[4] The Board of Directors was unimpressed.

The formal statement of opposition from the Board was long but unambiguous. Their recommendation was unanimously in favor of a vote against the proposal and they backed that up with a full page of reasons including the usual national energy security argument. They next cited the Naughton power plant upgrade as an example of holistic attention to environmental responsibility. That was followed by a reference to the company's tinkering with drill rig fuel alternatives involving bi-fuel and catalytic converters. Lastly, it cited movement toward meeting requirements for tier-2 engines. Flare-less completion technology was also cited.[5]

The company's explanation went on to cite corporate obligations to improve stockholders value and balance that against costs of environmental compliance. It characterized environmental laws as risks which along with the costs of compliance were topics worthy of stockholder/Company discussion. It expressed clear awareness of national focus on emissions and climate change and assured stockholders it took all of this seriously.[6]

Not to worry though because the Company cited environmental awards received from the EPA, Wyoming Game and Fish Department, and the Wyoming Wildlife Federation. All of this demonstrated that reporting in the manner requested by the stockholder proponents would not reduce emissions or improve environmental performance of the Company. The Board of Directors, therefore, recommended a vote against the proposal.[7]

Questar

This operator was the one with whom I interacted almost as routinely as I did with Shell. Mike Golas was their contact and as I have explained, he was their representative on the AQTG. Mike was a courteous guy who seemed genuinely interested in limiting the impact of gas development being created by his firm.

Early on he offered discouraging advice to the AQTG that the window of opportunity for environmental actions was all but closed. He explained that industry operates on a simple formula: the beginning of a project is the time for improvements in methods to be incorporated. As time goes by, that window of opportunity closes at an exponential rate and any improvements increasingly fall into the realm of "retrofit" which becomes objectionably costly in the eyes of shareholders.

He further explained that on the Anticline, we were already at the bottom of the curve. In other words we were about to lose any opportunity to influence the environmental outcome of both Jonah and the Anticline.[8] This was an example of how his participation on the AQTG could be reasoned and measured. He often offered an open invitation for AQTG members to accompany him around the Anticline so we could be instructed in the ways of Questar. I took him up on the invitation several times.

I followed him around in his corporate office in Pinedale and met several engineers who were happy to fill me in on their job description and activities on the Anticline. That said, I was always aware of a tangible cautious attitude when they spoke to me. I suspected they were always on the alert to not say too much while appearing to be fully cooperative. I took this as to be expected and did not hold it against them.

One field trip I went on with Mike was particularly instructive and actually fascinating. He took me to a well pad where a fracking operation was being

By the Rasping in My Lungs

conducted. Upon arrival, we drove around the pad using the road established for big rig truck support and he complained about restrictions. I gathered from his explanation that the state or BLM or both had placed a tight envelope around the pad perimeter that imposed difficulties on drivers to negotiate turns necessary to stay inside pad boundaries.

We next walked the area around the wellhead being fracked. That was impressive and dangerous. Impressive because fracking support pump trucks and fluids tank trucks were backed up to the well head almost too tight to walk between. They circled the well like automotive flower pedals and their engine noise was deafening. Guys dressed in their corporate-colored coveralls and saucer-shaped hard hats scurried around in a well-rehearsed industrial dance.

Our walk was dangerous because as we wandered around the wellhead I realized we were stepping over a festoon of very thick rubber hoses equipped with specialty high pressure connectors. This made me retro to my Air Force training with 4000 psi activated aircraft weapons bay doors and the cautions that went with that. I asked the engineer accompanying us about the pressure being carried by those hoses and suggested we might be better off not being there. He seemed to show a sudden awareness, told me the pressure was 3000 psi and indeed we might want to vacate.

From there we entered the instrument trailer where the fracking operation was being controlled and monitored. This was the space age in a muddy laboratory. On one side was a bank of computer screens displaying a myriad of graphs and data readouts. Operators were glued to them and guiding workers via intercom links. As I watched and listened, I had to feel admiration for what was going on because the scientist and engineer in me recognized the level of hard learned sophistication at work. At the same time I wished they would better address the consequences taking place that were collateral to improving gas flow by fracturing the strata below.

As for that strata breakup going on, I was actually seeing it happen. The stress/strain graphs on the display screens were a colorful moving dance of lines. I watched curves slowly climb up the screen and suddenly dive down as a section of strata yielded to the fluids being rammed into their nooks and crannies. An operator would command someone controlling the pressure pumper and adjustments were made. This had been going on for hours and would continue for more hours. Finally after we had seen what there was to see, we left and returned to Pinedale.

Questar was generally aware of the threat to operations if air quality issues were not brought under control. It tried some approaches to the problem using combustors to burn tank flash gases but realized this was marginally effective. The company concluded a better way would be to eliminate the origin of emissions as much as possible.

In an AQTG meeting Golas announce his company's plan to initiate a fluids gathering pipeline system. It would reduce NOx and VOC emissions through elimination of storage tank batteries at each well. I saw this as promising and wanted to study it in action.

Accordingly, I asked to be permitted to do a research study on just how many tons per year would in fact be eliminated, expecting the results to be impressive. Mike simply smiled. He claimed the company planned its own study. I cautioned that if I did it independently, there would be more credibility. He agreed that my caution had merit but he would not yield. He claimed an engineer with the company had been tasked to conduct the study I was advocating and results would be reported to the AQTG.

That never happened.

EnCana and Engineered Concepts, LLC

Without doubt, my most bitter experience with establishment group think involved my facilitation of a test project between EnCana and a small but dynamic oil and gas production support company. I had naively decided I could develop a better solution to traditional gas dehydration technology and apply to the federal Small Business Administration for funding to develop the idea. I began a research effort to set my direction when I discovered Engineered Concepts in Farmington, New Mexico.

As I read about this company I realized they had the solution I was looking for and also, I had been silly to believe even for a moment that I could start from zero understanding and zero resources to invent an alternative technology. I contacted the owner by email and to my amazement I received a response only a day later. Thus began my excellent adventure with the Rodney Heath and his sons who owned the company.

I explained my belief that emissions from gas burned by on-pad process heaters, combustors, and burners were a large reason for our new problem with ozone and visibility. I wanted a solution that would reduce or even eliminate such emissions. Rodney agreed with me completely and described his solution he called a state-of the-art gas dehydration process dubbed the Quantum Leap Dehydrator or QLD.

He further explained that this technology had been field tested in Colorado by the EPA, which issued a very positive report on the test results.[9] As I have stated many times in this book, the technology effectively reduced emissions of NOX, VOCs, HAPs and CO2 and a few other species from hundreds of *tons* per year to a few tens of *pounds* per year.

We began a dialog that ultimately led to my facilitation of a meeting between the chief engineer for EnCana's Jonah operations and Rodney and one of his sons. This came about after they traveled to Pinedale to talk to me about what they felt they had to offer and to familiarize themselves with the situation on the Anticline. Elated by what they told me, I next contacted the EnCana engineer and proposed we all meet for a discussion of the technology and what it could promise. That meeting produced an agreement to proceed.

EnCana initiated a field test of three Quantum Leap units in the beginning of 2006. The EnCana engineer was quite certain that air quality impacts from the Jonah field had the potential of becoming a showstopper; so he was keen on finding a solution. Over the following months the test program proceeded with great success, albeit with usual start-up problems that haunt new technologies. He understood this and was quite tolerant of these problems. Soon, field personnel were expressing their pleasure over the performance of the technology and he advised me of his intent to recommend the purchase of 50 units.

Then I erred by inviting WYDEQ to become involved. Seeing the success of the test program, I began informing WYDEQ staff and officials of the field testing and urged John Corra to participate. This resulted in Rodney being invited to WYDEQ headquarters in Cheyenne to present the details of his technology. I later learned that during the meeting, Cynthia Madison, a staffer within WYDEQ-AQD challenged the Quantum Leap technology as being no cleaner than current dehydration techniques.

Her challenge centered on the small four-cylinder engine used to drive pumps in the dehy unit and its supposed creation of NOx emissions in amounts that reversed the gains from the overall operation. When requested to provide quantitative backup, she declined on the grounds that she had failed to bring her data with her from her Casper office.

This frustrating development forced Engineered Concepts to go home and develop a rebuttal without benefit of knowing the exact methods and/or analysis used by Madison. The one point of information that did emerge was that she used EPA protocol AP-42 for calculating combustion emissions. As I have previously related, I contacted Cindy Beeler at Denver EPA headquarters and inquired into the appropriateness of her use of that protocol. I was told it should be used with the realization that it was useful as a tool to create a general approximation but not an accurate depiction of any specific piece of field equipment.

EnCana and Engineered Concepts embarked upon a self-funded program of analysis of emissions from this four-cylinder engine that had by now become the showstopper. Lab quality analyses were conducted and ultimately the results refuted Madison's assertions. However, by now WYDEQ had circled its wagons in a defensive posture and never responded or acknowledged the results of the EnCana/Engineered Concepts effort. At this point EnCana terminated the process of purchasing the planned block of 50 dehy units because, understandably, it was not eager to invest large sums in a technology that WYDEQ was refusing to acknowledge as being effective.

Over the next many months, I lobbied the John Corra and even the governor, literally begging them to stop blocking the Quantum Leap Dehydrator approach. I pointed out that the units captured almost all of the natural gas that was otherwise burned and directed it into the market stream, which in addition to its environmental benefit, would increase tax revenues to the state. All was to no avail. WYDEQ had committed to its position and would not budge

Furthermore, our EnCana engineer was reassigned to the company's Calgary, Canada home offices and replaced by another engineer coming up through the ranks. He refused to work with us any further and perfunctorily

By the Rasping in My Lungs

ended the test program declaring he had no interest in pursuing it. Quiet scuttlebutt from field workers involved in the project was that our engineer had become too "environmental."

A suitable irony took place in 2008. EPA conducts regular conferences called Gas STAR where it and industry operators gather to discuss new state of the art technologies. At one of these conferences hosted down the road from the Jonah in Rock Springs, a local EnCana team described their unsuccessful efforts to solve the emissions problem through a variety of self-developed approaches.

An attendee later informed me that EnCana engineers from the Colorado Piceance Basin gas field who were testing an improved version of the QLD in this and another field in Colorado stated that the dehydrator was the best they had seen in 20 years of practice. They were thus dumbfounded by their Jonah cousins' consternation. At one point they turned to my contact and declared incredulously: "We just handed them the solution!"

Thus, a promising technology that very likely would have relieved Sublette County of emissions volumes by reducing them 1000 percent died on the vine. I invested three years of my time into the facilitation of this test program. I believe to this day that adoption of the QLD by EnCana would have compelled the other operators to follow suit and Pinedale would not be experiencing the ozone problem that still exists.[10]

A final post-script to this tale is that I was provided a not surprising piece of information a few years later and I confirmed. Cynthia Madison had left her job with WYDEQ and hired on with an oil and gas consulting company that includes in its services consulting on air quality permitting.

• • • • •

There were other problems I encountered with EnCana. The company provided a significant irritant in the form of a public relations spokesman named Randy Teeuwen.

I became familiar with Teeuwen as the result of my public challenges to EnCana's air quality impacts that were resulting from the sheer volume of drilling with which it was pressing ahead. Not long after he arrived in the area, he contacted me and invited my wife and myself to breakfast. I was suspicious of this but I wanted to measure his sincerity and this seemed a good opportunity to do that.

It didn't take long to assess him for what he was, a foot soldier sent ahead to calm the citizenry. He was very friendly and very assuring about the honorable intentions of his employer. What was lacking was substance. I never heard direct comments recognizing our concerns nor were there direct comments addressing specific engineering solutions to be applied to the problems.

Some months later I tried to interest EnCana and Shell in an approach to detecting methane leaks on well pads using optical technology. I proposed to offer my services of screening dehys for methane leakage using infrared imaging technology being introduced by the company FLIR Systems Inc. I was practicing real hutzpah here because I was proposing that EnCana and Shell fund me to buy an infrared camera that imaged methane leaks *and* pay me to do the screening. I fully expected both Teeuwen and Sewell to laugh me out of the county.

I was told by Teeuwen to write my proposal and give it to him because he and another employee in the Denver office were tasked to review field support proposals. After a five-month wait, he contacted me and requested a face-to-face meeting about the proposal. That was where he delivered the news that I expected but not for the reason I was anticipating.

He informed me that EnCana had decided to buy its own equipment and stated "… beside, we don't trust you." This was despite my written commitment to comply with their a priori gag rule that I not talk or write scientific papers about what I would see or measure.

I happened to have a neighbor who was an oil well trouble-shooter in the Alaskan fields who had warned me about this. I had mentioned to him my intent to submit my idea to the operators. He told me his story of having invented a down-hole retrieval tool only to discover that one of his clients had gotten ahold of it and reverse-engineered it. He concluded with the warning: "Watch what you reveal because those guys will steal your idea."

His tale rang in my ears afterward but I knew I had no grounds to claim the idea as my personal intellectual property. It was already being introduced to the industry by FLIR so it was inevitable that the operators would adopt it. I was simply a victim of bad fortune.

Rocky Mountain Shell

My first serious contact with the operators began with Rocky Mountain Shell. This branch of Royal Dutch Shell was moving into the Anticline in a big way at the same time as the other operators. I had become very vocal about my spectroscopy findings as well as my growing unhappiness with BLM and WYDEQ and that apparently got Shell's attention. It soon assigned a semi-permanent representative to our region who was tasked to address environmental rumblings from the citizenry. That person was Jim Sewell.

Shortly after he began periodically commuting to Sublette County from his Colorado based Shell office he reached out to me. We met one evening I believe for dinner. Over the course of the conversation he explored my perspective on the problems accompanying gas development in the Jonah and Anticline and I didn't hold back. Toward the end of our evening together he assured me of Shell's intentions to do the right thing and voiced a curious comment that really made me sit up straight.

He stated that he had been sent to Sublette County *because of me*. I revisited this often in my memory as months went by because I immediately realized there were two possible interpretations for this: either he was telling me Shell was attempting to address concerns I was being so loud about, *or* he was being assigned to run interference and minimize my ability to cause trouble that might threaten Shell's drilling schedule.

It didn't take long for me to conclude the latter was the correct interpretation. He reinforced my suspicions later when he asked if I believed him with regard to his assurances of Shell sincerity. It sounded like a test of whether I was being successfully deflected. My response was to nod my head approvingly while recognizing I was being treated to little of substance and a lot of visuals.

The visuals came in the form of an invitation to visit him in his Pinedale office and to travel into the field with him as he made his rounds to Shell wells on inspection runs. I returned the strategy with my own. I did visit him often in his office and I did spend a lot of time in the field with him, a few of those times running well after dark.

His approach probably would have been quite successful with someone not familiar with natural gas development and handling processes. However, I felt I had a hidden advantage in that I had spent a lot of time in an earlier life working in a natural gas compressor station and also in the field maintaining a pipeline. While this was not directly comparable knowledge, it did, nonetheless, give me a useful foundation toward understanding what I was seeing beyond what I was being told.

During one of my visits to his office he happened to be working on large three ring binders containing records of his company's wells. When I asked what he was doing he explained he was updating reports on production statistics and allowed me to look at examples. I thus learned he was also tasked to keep track of well emissions and report them to the home offices of Royal Dutch Shell in the Netherlands. I noticed the emissions volumes were tabulated in metric tons per year and I asked why. The reply was a reminder that the Netherlands was on the metric system.

Because this was early in our professional relationship, I failed to exercise critical thinking and let the comment go by. Weeks later it dawned on me that I should have explicitly asked if he was converting these numbers to standard tons per year for reporting to WYDEQ. The difference would have been significant. In truth, this detail certainly had to be addressed in his reporting and I had no reason then or today to doubt the conversion and correct reporting were being accomplished. That being said, the question has always remained in the back of my mind.

On the occasion of another visit, the infamous Anschutz emergency event had taken place a few days earlier. The incident was the upset which I talk more about in the next chapter. Details were never forthcoming beyond a cursory explanation from WYDEQ that an emergency pressure event took place necessitating an immediate dump of the well's pressure into the atmosphere.

In early 2004 I was in the hallway when a couple of Shell field engineers passed by and saw Sewell and me in his office. Somehow the Anschutz event entered into the conversation and one of the engineers commented that when he saw the black mushroom cloud rising into the sky, his reaction had been,

"Oh, why did they do THAT?!" It was instantly clear to me that these guys understood the razor thin margin of trust existing between the operators and the public.

Jim was also the man I had to go to for consideration of my FLIR imaging proposal I described earlier. I submitted it in writing and waited for many weeks to hear the outcome. Finally one evening in his office I pressed him for a decision. His reply was: "How will your idea increase production?" In that instant I realized any improvement on Shell's emissions performance had to be secondary to increasing output. I tried to bluff with the suggestion that failure to control emissions could in fact cause regulatory reduction of production. That fell flat as did my proposal.

He often commented to the local news media about Shell initiatives to address air quality issues. On one occasion he argued that an 80% reduction in emissions of NOx was a "goal" of a current approach by the company. This utilized technologies I described in the chapter about the AQTG meetings. Remember this involved two drill rig emissions reduction initiatives, catalytic converters and bi-fuels on older rigs that Shell, Ultra, and Anschutz were pursuing.

Technical difficulties developed and to outward appearances Shell had only been dithering with fixing the problems. As months passed, it looked to me like Shell was not committing the same level of resources toward fielding either technology with the same determination it had concentrated into its effort at forcing acceptance of year-round drilling upon us. As a result, his public claim that more drilling wouldn't mean more emissions just did not ring true.

• • • • •

After all the efforts and conflicts and challenges and frustrations, operators in the Jonah and Anticline demonstrated the ease with which they could fold their tents and move on to the next big gas bonanza.

In March 2014, EnCana announced it was selling its Jonah Field holdings of 24,000 acres and over 1500 wells to TPG Capitol, a Fort Worth and San

Francisco-based private equity firm for $1.8 billion. The sale included the 100,000 undeveloped acres of the Normally Pressurized Lance Field south of the Jonah.[11]

Furthermore, in August 2014, Royal Dutch Shell announced its agreement with Ultra Petroleum Corp to swap 155,000 Marcellus Shale acres and pay $925 million for 19,000 acres in the Pinedale gas fields.[12] This strengthened Ultra's dominant position there and expanded its operation to 1,577 wells accounting for 68 percent of the field. I also learned through Terry Svalberg that Jim Sewell had informed him of Shell electing to sell out because it felt there was more money to be made in oil. In a twist of market fate, however, the price of oil went into a dive through 2015 and 2016.

This market reversal blew back on Ultra. On May 1, 2016 Bloomberg reported that the company had filed for bankruptcy protection due to missed payments owed to lenders and bondholders.[13] This had happened even though Ultra was described as holding a primary asset in the form of "gas producing properties" in Wyoming.

This overall scenario was just what BLM field employees told me privately in 2008 was their recurring worry. They feared a resulting "offload" of liabilities by current operators onto new operators who might exhibit even less willingness and financial ability to address environmental problems. My query to the newest director of WYDEQ-AQD confirmed that indeed new owners assumed all liabilities.

It remains to be seen how this plays out.

Chapter 9

The Enviros and the Media

The Enviros

When I am in conversations about my years of challenging the gas guys to clean up their operations, I often start out by noting that in the course of that effort, I saw the dark side of all the players. That includes the environmental groups to whom I have been referring as the enviros. I am talking about the professional groups and in particular the Wyoming Outdoor Council (WOC), the Greater Yellowstone Coalition (GYC), the Sierra Club, the Wilderness Society, and Ducks Unlimited. These are the groups with which I had varying levels of contact.

Early on, I came to the conclusion that they were their own worst enemy. I found them to be almost fixated upon the single objective of winning financial contributions from their memberships to the detriment (I believe) of their supposed primary goal of advocating for the environment. During my involvement with them, this was exhibited by their taking credit for things that were of dubious ownership by them. I blame this behavior in part on their need to stand out from the competition in order to win funding from their respective members.

Furthermore, I witnessed moves by these entities to attempt a takeover of the hard work being done by local Pinedale activists. There were instances when they swooped into town to hold high profile public meetings or conducted field trips, all designed to make them look proactive. The truth was, however, they had been invisible during the first seven or eight years that the rest of us had been doing the heavy lifting to make operators and regulators clean up their air pollution.

Adding to the insult, they didn't even attempt to sit down with us in any effort to learn what we had already learned from our schooling of hard knocks. Instead, they stood before their audiences and declared their notion of what had to be done to protect air quality. In so doing they always clearly displayed ignorance of what had gone on before and a belief that they would somehow be immune to the obstacles we had experienced. One classic demonstration of this attitude was evident in their attempt to insert themselves into the PAWG and AQTG processes as I have already described.

The Wyoming Outdoor Council (WOC)

My contact with enviro groups started in an improbable way. By 2003 I had witnessed gas development in Sublette County assuming proportions of an invasion. Heavy equipment was busy in road construction, scraping off drill rig pads, and erecting rigs everywhere. I soon feared for the safety of everything being assaulted including the prairie surface, antelope and deer herds, and our clean air. During the early months, I realized the need for powerful voices to challenge all of this.

Thus, my personal saga with environmental groups started with a naive attempt to obtain what I thought would be "big guy" assistance. By late 2003 I had become particularly discouraged over the raid on air quality I felt was being perpetrated by the gas developers. I had written a few comments in letters to editors and impotently crossed swords with WYDEQ. Those events began to reveal that my single voice would have no effect. I concluded that a personal recruitment letter to the Sierra Club and the Wyoming Outdoor Council was the most effective next step for me to take. The results were mixed and an indicator of what would come.

I never received a reply from the Sierra Club but I did receive a very prompt response from the WOC. Its director, Dan Heilig, immediately reached out to me and expressed interest in learning more about what I had been doing with my fieldwork. I provided information about my activities and the findings I was accumulating. Some months went by and then in the winter of late 2004 (if memory serves) he invited me to accompany him and a former Pinedale BLM field inspector on a field trip into the Jonah.

I met them at the appointed time, climbed into his SUV and we set out. It was then I learned the purpose of the trip. The BLM fellow had recently retired and was very upset with what he was seeing on the pads. Specifically, he led us to several well pads owned and operated by British Petroleum. There he pointed to signs suspended across stairways leading to the top of produced water and condensate storage tanks.

The signs were unequivocal. They warned of the presence of poisonous gases that could threaten the health and lives of anyone going to the top of the tanks. Our BLM guide wanted to make the point that these gases were present by operator admission and they were being allowed to drift across the open countryside. Here was my first exhibit showing that operators also believed that dilution was the solution to toxic air emissions.

Indeed, once aware of this I personally traveled around the gas field and found a number of storage tanks that had open tops covered with a course wire mesh. This was to keep small animals and birds out while freely venting flash gases into the atmosphere. I mentioned this in a previous chapter. I hammered WYDEQ about this as did other citizens. Eventually, the agency promulgated new rules prohibiting such tankage. A small and rare improvement had been won.

In June 2004, the Wyoming Outdoor Council, the Jackson Hole Conservation Alliance, and the Biodiversity Conservation Alliance issued a joint press release warning against the leasing a portion of the Bridger-Teton Forest of the Wyoming Range. This was the same tract that I have spoken of so often in this book. The release cited several environmental concerns including roads, tree clearing, power lines, and industrial infrastructure.

They cited concerns that this area was close by the Pinedale Resource Area containing the Jonah and Anticline fields where 7,000-10,000 new wells were to be drilled in the next 15 years. The 2,000 wells already completed in

R. Perry Walker

Sublette County to date had produced impacts to wildlife, air quality and water quality. The release then cited my work:

> *Perry Walker, a retired Air Force physicist living in Daniel began an independent research effort a year ago, designed to identify air quality impacts by the gas development activity on Sublette county air quality. His efforts have so far revealed three previously unreported chemical elements in plumes from well completion flares which he feels may deserve investigation for potential health hazard impacts upon the area. For this and other reasons, he is concerned over the specter of such flares in the Bridger-Teton lease area. He feels that a statement in a U.S. Dept. of Energy study of natural gas development in the Rocky Mountain States sums it up: "Small changes in air quality can have noticeable effects, especially on visibility. The region is valued for its striking vistas and scenery, but the ability to see these sights over long distances has degraded over time as air emissions have increased from a number of sources, including traffic, urban development, and industrial activities. Natural gas development is one of these activities."*[1]

Over the following few years, my association with the WOC produced a professional relationship with one of its environmental lawyers. Initially I regarded him as being somewhat effective but that estimate slowly eroded over time. He seemed to be very active in pursuing legal remedies designed to limit drilling activity until environmental impacts could be better understood and he was often cited in regional news media.

For instance, in May of 2005, he weighed in to defend big game herds being pressed by gas development in the Red Desert. He and the WOC had participated with another group in an appeal to the Interior Board of Land Appeals in Washington, D.C., over nine parcels leased for oil and gas exploration on what his group characterized as "crucial" winter range in the Red Desert. They argued that the BLM's decision to lease these parcels did not address ongoing impacts from oil and gas development that could last for many years.[2]

In a worthwhile opinion editorial to the Casper Star, he argued against "categorical exclusions." This is a procedure allowing federal agencies to avoid analyzing the environmental impacts of an activity when the impacts are deemed minimal such as for building maintenance. But he argued that federal land management agencies, specifically the BLM and Forest Service, were using categorical exclusions to exclude the public from decision-making.[3]

This resonated with us in the Pinedale area. We were experiencing just such action from BLM as they applied the practice to circumvent environmental analyses of proposed well pad operational changes by the operators. I recall Shell specifically asking for and receiving this consideration regarding a drilling protocol change that, otherwise, would have required a new environmental assessment.

In 2005, we attended a meeting convened by EnCana for the advertised purpose of receiving input from Pinedale citizens about their concerns over gas development. An EnCana public relations officer led the meeting that produced the usual list of concerns from us. I recall in particular, the WOC lawyer spoke up and stressed almost in ultimatum fashion that above all, we were determined that air quality in the Class I wilderness areas must be protected at all costs. This impressed me at the time but months slipped by and neither he nor EnCana ever made good on the issue.

A more-bitter event occurred somewhat later. He and another enviro lawyer from another organization stood before a group of us who had been pressing the BLM and operators to implement real measures that would address the growing loss of local air quality. They urged us to form into an official group complete with officers so that we could qualify as a formal entity with legal standing. The stated purpose was to become a stakeholder they could represent in a contemplated court challenge aimed at the BLM for failure to execute air quality protections contained under federal law.

This sounded worthwhile and it gave us hope that we would at last be in a position to demonstrate BLM management failures. Months went by and no legal action was ever mounted. Ultimately we learned that he and his cohort had apparently developed second thoughts. They now argued to us that such court action was doomed to failure because the Anticline and Jonah were not far enough along in development for air quality impacts to be demonstrated. They decided that absent concrete evidence of air quality degradation, they would be thrown out of court. This revived in my mind the thought problem, what kind of evidence and how much of it would ever suffice to motivate action?

Over these few years of my working with the WOC, things changed for the worse. In 2009, the WOC chairman, Dan Heilig approached me to ask if I would be willing assist him in developing his legal challenge of BLM's adherence to federal information quality guidelines. He informed me that he was working up a plan to challenge the BLM in court over its environmental assessment practices and specifically the data the agency used to make its decisions allowing drilling. He wanted to invoke a recent law called the Information Quality Act, which was contained as a two-sentence rider in the 2001 federal appropriations bill.

The purpose of the act seemed reasonable. The title appeared to be a laudable effort to insure that qualified, sound investigative procedures had produced information upon which government decisions would be made. Critics argued that the hidden purpose was less noble. In fact the Act could be applied in a way to deny credibility to research data that ran contrary to what the government and in particular, the Bush Administration desired to support.[4] Backing this up in March of 2004, the Union of Concerned Scientists published a report, *Scientific Integrity in Policymaking*, criticizing the unprecedented "manipulation, suppression, and misrepresentation of science by the Bush administration."[5]

Heilig's plan was to turn the provisions of the Act against Administration appointees who had been so successful at blunting environmental challenges against the gas and oil industry. I considered this to be brilliant. I could see no better way to trip up the bureaucracies than to catch them in the web of their own making. I set to work and after two weeks of night and day research and writing, I forwarded to him what I thought he wanted and did so by the deadline he had imposed upon me.

After several months had passed with no word as to the success or failure of his idea, I contacted Dan and asked what had happened. I couldn't believe my ears. He offhandedly declared that he had decided the idea would be a waste of his time and effort so he dropped it. To describe my reaction

By the Rasping in My Lungs

as chagrined would be an understatement. Clearly, my time was of no value but his was.

On another occasion a few years later, I was reading through a publication issued by WOC to its members. It contained brief summaries of environmental actions being undertaken by the organization and listed some accomplishments. Of interest to me was a statement that WOC had been responsible for a new well green completion requirement in the Jonah and Anticline gas fields. This was not entirely true.

In fact, green completion had been the direct result of my actions of photographically documenting the Anschutz well venting event that covered Sublette County with thick haze. That event was the upset.

It produced a spectacular black mushroom cloud that climbed well over 1000 feet into the sky, bumped up against an inversion layer, and spread out to fill the region with a dense particulate haze. I took a series of photographs from different locations around the site illustrating the deep haze that formed in all directions. I also received reporting from friends who had been driving in various locations around the county stating they had experienced the haze. From this information I calculated that the event had immersed approximately 1200 square miles in a hydrocarbon fog.

I photographed the haze from different vantage points and made them a part of a formal written complaint to WYDEQ. I info copied the Wyoming Outdoor Council and to my surprise, their lead legal counsel hit WYDEQ with a request for rule making. In a truly rare experience with WYDEQ, I received a phone call from director John Corra who stated "You have our attention and we will act on this."

Thus, pressure on WYDEQ to act was actually a joint effort. The Wyoming Outdoor Council used my documentation to present WYDEQ with a petition for rule making. Subsequently, WYDEQ issued its requirement for green completion. Together we double-teamed WYDEQ into admitting a problem existed and the rule was established.[6]

I contacted the newest director of WOC and told her the story. I warned that if her group regarded the accomplishment as being their property by reason of my having been a member of WOC, then they needed to rethink that and give credit where it was due. She confessed she had known none of this. I

345

thus came to realize there is a short corporate memory in these organizations, perhaps due to frequent turnover of staff.

WOC interest in my activities waned after Heilig's tenure as WOC's director. I continued to try keeping them informed of my work results but lack of return contact ultimately convinced me it was time to concentrate my attentions elsewhere.

• • • • •

This narrative about the WOC concludes with a much earlier noteworthy episode connected to Ultra Petroleum and the Naughton Power Plant. WOC was prominent in the story line I related earlier in the chapter about the operators because it was specifically cited as a lead group that had to be appeased. It was seen as an important factor for Ultra's consultant as an entity that had to be brought on board. This effort apparently was successful because the WOC sided with Ultra in the court challenge to its EIS by Amoco and McMurry Oil. The same consultant commented to me that the EIS was the first major oil and gas EIS for a new field in Wyoming that was not opposed by enviros.[7]

Sadly, subsequent failures by BLM and WYDEQ to monitor and act upon air quality losses taking place for three years after approval of the EIS in year 2000 opened the door to emissions that far exceeded the air quality threshold set by BLM. Thus, the expensive mitigation requirements threatened by BLM had to be implemented anyway but even they have proven inadequate. WOC, if it has any remaining corporate memory of the affair can count itself among those who were fooled by BLM and the operators.

The Upper Green River Valley Coalition/Alliance (UGRVC/A)

The name sounds imposing but it was actually the moniker for Linda Baker who is a dynamo individual. I first became aware of her as the result of a simple notice in the local Pinedale Roundup newspaper. It carried a photo of her and a brief statement to the effect that she was setting up shop as an environmental activist to take on issues of concern to area residents resulting from gas development.

I immediately contacted her to offer my services and explained my background and motivations. This simple beginning flowered into an excellent professional relationship. We found ourselves mutually supporting each other not so much through intent as coincidence of motivation and individual resources. She knew how to organize the public and make contact with the media and I had background in a scientific approach that was new to the game. Synergistically, I kept her aware of my field work and she directed media reporters, one after another, in my direction.

In 2005, Linda took on the task of serving as chairperson of the PAWG and almost immediately found herself in a minefield of controversies. She and I were very much in sync over a number of topics that angered BLM and Prill Mecham in particular as well as many citizens including myself. Chief among those was the issue of BLM (Prill) interfering in what the PAWG would be allowed to address.

For instance, Linda and I felt that BLM attempts to control discussions under the nonsensical notion of "pre-decisional" subject matter was a classic oxymoron. The PAWG and its task groups were supposedly created to look out for negative impacts from gas development, alert BLM, and offer possible solutions. We warned that the record clearly showed there was no turning back once BLM rendered its decisions and even proposals offering adjustments were rejected. Thus, the pre-decisional caveat was actually being used to firewall the agency and operators from any inconvenient delays.

We both pressed BLM on its concepts of good management. We challenged the popular notion advanced by BLM and the operators that off-site mitigation was an effective counter to destruction of the Anticline winter deer habitat. We questioned why they thought a wildlife species genetically conditioned over thousands of years to migrate to the Anticline would suddenly recode their genetic memory and go to another location selected by BLM.

Outcome Based Management was a favorite assurance by BLM but Linda pointed out and I too emphasized there was no timeline attached to such management practices. BLM and the operators would be free to promise and maybe marginally implement but absent a formal objective and deadline, no material improvement could be guaranteed. A variation of this was Adaptive Management. I described Prill's sidestep of that promise in the chapter about the BLM. Linda weighed in as well by pointing out that BLM had routinely rejected her PAWG warnings. Drilling was damaging wildlife and BLM continued to ignore PAWG advice about ways to mitigate those damages.

Linda was instrumental in assisting the local environmental protection group CURED (Citizens United for Responsible Energy Development) in creating a pamphlet guide designed to inform readers about issues surrounding the Anticline. That document was titled *A Citizen's Guide to the Pinedale Anticline*. It presented a checklist of commitments contained in the Anticline ROD as well as information about expectations of completion and points of contact to whom questions should be directed. The purpose of the guide was to give citizens a tool to use for monitoring BLM success in meeting its environmental commitments.[8,9]

On one occasion, she took up the issue of BLM's frequent use of categorical exclusion to sidestep any need to execute an environmental assessment of an operator's proposed operation modification. Recall that operators frequently decided in the course of drilling a well that methods or processes had to be changed. If those posed a serious prospect of altering the environmental impacts already accepted in original permitting, new assessments were in order. However, BLM preferred to define such changes as minor variations within the existing permit, thereby allowing exclusion from new environmental evaluation.

Her activity won her public criticism from the County Commissioners. They insisted that existing environmental analyses and mitigation determinations were sufficient to justify categorical exclusion. To back this up they accused her of invoking unspecified non-factual information. I knew better because my association with her had revealed she was well aware of the consequences that would result if she were to be caught in falsehoods or misstatements. She was, therefore, meticulous about such things.

I thus wrote to the *Pinedale Roundup* in her defense. I countered that contrary to their assertion, she and environmental groups had not fought energy development per se. Instead we were attempting to insure transparency, compliance, flexibility, and responsiveness by operators and regulators. In so doing we were actually seeking balance between the national imperative of pulling gas out of the ground while simultaneously insuring attention toward our air, water, dirt, and wildlife.

Possibly our most successful joint project was our effort to independently monitor ozone behavior around the Anticline and Jonah fields.

In March of 2008, Linda and I pondered how to cost-effectively conduct point measurements of ozone levels. She suggested badges made of chemically treated paper that changed color according to ozone intensity exposure. As a result, a pilot project was undertaken with the help of ten individuals who volunteered to participate.

I developed a deployment protocol and put it into writing as an instruction sheet. That instruction sheet and sufficient badges for about 12 days of testing were distributed and at the end of a week, the badges were collected and read. Results were encouraging because meaningful results were successfully extracted from the badges along with discovery of inefficiencies hidden in our process.

A few weeks later, a citizens group gathered to discuss what to do about a series of regional ozone alerts that had recently taken place. I told this group about the pilot project and they immediately asked for an estimate of cost to conduct an expanded version of the project. In short order the project was undertaken with about 40 volunteers over a period of six weeks.

Results were provided to the Wyoming Department of Health and WYDEQ at their request but as was so common about these agencies a response was never forthcoming. A detailed discussion of findings is part of the last chapter of this book.

Sadly, as controversy grew over the impacts that drilling was bringing to Sublette County, other professional environmental groups decided to move in for some face time. There were a few representatives from these groups who swept into town to set up public meetings and get themselves appointed to the PAWG. They had no real clue about the history of our battle and they exerted no effort to learn. Instead, they shouldered Linda aside and tried to take over.

They also tried to take over the air quality battle and issued a lot of public commentary alleging their concern by sounding off in assertive tones. I sat in on one of their public forums and after listening to a lot of superficial chatter from the organizer I asked if she had any clue about what Terry Svalberg, my contacts in GYC and UGRVC, and I had accomplished. She was staggered momentarily but quickly recovered to say she was eager to learn about that. However, I never heard from her again and she and others carried on with their attempt at takeover.

My contacts with Linda as I wrote this book revealed some scars. She conveyed a degree of justifiable bitterness and reluctance to revisit this period of her life. Such was the toll taken on us who were simply trying to protect a pristine region from undergoing total destruction. I coined a phrase to describe the gas guys she and I had battled for so long: "petro braceros," meaning energy prospectors with no allegiance to the region and determined to move on to the next lucrative destination after razing this one.

The Greater Yellowstone Coalition

This organization, based in Jackson, Wyoming, ultimately became the one group I most often worked with. This happened because of one individual, Lloyd Dorsey. He was a GYC field representative also based in Jackson. I can confidently state that Lloyd was the only individual in the entire GYC organization who grasped the long-term ramifications of ruined air quality in Sublette County.

GYC's area of concern included the Wind River Mountains and the Wyoming Range but they often ranked a distant second place behind GYC emphasis on issues within the Jackson Hole valley. There were environmental concerns over mining impacts in nearby Idaho and Montana that received energetic effort from them but there never seemed to be similar interest in the two mountain range ecosystems to their south.

By the Rasping in My Lungs

Over four or five years, I kept Lloyd in my loop and aware of my field findings resulting from my spectroscopy and statistics research. In return, he tried to elevate my efforts to the level of the GYC Board of Directors but that too always yielded minor results. He pursued these actions because he realized that air pollution invading the far reaches of the GYC region of preferred interest would, if left unchecked, eventually flow like water into the Jackson Hole valley. Here, visibility was already being degraded by vehicle emissions.

On a few occasions, Lloyd arranged for a short session with some of his directors for the purpose of having me explain air quality impacts I was tracking. This approach failed to get much attention until the subject of ozone began to make headlines in the Pinedale area. Then, some increased attention started to come in my direction.

I informed Lloyd of my intent to establish an EPA compliant ozone monitoring station on my property and I asked if he thought there was any hope of obtaining a modest funding grant from GCY to help pay for it. Over the next few years, he included a budget line item in his annual submissions to his headquarters recommending a few hundred dollars donation to my effort. That didn't gain traction at first but he proved persuasive and I did in fact receive $1000 that was tremendously helpful.

In gratitude I invited him to assist me in erecting the tower I had built to hold my meteorology instruments and ozone sampler. This was useful for both of us because I needed some help with the heavy lifting. For his part, I suspected his superiors would be more supportive if they saw him involved in actual physical support of my site and thereby be assured I was legitimately pursuing an issue of relevance.

By chance, I had been recently interviewed and video recorded by a team of environment documentary investigators out of Jackson. They learned of the tower erecting party I had planned so the videographer returned for the purpose of collecting some footage for a planned documentary. Thus, we were chronicled on a video record that backed up what I was doing.

Lloyd also introduced me to a GYC board member at an annual GYC conference in Jackson. I immediately seized on that opportunity to explain my full range of efforts to track well emissions and ozone events, believing full well this would be my only chance to win his heart and mind. He did indeed seem very interested in seeing his fellow board members get on board with my project and requested that I keep him advised. We went our separate ways with his parting observation that he felt my topic needed attention.[10]

Over the following year or so, I tried to keep that same board member aware of my latest developments. This was by email and once by telephone and in what was our last conversation, he told me again to keep him informed. Not long thereafter, I drifted away from further contact because I had seen no consequence, good or bad, resulting from keeping him informed. I needed to focus my time where it might yield better results.

On another occasion, I attended a social gathering of GYC members at private residence north of Jackson. The event was a pleasant treat of good food and excellent scenery. At one point, the director of GYC gave an impromptu speech outlining what he viewed as notable achievements. I noticed he failed to mention funding support of my ozone monitor involvement, so I asked for the floor.

Actually, as I recall, I interjected myself and took over long enough to tell the audience about the monitor. Seeking to make them feel like they were active players, I invoked the old tag line from the 1960s TV science fiction program "Outer Limits." I quickly summarized what I was doing and why. I ended with "You are about to participate in an outstanding adventure." That elicited comments from the audience: "Cool," "Wonderful!"

Up to that moment, the director of GYC had displayed cool disinterest in what Lloyd and I were doing but with the audience reactions ringing in his ear, he went into PR mode. He quickly came over to me with assurances I could count on his and GYC continued support. However, not long thereafter, he left GYC for the Hewlett Foundation and the promised support went dormant.

My last effort involved my attempt to bring GYC into the space-age. I had met a young staffer through Lloyd whom I was led to believe had been tasked to expand GYC's horizons. I suggested that he lead an effort to contact NASA and build a partnership that supported NASA's environmental monitoring satellite program. I argued that this most likely had not been tried by any environmental group; so he could make GYC a pathfinder.

I put together a presentation for him complete with a three-ring binder of visual aids. I urged him to determine what satellites of the NASA "A-Train" could serve GYC charter objectives and work with the agency in publically advocating support in Congress to that end. Of course I emphasized the applications of air quality through pollution monitoring of Sublette County as a direct example. Months went by and I never heard from that staffer again. Lloyd never received any word if my ideas had been explored either.

By now I had become frustrated over the limited success I was having in winning the heart and mind of GYC. I Sat down with Lloyd one day and asked why I was spinning my wheels. I learned important points I should have realized early on. Specifically, I was tutored that such groups are critically dependent upon membership support through funding donations and perceptions of environmental protection success. These perceptions crucially depended upon visuals.

A protected stream that was once threatened has impact. Wild game continuing to roam freely is emotionally satisfying. Unfortunately, clear air is always there until it is suddenly not there and by then it is too late. Conversely, I was hard pressed to cite much if any research exploring the consequences of long-term exposure to air pollution by large and small game. Much is known about such effects on alpine lakes and pine forests but those effects are so gradual they are missed until, again, too late.

Sadly for me, GYC initiated one of its internal reorganizations in 2015 and Lloyd was let go. Many times I had made a point of telling the top managers of the organization they had a valuable resource in Lloyd and should not lose sight of that. I did this at every opportunity because I was concerned that he was twice the age of all the others in the organization.

That can be a liability in a business where youth, energy and obedience are premium commodities. I saw evidence here and in other environmental NGOs that groupthink is expected. Lloyd had years of experience over all of his peers and I remain suspicious to this day he was let go because of his senior reluctance to play that game.

The Wilderness Society and Trout Unlimited

My contact with these two environmental advocacy groups was limited but somewhat indicative of my broader experience with such groups.

Trout Unlimited appeared in the press occasionally with criticism about oil and gas drilling methods in the West. Like me it suggested that baseline data was lacking in defining the impact on fish and wildlife. It too criticized the use of exemptions by regulators that freed operators' hands although this criticism was aimed toward protection of clean water, whereas mine was aimed at air quality exemptions.[11] Still, we were on the same page.

I encountered Trout Unlimited during a meeting with the director of EPA Region 8 and his staff at a private residence in the upper Green River valley in May 2006. I talked about that meeting in the chapter about the EPA. There was a middle level executive in attendance from Trout and initially we had no interaction. After I had briefed the Director and his staff about my air quality work, the Trout guy seemed to find me interesting.

He inquired into my work so I gave him an extended explanation of my spectroscopy activities. At that time, my spectrometer work was still my dominant theme because I wanted to showcase it as a new and different look at the problem. He responded with interest and when I finished, he made comments promising to take what he had just heard back to the home office.

He also seemed to indicate interest in advocating support for my activities and I took hope that I was hearing words pointing toward funding assistance. Evidently I was mistaken in my interpretation because I never received further contact from him or Trout Unlimited.

My relationship with the Wilderness Society was a little more encouraging and supportive. Again, Lloyd Dorsey was my advocate. He personally knew the regional director Peter Angst and recommended that he lend support to my effort of building my EPA compliant ozone monitoring station. This support actually materialized in the form of a funding grant, which in combination with the funds provided by GYC was a great help. That enabled my acquisition of an ozone calibration source and environment controls for the instrument enclosure.

Peter and I had further exchanges during subsequent months. He would query me about the nature of ozone related hazards and also sought clarification about precursor conditions that lead to visible haze. He seemed to recognize the long-term issue of NOx and VOC contributions to both problems but lacked the science background to decipher the atmospheric chemistry details. On one occasion following his questions to me, I tried to assist him by assembling a short and simple tutorial explaining the relevant basics.

On a few occasions, I recall Peter asking me for comments that could be used in press releases addressing air quality concerns about the proposed drilling projects in the Wyoming Range. The quote I cited earlier in the section about the WOC is an example. To its credit, his branch of the Wilderness Society was early out the gate with public warnings about the dangers presented by these projects and often weighed in against them.

The News Media

Between 2004 and 2008 I was contacted or visited by reporters from no less than 16 newspapers, magazines, and TV outlets. All of them were sent my way by Linda Baker. So much time has passed since then I don't recall the conversations I had with many of them but some do still stand out in my memory. It was common to be interviewed by local news groups like the *Pinedale Examiner* and the *Pinedale Roundup* as well as *Planet Jackson Hole* and the *Casper Star-Tribune*.

However, I was surprised by contacts from other organizations far from home. These included the *New York Times, Washington Post, Los Angeles Weekly, Los Angeles Times, Wall Street Journal, High Country News, Ventura County Star, National Geographic, American Legion Magazine, and the Denver Post*. I was also interviewed by CNBC TV News and most surprisingly, a news crew from a German TV organization. Lastly, WOC produced a video news piece that ran on Casper TV.

Initially, I felt a little inflated by so much attention but as time passed I learned much that jaded me about that industry. I found it nearly impossible to get them to focus on my work and its findings. Instead, with one exception, all chose to showcase my astronomical observatory with its big telescope and segue off that into their own superficial story lines. The attention given to my observatory was to some degree my fault because I always started my story by explaining the loss of clarity in the night sky and my declining ability to see the Sombrero Galaxy as my personal tipping point.

Occasionally a writer would exaggerate odd details to set his stage. In the case of the *Ventura County Star*, the writer did a pretty decent job of summarizing in a few sentences my field research and the resultant spectroscopy findings. However, in his introduction he characterized me as living on a hilltop in a mobile home full of cats and Air Force memorabilia.[12] I could grant the memorabilia detail but the house was a nice manufactured home, not a mobile home, and I was until then unaware that three cats constituted a house-full. I wondered how this description might have colored a reader's subsequent assessment of my credibility.

I learned also that news outlets have short interest spans. I did an interview with a reporter from the *Los Angeles Times* that led me to this realization. I had filled her in on details of likely violations of the air quality terms of the Anticline EIS and subsequent ROD. A year later, events had moved in a way that further supported those concerns; so I called the reporter and advised that her story of the previous year could be given some very interesting updates.

I was disappointed in her response. She explained her doubt that she could interest her editor in any kind of follow up. The subject had received sufficient attention that her editorial staff felt was warranted. Also, her sense was that the readers would not possess an attention span justifying a revisit of the matter.

Perhaps my most unexpected lesson was that sometimes the smallest can be the best. I refer to the story done on my activities by *High Country News*. Until one of its reporters contacted me for an interview, I had not heard of that newspaper. The reporter came to my home (the same mobile home full of cats) and spent several hours listening and writing. The final product was impressive.

The story was well done with an excellent photograph of me in front of my computer displaying a spectrometer plot of a gas well source I had sampled. My opening comment about the Sombrero Galaxy was included but only as the starting point for a story that did an excellent job of capturing my fieldwork, my political lobbying, and my local reputation. The latter was well summarized at the end of the story by a quote from Dan Heilig the former WOC director who described me as "the ideal citizen activist ... [who] shows how a single individual can make a difference."[13]

A similarly good job came from a reporter with the *Planet Jackson Hole* also unknown to me until the interview request. The young reporter spent an afternoon with me attentively scribbling notes of my adventures in fieldwork and politics. Here again the published product was a satisfying compilation in a few paragraphs.

About four years later I was surprised to receive a call from that reporter inquiring about my activities subsequent to our interview. After answering her question I told her I felt her original story ranked at the top of the many done about me. She offered heartfelt thanks, adding that my timing was perfect because she was thinking about leaving the reporting profession out of discouragement. I have always wondered if she stayed with it. I hope so.

The rest of the print media reporters proved lackluster. They all said little beyond a brief commentary about my observatory telescopes. Eventually, I began to resist further interviews because they seemed a waste of time. I tried to direct reporters' focus by stressing my air quality science and political lobbying but I never seemed to succeed.

The TV interviews were no better. A crew from Germany settled in for an afternoon and worked hard to acquire just the right camera angles inside the observatory buildings and taped some comments from me about sky transparency. Then they were gone.

My interview with CNBC TV was by far my most infuriating experience and taught me *never* to assume honorable intentions by such groups. The crew spent a full afternoon taping me in front of my observatory while asking a long series of questions about my fieldwork, my findings, and my frustrations with the regulators. After it was all done the interviewing reporter assured me he had not come to embarrass me or play "gotcha!" That should have warned me but I missed it because of my innocence.

About three months later the interview was run on a nationally broadcast program and when it was over I was furious. During the time of question and answer with the reporter, I had been asked how I got started in my activities. I explained I had begun to realize a loss in night sky transparency that motivated me to undertake my spectrometry study of gas field sources of pollution.

I went on to explain my faith in Dick Cheney (who was our Vice President at that time) being a son of Wyoming and surely interested in preserving our values. I wrote to him inviting him to a personally conducted tour of the region so I could show what I was discovering.

That was careless on my part for two reasons. First, I never should have said it, and second, I had no idea Cheney was maneuvering in the manner I described in detail in a previous chapter so I had wasted my time. My comment about writing to him came back to bite me in a big way.

The start of the interview was when I mentioned my realization that the night sky was losing clarity and why. Toward the end of the interview, I made the comment about writing to Cheney. The CNBC editing department then got inventive. They spliced the night sky comment to the wrote-to-Cheney comment absent all the commentary in between. I was thus portrayed as an old fool who griped to Dick Cheney that I couldn't see my stars anymore!

In a fuming state, I shot an email to the editing department of CNBC informing them they best not ever contemplate a follow-up interview because they would not be allowed to set foot on my property ever again.

• • • • •

I had other experiences with other environmental and press groups both professional and citizen based that further reinforced the patterns I have described here. They produced similar outcomes and usually displayed their own variety of dysfunctional behavior and institutional failure. To this day, when I read or hear individuals' comments in the media either personally or as a representative of a group, I feel great skepticism that I am receiving an accurate account free of artistic license.

The Science

Chapter 10

Real vs. Virtual

My long history of challenging the BLM approach to environmental impact assessment began as the result of a follow-on drilling project in the Jonah Field. The Jonah had already been under development for a few years and was proving sufficiently lucrative for the operators to press for even more drilling. In compliance with NEPA and other federal requirements, BLM proceeded to undertake an environmental impact assessment and then to publish an environmental impact statement. If the Agency concluded that all environmental concerns had been addressed, it would then issue a formal Finding of No Significant Impact (FONSI), possibly containing environmental restrictions on operators.

Recall from an earlier chapter that on November 13, 2003, BLM-Pinedale hosted a Jonah Infill public meeting that included disturbing assertions. It had made the determination that it had jurisdiction over surface disturbance only and would confine its attention to that topic. It was thus abdicating air and water quality issues to state and federal agencies.

I challenged this notion by pointing out that BLM not only had jurisdiction but a chartered responsibility to execute stewardship over air quality. Were this not so, one had to ask why the BLM had on its staff in Cheyenne a position titled "Air Quality Scientist?" Why was that position tasked with assessing potential air quality impacts from energy development projects on government lands throughout the entire state of Wyoming? The BLM disclaimer of jurisdiction over air quality had to be taken to mean that BLM-Pinedale had no interest in pursuing air quality stewardship.

I insisted that state and federal agencies charged with protection of the environment needed to take notice and revisit criteria and guidelines under which they had thus far been operating. I noted that assessments by those same state and federal agencies regularly contained vague language to the effect that current rates of degradation in air quality fell "… within acceptable limits …." I challenged this notion on the grounds that these agencies had failed to perform sound scientific investigations into the problems. I backed this up by advising that in my private conversations with air quality scientists in those same agencies, they admitted there were no properly sited or designed air quality monitoring stations to measure the specific impact of Sublette area energy field activities on the local air mass.

Lastly, I warned that The Wind River Range designation as a federal Class I protection zone had to be heeded in a less loosely interpreted manner than seemed to be happening. Of equal importance, attention had to be focused on the possibility that emissions from the Sublette region might be migrating toward the Jackson Hole area where locally generated haze pollution was already becoming obvious. Sublette County gas field emissions added to the local emissions would combine and soon threaten transparency of one of America's premier tourist attractions, namely the Teton Mountain vista.

This BLM announcement became a flashing red light and a call to battle stations for me. With my suspicions aroused, I looked for ways to defeat this dangerous decision. I had to establish that transparency of the sky was in fact changing. The weather monitoring system I was operating at my home included an ultraviolet (UV) sensor that offered a quantitative start to assessing the suspected changes. I had been collecting daily measurements as well as routine temperature and wind statistics so I had a database of UV statistics extending back to 1999.

Upon graphing these UV data, I was surprised and intrigued by evidence that the UV signal had declined with the maximum decrease of 30 percent taking place in the first half of that four-year period. Notably, this decrease seemed to coincide with the uptick in drilling taking place in the Jonah field. However, a complicating factor was the fact that the area was entering a period of drought that could have been introducing dust into the atmosphere. Nevertheless, when BLM issued its notice for the EIS public review and comment period concerning this newest Jonah project, I began my comments with my UV observations.[1]

I further argued that the region bounded by the Wyoming Mountain Range to the west, by the Wind River Mountain Range to the East and by a line connecting Evanston and Rock Springs to the south had suffered measurable degradation from human caused visual haze and pollutant emissions originating from extreme drilling activity in the Jonah Field.

- I based my arguments on multiple lines of evidence:

- Measurements of solar ultraviolet radiation I just described.

- Less empirical but, nevertheless, useful observational evidence derived from my amateur astronomy based familiarity with the clarity of the night sky.

- Photo chronicled events of visible haze that were coincident with primary gas field operations.

- Cataloged wind patterns in the region between the Wyoming and Wind River mountain ranges bounding the Sublette County gas fields.

- Spectroscopic field measurements of well completion flaring and on-pad production facility exhaust emissions.

- Infrared imaging of storage tank vents and tank vapor combustors.

- Analysis of state and operator reported well production and emissions volumes statistics.

R. Perry Walker

These lines of challenge ultimately led me on an odyssey of air quality activism I never could have anticipated.

Visible Haze

As I have stated more than a few times in this book, by 2004 the problem of air quality did in fact escalate to a front-and-center public concern because of increasing haze. It was obscuring once pristine views of the Wind River Mountain Range. Local citizenry complained to BLM, WYDEQ, and the operators demanding that efforts be focused on three sources of particulate emissions that were contributing to the problem. These were smoke and emissions from well completion flares, smoke and emissions from the banks of engines powering drill rigs and drill rig burn pits, and dust lofted by big rig truck traffic on the network of dirt roads in the gas fields.

By early 2003, I had already become certain that drilling operations were degrading visibility of the regional air mass. As an amateur astronomer, I had been observing the night sky in the Daniel region since 1993 through two large telescopes. One particular galaxy I observed every year in the spring was the Sombrero Galaxy. It had become my standard candle.

Beginning in 1998, I realized that a specific optical characteristic seemed less prominent. By spring of 2003 I further realized that the galaxy was altogether missing some additional characteristics I had considered to be constant aspects of its appearance in the eyepiece. This was not the result of intrinsic changes in appearance of the galaxy. The changes were definitely due to intervening particulate obscuration in the local air mass. I knew full well I would convince no one of the loss of sky transparency by invoking esoteric sky gazing as my sole rationale. I had to come down to a lower altitude.

Thus, to prove my claims of visibility impairment, I elected to address the issue of air quality at a personal scientific level. This began in the form of simple methodical photo documentation of visible haze developing along the Class I protection zone of the Wind River Mountain Range as well as other areas of the Sublette Flats region. The visible haze phenomenon had been growing in parallel with increasing activity in the Jonah, Big Piney, and Anticline gas fields. On several occasions I provided this photo documentation to WYDEQ air quality personnel, and to the BLM air quality scientist Susan Caplan in Cheyenne but no response was ever forthcoming.

My initial approach was deliberately simple because it invoked an old method of observing sunlight scattered by suspended particulates. I positioned a camera in the shade of a narrow post supporting the front porch of my home. This blocked the sun disk while revealing the light being scattered around it. As a quality check, I always monitored regional weather to rule out pollution being transported from afar. For example, there were instances when the jet stream had carried dust from Mongolian sandstorms that created significant obscuration in the Sublette region.

Not surprisingly, hazy days nicely coincided with previous night's well completion flaring events. I spent many late nights driving the roads in the Anticline and Jonah fields documenting these events and recording wind behavior taking place at the same time. The next day I documented visible haze obscuration of the Wind River Mountain Range. There was a definite correlation between the two when winds traveled from the flare pits toward the mountain range.

As I realized this pattern, I set about assembling photographic evidence of example events. There were many times when the mountain view was degraded by a milky white curtain. A curious aspect about these flaring operations was that they nearly always took place under the cover of night. The fire was always impressive and even attractive but the night effectively hid dense smoke plumes rising high into the atmosphere. I continued this project through 2006 and built up a large record of haze events. I cited these in my EIS comment documents to challenge BLM assertions that modeled haze events would occur only a handful of days per year. BLM and WYDEQ ignored all of it.

I assembled a statistical count of the number of completion flares I could see being performed per month and identified them by the operators doing them. Although my home was located on a hill top 30 and 50 miles north from the gas fields, I was able to observe flaring events at night because of the scattering of light from the upper atmosphere above the flaring pit. To improve on my ability to see these events, I constructed a small wide field telescope equipped with a night vision enhancement eyepiece. I was able to distinguish flaring events from well site floodlights because flare fire intensity grew and dropped just like a camp fire increases and decreases in intensity with the size of the flames.

I recorded my observations from 2005 through 2006 and kept a count of flare events. I counted scores of them in the first half of this period as well as a decline in their number toward the end of the period. I was thus satisfied that the limits on this activity in the form of green completion recently mandated by WYDEQ was apparently having an effect.

I also collected statistics reported by WYDEQ for the period that coincided with the height of flaring activity. This period stretched from July 2005 through June 2006. Only statistics were available for EnCana, Ultra, BP America, and Questar. They came from the EnCana drilling permit units called Studhorse Butte, Yellow Point, Caprito, Sugarloaf, Corona and Jonah Federal; Ultra units Mesa, Riverside, Warbonnet, Studhorse Butte, and Boulder; BP America units Cabrito, Corona, Rainbow, Studhorse Butte, and Antelope; and Questar units Stewart Point, and Mesa.

In this period, EnCana completed 116 wells representing 52.3 percent of total wells completed, Ultra completed 26 wells representing 11.7 percent of totals, BP America completed 29 wells representing 13.1 percent of totals, and Questar completed 51 wells representing 22.9 percent of totals. In all, EnCana, BP America, and Ultra reported flaring a little over 597 million cubic feet of natural gas from their 171 well completions. Together, they were responsible for total emissions of 527.4 tons of VOCs, 379.1 tons of HAPs, 389.6 tons of NOx, and 376.3 tons of CO. Questar stood out for its clean operation because it was the first to initiate green completion and centralized fluids gathering technologies. As a result, its 51 wells flared no gas and were credited with emitting no VOCs, HAPs, NOx, or CO.

EnCana, because of the sheer numbers of well completions, was responsible for 76.7 percent of flared gas, 82.4% of emitted VOCs, 98.9 percent of HAPs, 97.3 percent of NOx, and 99.3 percent of CO. This underscored why so much local attention by residents was being focused on the rate of gas field development and demands that BLM slow that rate.

There was another haze source that generally received only superficial discussion. That was the presence of a large number of drill rigs operating at

By the Rasping in My Lungs

any one time. The section in the Anticline EIS that discussed the BLM preferred alternative stated BLM would not regulate the number of wells or the pace of development under this alternative. Most sobering was that the preferred option projected development of 4400 new wells with a bottom hole (production zone) spacing of 10 acres per well through the year 2023.[2] Apparently, BLM had no interest in limiting drilling rates.

Previous projections envisioned rigs equipped with diesel engines in the 1000 to 2000 horsepower class. In fact, the rigs operating in the field over the previous year had frequently been in the 3000 to 5000 horsepower class. This constituted a significant increase in SOx and NOx emissions potential. Furthermore, regulators were invoking definitions of rigs as mobile sources not subject to more stringent fixed source regulation.

In a publication on visibility published by Colorado State University, sulfate and carbon species were declared to be the single largest contributor to visibility reduction.[3] In the book "Atmospheric Chemistry: Fundamentals and Experimental Techniques by Barbara Finlayson-Pitts, it is stated, *"The major source of SO_2 is the combustion of sulfur-containing fuels."*[4]

Since rig engine internal combustion processes also create NOx emissions, it seemed logical and appropriate for BLM to amend its list of conditions of approval (COAs). They needed to include a requirement for operators to rapidly transition to the use of diesel engines in compliance with EPA Tier 1-3 specifications and ultimately Tier 4 emission standards.[5]

I publically challenged BLM to apply all of its pre-approval leverage toward inducing operators to adopt these measures particularly because of the likelihood that there would be scores of rigs in operation simultaneously. The region impacted by such a high level and long duration of proposed Jonah development activity was considerable and demanded that BLM be proactive. However, BLM proved typically dismissive of such requests by declaring again that it had no authority to tell the operators how to conduct their business.

Drill rigs were a significant source of smoke from engines and from burn pits. There were times when black soot layers hung in the air across the prairie in miles-long ribbons. Recall from Chapter 6 that WYDEQ air quality director Dan Olsen had long been suspicious that rigs were a major contributor to visibility impairment.

Recall also my mention in an earlier chapter that I wondered if this type of soot could be adding to local climate heating because it did seem like summers here were becoming curiously intense.[6] However, a Lawrence Livermore scientist was gracious in rejecting my idea.

Eventually public unhappiness over haze motivated the operators to move toward cleaner rigs. EnCana chose to utilized rigs powered by natural gas fueled engines rather than diesel fuel and as their numbers increased, particulate emissions from them declined.

The third source of two kinds of particulates, big rig truck traffic, was very heavy. Observation indicated those trucks were badly maintained because of carbon soot exhaust plumes they commonly emitted. Also, the gas fields had become interlaced by literally hundreds of dirt roads and heavy traffic operated on them 24 hours a day every day, lofting large amounts of dust into the air.

That dust was very fine and could stay airborne for long periods particularly in strong winds that were common. I insisted that photographic evidence of the type I had submitted strongly indicated this dust was a major component of the haze phenomenon. Eventually, the public outcry forced operators to routinely water spray these roads which did indeed somewhat reduce the problem.

Winds and Models

A key element to the environmental impact assessments conducted by BLM and WYDEQ was an approach to assessing visibility impacts through the application of computer modeling using a code called CALPUFF. Early on I took issue with CALPUFF and any modeling approach that only used wind behavior data in terms of speed, direction, and frequency on an annual averaged basis. This was a crucial point because wind field behavior was a key input factor to CALPUFF.

BLM actually admitted this in its Anticline EIS by observing that "... *frequency and strength of winds greatly affect the transport and dispersion of air pollutants.*"[7] The importance of this admission had great bearing on the main purpose of modeling which was to estimate the inventory of certain criteria pollutants that could be expected to reach the adjacent Bridger-Teton National

Forest and Wyoming Range. In my mind, there were several flaws in the modeling approach, the foremost of which was wind data being used.

I visited a modeler at WYDEQ who privately agreed that the wind data available in the Sublette region was seriously inadequate for the task of producing a truly accurate model-derived wind field. Other conversations I had with various corporate members and government air scientists supported this assessment. Anticline preliminary air quality models demonstrated that predictions for the Sublette air mass rested primarily upon assumptions, generalizations, estimates, and extrapolations.

My data also showed the winter season experienced most of the wind movement toward the national forests. This same source of those month-to-month data on the strength, direction, and frequency of the winds was available to the modelers but it was not used. That source was the Pig Piney airport meteorological monitoring facility. It might have injected accuracy into the modeling for this EIS and later EIS exercises. Instead, annual averages of data were employed using a data set drawn from a meteorological station southeast of the Anticline, operated by the British Petroleum Corporation.[8]

Winter seasonal movement had the potential of becoming even more relevant and problematic because as the Anticline was being opened up, the drilling season was expanded from summertime to year round (recall that drilling had been limited to summer only to protect wintering big game in the region). A possible result would be under-estimated transport of the pollutants to national forest areas and under-estimation of ozone creation. Potential impact due to entrapment of NOx and VOC pollutants in the winter snow pack and subsequent concentration by summer melt runoff would also be underappreciated. Just what impact such transport was having on those forests remained undetermined.

Perhaps to establish cover, BLM in a later EIS cycle made an assertion that there was a scarcity of data "... *addressing temporal profiles of dominant important pollutant species.*"[9] This went right to my belief that had there been an adequate number of monitoring stations installed before the Jonah and Anticline fields were being developed, or even by 2004, there would not have been a scarcity of data.

The same EIS document also commented that contract modelers chose to fix the data scarcity problem by applying "... *simplifying assumptions* ..." regarding background chemistry quality and use of "... *prototypical assumed characterizations* ..." of such chemistries for rural and urban locations.[10] A great source of frustration to me was that these assumptions seemed to always be applied in a manner that resulted in a cumulative effect of biasing predictions toward smaller impacts.

The EIS further commented on quality and types of data applied in the modeling effort. The TSP noted that wind speed variations had the largest degree of impact upon ozone creation mechanisms and ozone precursor pollutant emissions.[11] The TSP further stated, "... *a need exists for accommodating variations in point source VOC speciation within the context of a screening analysis.*"[12] These issues could have been accurately addressed had there been sufficient air quality monitoring in place before gas field development began.

The same flawed approach was essentially followed throughout the Jonah and Anticline EIS process. Wind behavior in the local region was certainly far more complex than was being acknowledged by BLM and its contract modeler. Averaging was a wrong practice. Analysis of wind data recorded by the government meteorological station at the Big Piney airport showed just that.

For the Anticline EIS documents, modeling was done using a temporary distant met station record of previous year's wind history for one scenario; in other scenarios, wind history averaged across years between 1999 through 2004 was used; in still another modeling exercise, years 2001-2003 from a national mesoscale data set called MM5, plus data from the British Petroleum meteorological station were used.[13] It was very possible that seasonal cumulative impacts were lost because of modeling on the basis of simplified annual averages over so great a period.

• • • • •

It didn't take long for the BLM to realize it had to require Jonah and Anticline EIS analyses to address haze and ozone creation and transport. Visibility, human health impacts, and health of the nearby national forests were of concern first because of emissions of NOx and VOCs being the building blocks of visible haze and health threatening ozone. NOx and VOCs were also of concern because of their potential for damaging the forests in those mountains from mechanisms such as soil and water acidification. Also, ozone is known to have negative impacts on the health of pine forests.[14]

Beginning in 2003, I vigorously urged WYDEQ to establish a comprehensive set of air quality monitoring stations. These stations were needed so that regulators could construct a record of chemical emissions from the gas fields because at that point no record existed. I worried that it was rapidly becoming too late to establish a baseline of measurement data before the air shed was compromised to a degree that the original pristine character would be lost as an accurate reference point.

Sadly, WYDEQ initially resisted but then did install one and then two stations. However, one was in the middle of the Jonah field and was relocated a few years later due to EnCana drilling wells right next to it. To realize how meager an effort these two stations represented, bear in mind that the Jonah Field covers 30,000 acres and the Pinedale Anticline Field covers 198,000 acres for a total of 228,000 acres of natural gas development. Adding together the 1400 wells in the Jonah and Anticline, plus those in three other adjacent fields, there were a total of 2900 wells in operation in a region of approximately 468,000 acres by the end of 2005.[15]

Eventually during year 2005, actual valid data was collected but by then air quality had already been seriously degraded in terms of visibility and hydrocarbon pollutant content. Nevertheless, in 2006 during a closed door meeting involving, WYDEQ, BLM, USFS, and EPA, it was decided that 2005 would become the designated baseline year against which future gas field development impacts would be compared.[16] This became a central point of my technical criticisms of all subsequent EIS exercises done by BLM.

Understandably, I wanted to see more wind field measurement underway and I made specific recommendations in two comment documents responding to BLM environmental impact statements. I urged the BLM and WYDEQ to expand the existing pair of air quality monitoring stations to a total of ten, positioned to measure upwind, mid-field, and downwind conditions. At first, these recommendations didn't gain any traction with BLM or WYDEQ but by 2011

WYDEQ had for whatever reason installed a total of six stations that were close to my suggestion. Such a network was necessary because by the beginning of 2012, my data derived from WOGCC records showed a total of 3081 active wells in the Jonah and Anticline fields with more due to come online.[17]

Meanwhile to fill the void, I collected daily surface wind readings from the Big Piney/Marbleton airport Automated Weather Observing System (AWOS) station. The data revealed a pattern of activity with speeds commonly ranging from 10 to 25 mph coming from directions between compass headings of 145 degrees and 225 degrees, i.e., out of the southeast, south, and southwest.

My own meteorological station data showed surface winds following the same directional pattern but reaching speeds commonly in the range of 20 to 40mph. Wind data from a state DOT station ten miles north of my location indicated lower speeds but frequently changing directions because of its being situated at the end of a "V"-shaped valley where there was a high degree of turbulence. All of this pointed to reasonable probability that the air mass containing gas field emissions would run up against the mountain slopes and invade the higher peaks and valleys.

The ultimate outcome would be a seriously degraded air transparency index in those mountains. An additional collateral effect would be increased damage to mountain forests and watersheds from plume induced acidified alpine lakes. Five years later, this dynamic would be confirmed by official WYDEQ and USFS monitoring stations.

In light of the potential for the Jonah Infill to grow to as many as 3100 *new* wells, a state sponsored program of empirical measurement seemed essential if air quality was to be effectively protected. Such a large increase in activity over that first visualized in the initial EIS rendered all of the predicted air quality impacts to be no longer credible. The latest BLM document titled *"Technical Report – Pinedale Anticline Oil & Gas Exploration & Development Project Draft EIS"* was, therefore, of heightened importance. Unfortunately, it relied heavily upon the original outdated meteorological analysis and associated air quality assessment contained in the original Jonah development plan.

I considered this to be crucially flawed because that earlier analysis from the time period 1995-99 using wind data collected during 1995 was now questionable. My own observation records indicated a possible significant shift in those patterns since 1995. Furthermore, the analysis condensed the year's weather activity into overly concise generalizations that obscured and even ignored monthly specific patterns. I thus believed the assessment in the Technical Report had to be regarded as being no longer relevant.

I presented my case by explaining that I had been analyzing wind patterns beginning in April 2003 utilizing data from the federally operated Big Piney AWOS site. The above cited Anticline EIS Technical Support Document contained a contractor-authored analysis that distilled 1995 wind history into a single pie chart showing speed vs. direction and percentages of occurrence. Likewise, I also diagrammed my findings but was more detailed.

My records showed in 2003 there was a high degree of month- to-month variability in wind behavior revealed by clustering of wind direction versus speed in my own wind rose diagrams.[18] The year of 2004 was a much less variable wind direction year but even so, a graphing of the year's directional trends for winds of all speeds showed they traveled toward four Class I regions from 21% to 42% of any one month. Importantly, the higher sun angle period between April and August when solar flux conversion of haze precursors is most efficient was also a time of infrequent but non-trivial wind movement toward the class I areas.

I felt a key point to be emphasized was that a simple annualized approach to estimating haze impacts could lead to over simplified conclusions. Conveniently, such conclusions were more amenable for reasons of regulatory minimization. In fact, it seemed likely there were significant monthly variations deserving serious consideration as to true short term and long term impacts in the Class I areas.

I created wind rose diagrams depicted speed vs. direction on a monthly basis and broke down the prevalent direction of those winds by depicting the percentage of occurrence in each of eight 45-degree segments within the 360 degree compass rose. In addition, these diagrams depicted the direction *toward* which the winds blew rather than the meteorological convention of direction from which the winds blow. With each wind diagram, I included a pie chart showing the percentage of the month that winds of all speeds blew towards each compass quadrant. I did this because I hoped readers not versed in meteorology could more directly appreciate possible transport of emissions from Jonah and the Anticline into the Wind River Range and the Wyoming Range.[19]

The first and most obvious character of the wind trends was that those above about 10 mph tended to cluster with respect to direction. This appeared on my diagrams as lobes of concentration toward specific quadrants. In all but one of the months charted, there was substantial clustering directly toward the Wind River Range as well as a substantial cluster that tracked toward the southeast end of that range.

While the duration of these clusters was unclear at the time, their presence indicated likelihood this was promoting transportation of haze precursor and air acidification gases from the Jonah and the Anticline into the mountain range. I also noted there were months when the winds concentrated toward the western Wyoming Range thus indicating a possibility of air quality impacts in that ecosystem as well.

For direct comparison, I converted the contractor's wind diagram to show its wind direction trend in the same non-conventional format I was using. A noticeable difference stood out. The EIS Technical Support Document depicted a major directional concentration toward the southwest with a minor secondary concentration pointing toward the southeast both of which were at odds with my own wind trend diagrams.

For seven years from 2004 through 2010 I logged daily Big Piney meteorological measurements. I eventually converted these data to tabular and graphic form and presented results to WYDEQ and BLM in my EIS comments. The graphs were in two formats. The first displayed monthly wind direction versus percentage of occurrence toward each of the two mountain ranges. The second format displayed wind direction compass heading toward which the winds were traveling versus the percentage of occurrence toward each of the two mountain ranges. Emphasis on the latter was added by marking the compass headings that coincided with north and south limits of those ranges.

These graphs revealed summer wind travel toward the south that did often spare both mountain ranges. However, the graphs also revealed the presence of wind directional travel toward the two ranges in winter months that were significant. This winter effect presented the specter of pollutants being carried over and into the snowpack where they would become concentrated during spring melt and runoff. This appeared to support USFS concern that gas field operations would foster alpine lake acidification.

All of this served to convince me that BLM's modeling contractor was under-accessing wind behavior outside of the summer season and consequently failing to sufficiently factor in winter wind transport into the mountains. Lastly, modeling assertions of constant annual wind behavior were contradicted by the large degree of year-to-year variation shown in the graphs.

• • • • •

A provision in the federal CAA places upon the operator the responsibility of demonstrating *"that emissions ... will not cause or contribute to concentrations which exceed the maximum allowable increases for a class I area."*[20] As I have earlier stated, BLM was the entity actually engaging in such demonstrations. This seemed a conflict of interest and a compromise of BLM's land stewardship responsibilities in that such activity placed it in the role of energy industry advocate. It was doing so by contracting private firms to conduct air quality modeling and presenting the results in an advocacy role on behalf of industry.

Based upon those model results BLM acknowledged that impacts *"may"* indeed occur but at a rate it judged to be an acceptable level. Such judgments of acceptability arguably violated the intent of the CAA and seemed to conflict with the fact that local BLM officials regularly disavowed any authority to address air quality issues.

The project Technical support Document and its discussion of VOC/NOx point source screening tables addressed in some detail the issues of model input considerations. Acknowledgement was given to sensitivities of the model to ambient air quality and the scarcity of data addressing temporal profiles of dominant important species. The discussion went on to cite *"simplifying assumptions regarding background chemistry quality,"* and use of *"prototypical assumed characterizations"* of such chemistries for rural and urban locations.[21]

The discussion continued with the development of *"reasonable"* worst-case model inputs (rural) and recognized *"counter-intuitive results."* These results came from running various emission mixes ... *"with rural and urban background concentrations that produced outcomes of greater ozone increments in rural settings* under equivalent emission rates."[22] Regarding meteorological input parameters, it stated that: "*A true ... analysis of all possible combinations of wind speed plume dimensions, starting time and temperature was not performed because of the range, continuous nature and number of variables involved.*"[23]

Additionally, the Technical Support Document stated that "*Wind speed variations impart the greatest degree of sensitivity on maximum ozone increments*" and a dilution effect due to increased dispersion near the source accompanies elevated wind speeds.[24] In the conclusions it was stated that "*these reactivity-sensitivity simulations suggest that background chemistry is a limiting factor in determining ozone increments due to ozone precursor emissions.*" Also, "*a single scale-up factor as used for the rural (input) table, is not adequate.*" Finally, it was stated that "*... a need exists for accommodating variations in point source VOC speciation within the context of a screening analysis.*"[25]

The Technical Support Document harbored many claims that seemed to warrant serious skepticism. Section 3.0 "Near-Field Modeling Analysis" was an impressive description of the modeling work but the results were, nevertheless, model results, not reality. Section 3.1 stated, "*One year of surface meteorological data, collected in the JIDPA from January 1999 through January 2000 was used in the analysis.*"[26] The wind rose representing this period generally agreed with my own research from 2003 through 2004 but again I cautioned against total reliance upon measurements from this single location so far in the past.[27]

By then my own database had grown as a result of my two years of hourly measurements from the Big Piney AWOS station as well as seven years of daily observations at my home north of the Jonah field.[28] I offered these data in various EIS comments in graphical form to challenge the TSP.

Also, I had begun collecting mid-day values from the recently activated Jonah-EnCana and Boulder-Shell stations started up by WYDEQ. Unfortunately these observations were fragmentary due to the fact that the readings were, according to WYDEQ and an EnCana representative I spoke with, not being archived for batch access.

This made it impractical to log into these sites each hour as would be necessary. Furthermore, both stations went through an initial period of poor reliability regarding uninterrupted operation. Consequently, initial comparisons of the two stations showed worrisome departures from each other and from both the Big Piney and Pinedale airport monitoring stations nearby.

Section 4.0 "Mid Field and Far-Field Analysis" maintained that far-field results modeling "... *indicate that neither direct impacts nor cumulative source impacts would exceed any ambient air quality standards.*"[29] I was suspicious of the use of the word "*indicate.*" It was important not to lose sight of the results being the product of modeling rather than empirical measurement. For this reason, I argued that the extensive set of supporting tables depicting far-field visibility impacts should be regarded as an academic treatment requiring empirical measurement-based validation.[30] The accuracy of the conclusion that JIDPA emissions would be of no serious consequence to air quality in the far-field air mass was too important to address only with modeling

The air quality modeling-domain to which far field modeling was applied encompassed most of western Wyoming, northwest Colorado, northeastern Utah, and a sizeable portion of eastern Idaho. While this may have been useful and perhaps even necessary to account for influxes from those states into the Sublette region there remained the credible question of just how useful this was in the absence of modeling on a tighter scale of the JIDPA and nearby Class I areas.

My own wind studies indicated that wind transport on a far-field scale from the south (Colorado, Utah) and from the west (Idaho) was not common. More often, wind transport occurred from the northwest. This assessment might seem to contradict my earlier observations citing wind predominance from the southeast, south, and southwest. In fact, this was evidence of an important difference between events on the far-field scale and the near-field scale.

I maintained that the nature and behavior of wind fields on a near-field scale across the Sublette Flats and certainly along the foothills of both the Wind River Range and the Wyoming Range were not well characterized and not understood. I thus insisted upon collection of sufficient data across that local domain and modeling of the close-in region of the JIDPA.

I also insisted upon parallel collection of empirical pollutant measurement data for validation of the model results. I urged a much more comprehensive network of sites to monitor O3, SO2, and NO2 if regulators were serious about comprehensively gauging outside regional emission flows toward our

Class I areas. I tried to be specific by submitting a diagram of my notional concept of what might constitute a useful network.

I very much wanted to see local air quality monitoring incorporate considerably more methodology. At the time, three planned WYDEQ operated monitoring stations seemed an inadequate response to the Jonah and Anticline activity. They were intended to monitor only NO2 and O3 plus some meteorological parameters.

However, as was earlier mentioned, the two stations then in operation were unreliable due to down time and were strangely out of agreement with each other regarding wind, temperature, and humidity parameters. They were also at odds with the Big Piney AWOS and the Pinedale airport monitoring system located only a few short miles away from the Boulder Shell monitoring station.

I had been long convinced of the correctness of my arguments in favor of a better monitoring network because of a statement in the Finlayson-Pits textbook referenced in an earlier paragraph: validation of … models' predictions *must* be done by comparing predictions with "*observed concentrations measured in appropriate ambient air monitoring programs..*"[31] I felt that a network such as the one I suggested should prove more supportive of accurate modeling efforts.

I took particular exception to a JIDPA assertion in its cumulative impacts subsection about mid and far-field impacts. Here it was offhandedly stated, "*Visibility impacts are predicted to be above the 'just noticeable visibility range.'*"[32] I had previously submitted photo documentation in comment submissions that illustrated a multiple deciview reduction in transparency of the airshed toward the Wind River Mountains. I was certain this benign prediction was already invalid.

As BLM and WYDEQ continued to assert these opinions in subsequent EIS documents, my frustration grew to the point where I began to demand publically that they stop modeling and simply look out the window.

In 2005 I happened to be in Laramie where the offices of BLM's modeling contractor were located. I paid an unannounced visit hoping but not expecting to speak with someone involved in the modeling effort. I was connected briefly with one such person who came to the lobby. We had a short conversation about where the modeling effort was in terms of completion and that was the end of it. Luckily, I paused to ponder what to do after this uninformative chat because soon after the person with whom I spoke had left, another came rushing into the lobby to shake my hand.

He introduced himself as the director of the modeling effort and excitedly noted that my name had become respected in the back rooms. He elaborated by explaining that I had presented them with wind arguments and data that, as he put it, made them realize that far field modeling was not enough and that near field modeling would now have to be added into their contract with BLM. We talked a bit longer in better detail until my schedule dictated that I take my leave. As we parted, he commented that my formal EIS comments had demonstrated the value of public input.

His assessment reversed somewhat many months later when I was vigorously challenging the model in a BLM public meeting. The fellow who had sung my praises in Laramie was now in the employ of one of the operators and heatedly counter-challenged my challenges. I thus received my first lesson (not to be the last) about how easily the operators could turn the tables through the application of more lucrative employment offers.

Optical Spectroscopy

Seeking more empirical methods of identifying gas development-generated contributions to the area's haze and ozone problems, I elected to employ the tool of optical spectroscopy. I chose this option because I had recently retired from the U.S. Air Force where I had performed a variety of duties as a physicist and engineer, including atmospheric spectroscopy. I was thus somewhat familiar with the physical and optical behavior of the atmosphere.

I began by purchasing a state-of-the-art portable spectrometer manufactured by Ocean Optics Inc. in Dunedin, Florida. This instrument was a laptop computer compatible unit optimized for the ultraviolet-to-visible part of the spectrum. Crucial to me was the fact that it was affordable because I was entirely self-funded. I soon assembled it into a field portable arrangement that consisted of a laptop computer, the sensor mounted on a small rifle-scope for precise pointing, and a cold pack around the spectrometer to increase its sensitivity.

I reasoned that by electing to look for signatures of pollutants obscuring the visible part of the spectrum, I would hopefully gain the advantage of having an easier time convincing a less trained audience of findings they could actually see in the sky. Furthermore, by addressing the visible spectrum, I could point to the presence of increasing haze in the region and leverage my findings by invoking parts of the Federal Clean Air Act which address visibility.

Beginning in the summer of 2003 I collected optical spectra at various locations in the local area, including drilling sites during times when flaring was underway. My work had to be nocturnal because as I mentioned earlier, the practice took place primarily at night. An added bonus of nighttime observation was the absence of complicating sunlight that would have to be subtracted from the signal.

Between 2003 and 2005 I observed scores of well completion flares, collecting both spectra and photo documentation. These events lofted huge amounts of smoke into the air, generally over a period of at least twelve continuous hours of burning at each well. The spectra obtained from them immediately revealed the presence of signature peaks associated with two obvious elements, sodium (588 nm) and potassium (766 and 769 nm), and a third not so obvious, lithium (670 nm).[33] The abbreviation (nm) is the wavelength measurement, in nanometers or billionths of a meter.

To help place this in perspective, this sodium signal occurs in the yellow portion of the visible spectrum, lithium occurs in the red portion and potassium occurs in the far-red portion. In addition to these element signatures, other combustion products were revealed but they were a source of frustration because I was unable to identify many of them due to a lack of reference research available to me.

This was largely a result of the fact that spectroscopy of the type I was applying was in the visible part of the spectrum. Europe seemed to be interested in the ultraviolet end of the spectrum while in the U.S., the infrared portion was preferred. I had no contacts in the academic world who could help and the Internet proved devoid of visible spectrum information about hydrocarbon combustion chemicals. Thus, I could find little identification data to help me determine what combustion product spectral peaks I was detecting.

I made known my findings and advised that the unidentified peaks were, nevertheless, evidence of combustion pollutants warranting attention. In my various EIS comments I warned that I found strong evidence indicating that sodium, potassium, and lithium originated with the constituents of fracking fluids being used. I conceded it was probable that a minor part of the sodium signature originated from the ground water being flushed out of the borehole but there remained the high likelihood that fracking fluids were the predominant source.

I set about clarifying the fracking chemistry issue by undertaking a six-month search of fracking-related publications and patent documents. Ultimately, I assembled a list of fracking-specific compounds revealing a prominent presence of sodium, potassium, and lithium in them.

With these data in hand, I tried in vain to persuade BLM and WYDEQ to initiate rules curtailing this practice. Despite the spectrographic and photographic evidence I had accumulated, both agencies either dismissed the issue or, in the case of the BLM, again argued that it had no authority to dictate to the operators how they conducted their field operations. As for WYDEQ, it showed no interest in my spectroscopic and photographic evidence when I tried to present it to back up my calls for restrictions on flaring. EPA officials at the Region Eight offices in Denver opined only that my results were "interesting."

I continued to argue that all operators should be compelled to adopt green completion methods with only few safety related exceptions being allowed. Ultimately, WYDEQ did adopt this concept to regulate flaring. However, as I explained in earlier pages the motivation for this change was the upset event experienced by Anschutz and not explicit concern about flaring.

By the end of 2004, I had applied my spectroscopic approach to well head production facilities, i.e., tank vapor combustors and dehydrator reboiler burners. I did so because I was challenging industry and WYDEQ insistence that standard production emission controls were 98% effective. An EPA study had indicated this figure to be critically dependent upon "... *the rate and extent of fuel-air mixing and on the flame temperatures achieved and maintained.*"[34]

Logic seemed to dictate that such a high level of efficiency would require, at a minimum, an intense level of maintenance of the combustion devices and the feedstock controls but during my fieldwork I had seen no evidence to indicate this was happening. Instead, the units were installed and periodically visited by an industry field worker to log the combustion flame temperature on the assumption that this parameter if maintained above a minimum value assured proper operation.[35]

Tank vapor combustors produce a lot of NOx and a host of other hydrocarbon combustion products that go straight into the atmosphere. My own on-site observation of certification measurements of a British Petroleum combustor by an outside contractor supported this. Regarding VOCs, measurements of the combustor validated WYDEQ's expectation of 98% efficiency in reduction of VOCs. However, close questioning of the lead contractor technician produced the admission that they were still grappling unsuccessfully with those other combustion products. He also noted that his diagnostic instrument displays were showing increases in area VOCs despite the performance of combustors. He interpreted this to mean these combustors were being overwhelmed by the sheer rate of well development.

My spectroscopic samplings of just a few dehy reboiler exhaust stacks indicated a high degree of variability in the combustion process from unit to unit. Many glycol reboiler flames were bright yellow because they were rich in incandescent carbon particles. Others exhibited flames that were orange, deep blue, light blue and white. This signaled they were in various states of fuel-air mixture combinations and various states of incomplete combustion. As a result, they produced combustion gases yielding spectra that exhibited a variety of spectral peaks.

I was able to build a limited list of chemical species that emit energy in wavelengths matching some of those displayed in the dehydrator spectra. Those compounds were the result of methane combustion and a few were CAA criteria pollutants.[36] All of this was important because the glycol heaters vented directly to the atmosphere, were not regulated, and existed in great numbers already.

To satisfy opponents who objected to the rapid spread of individual well sites, regulators and industry undertook a field development architecture that sited several wells on one pad. Despite that, there still was a necessity for multiple tank emission combustors and glycol recycle heaters because one unit could process only so much gas. For instance, a pad on the Anticline supported 12 wells and 12 dehydrator combustion vent stacks. Thus, it was not unrea-

By the Rasping in My Lungs

sonable to foresee several thousand units contributing emissions to the atmosphere. This justified the high level of citizen concern and argued for much more restrictive control over their operation.

The prospect of so many emitting units argued strongly for abandonment of the method altogether and adoption of a totally new approach. Actually, the EIS declared that operators would be allowed to adopt new technology as it became available. This briefly seemed to open the door to the Quantum Leap Dehydrator technology I described in the chapter about the operators. It had been tested in the field, and was certified by EPA as being 99.74% effective and offered the promise of replacing the glycol approach that has been in use since World War II.

Sadly, this promising solution like so many other initiatives never happened.

Infrared Imaging

I applied a second technological approach to monitoring the gas fields I had learned while in the Air Force. This was the technology of infrared imaging. I realized that a likely guaranteed attention-getter from the public and regulators alike would be imagery of normally invisible natural gas leaks from production equipment. I also was certain there were plumes of invisible emissions vapors venting from the hundreds of storage tanks that were everywhere in the Jonah and Anticline.

I had discovered that the infrared imaging equipment manufacturer FLIR Systems Inc. had recently begun marketing a sophisticated camera that detected the signature of natural gas. The device was equipped with optical filtering which passed infrared energy in the bandpass that coincided with wavelengths of the emissions I wanted to expose. The price of the unit was far beyond my reach but I wanted to see it in action so I contacted the regional sales representative and requested a demonstration. To my surprise, he replied immediately and we arranged to meet at the entrance to the Jonah field.

In anticipation of this field imaging exercise, I had rigged a visible camera to record images from the infrared unit finder eyepiece for later review. I had to use this approach because even though the IR unit recorded the scene, I had no way of downloading it and in light of its $20,000 price tag, I would not be taking it home to download there. I, therefore, kept the sales rep almost a prisoner for the afternoon so I could collect images of all the class of emission sources I felt I needed to observe.

We spent an afternoon driving around the field, stopping at various well pads and viewing dehydrator piping, tank vent stacks, and flash gas combustors. My hopes were fulfilled beyond expectation. By the time we had viewed only a few dehydrators it was clear that piping from reboiler units was leaking natural gas from many fittings. Sampling at more and more pads confirmed this. Clearly, methane was venting into the local air mass in notable volumes. These results were sensational to many and unwelcome to many others.

Dehydrator re-boilers are festooned with metal tubing, valves and fittings. If not assembled very tightly and with thread sealants, they become sources of methane leaks. Such leaks are revealed in IR imagery as small dark plumes contrasted against the surrounding background. The same happens with external piping carrying gas to various pieces of equipment on the well pad.

Tank storage of fluids had been the standard before some Jonah and Anticline operators converted their fluids recovery handling methods to collection and pipeline transport off pad. These fluids have been removed from the wellhead gas before it goes to market. Such tanks can be very large in order to hold thousands of gallons of condensate gasoline and produced water. As produced water is pumped into tanks, dissolved gas and vapors bubble out of the water and vent stacks on the top of the tanks keep pressure in those tanks under control.

One could look up at vent stacks protruding from the tanks and see nothing but when viewed in the infrared, emission plumes were coming from every one. Such venting resembled dense dark fog streaming away in the wind. Tank emissions combustors produced less discernable output thereby indicating they were doing their job of reducing raw emissions to the atmosphere but they too showed thin plumes. Overall, one could conclude that total emissions volumes pouring into the atmosphere were anything but trivial.

As I processed the IR images, I began to realize there might be a way to apply photogrammetric techniques to extract information about actual volumes of emissions being viewed. If that could be done, it would be a simple extension to quantify total volumes emitted per year. I had interest in this possibility because I was pretty sure that pictures alone would have only limited influence on WYDEQ and BLM.

I sat down with the WYDEQ director John Corra and described the infrared camera, what it could do, and why I thought he should acquire one. I suggested that by so doing, he could screen production facilities on the well pads, identify the leak sources and present the information to the operators for their action. I further suggested this would be win-win; he would be actively reducing pollution of our atmosphere and he would be enabling those same operators to eliminate monetary losses represented by lost gas not going to market. Both parties stood to gain something.

Director Corra's reply was a terse observation that he saw no regulatory value in such a device. That signaled I would be wasting time pushing this idea. I set this to the side because I was just one person being overwhelmed with data and discoveries of ways to apply it. I often lamented that had I been wealthy or even outside funded, I could have kept as many as ten researchers busy developing lines of evidence-based revelations about regulatory failures.

About three years later I learned that WYDEQ did in fact purchase the camera and GPS position recording equipment. Their stated purpose was that they were applying it to screening pad facilities and recording GPS coordinates for the record. I was only marginally impressed because this sounded like just another paper building exercise to fill file cabinets.

Ozone

As I stated in Chapter 6, the winter of 2005 proved to be a watershed period for WYDEQ, the operators, and concerned citizens in the region. That was the first sign of a new experience being brought to the region by gas development in the form of ozone pollution. The issue became impossible to dismiss when ozone levels frequently spiked well above the EPA recommended 8-hour exposure level of 75 ppb.[37]

In December 2007, a study of the problem done for BP America by Argonne National Laboratory was released. The study used a box photochemical model that implemented two different VOC model-lumping schemes and drew upon meteorological and pollutant concentration data provided by WYDEQ. Study calculations indicated that ozone mixing ratios are sensitive to (a) surface reflectivity (also called "albedo"), (b) column ozone (relating to ozone levels normally expected to be present), (c) NOx mixing ratios, and (d) available terminal olefins.

The study went on to state, "... *if one assumes that measured VOCs are fairly representative of the conditions* [in Sublette County], *sufficient precursors might be available to produce ozone in the range of 60-80 ppb under the conditions modeled.*"[38] The fact that there were many days when actual ozone levels ranged much higher than what the model predicted suggested that first, real conditions were certainly "right" for the production of ozone, and second, the model could dramatically under-predict what happens in the real world.

The study contained numerous caveats and repeated admissions that *actual* VOC measurements were insufficient to allow a full quantitative analysis. Much detail was presented regarding the large number of variables that had to be factored in. These included meteorological conditions, photochemical activity, NOx conversion mechanisms and pathways, trace gas mixing behavior, and ground surface-cover conditions.

The authors also acknowledged significant differences between their two modeling methods. This had to do with model sensitivity of ozone production due to presence of propane and butane which were noted specifically as being well above expected levels for a continental site. Of particular note was the observation by the authors that benzene and toluene were also elevated. Had WYDEQ not dropped reporting of these hydrocarbons from operator permit applications as I explained in Chapter 6, I likely could have supported this finding.

The study seemed more of an academic exercise than an attempt to inform what specific actions might be designed to solve the ozone problem. As such, it was premature for WYDEQ and the operators to embark upon a campaign of arguing that controlling VOC emissions was more important than controlling NOx emissions in combating ozone formation. Nevertheless, WYDEQ and the operators took these findings and leveraged them into a blanket justification for the events that followed.

In late 2008, EnCana submitted a permit application for 32 wells containing an "offset" proposal letter as part of its package. That letter contained assertions that EnCana had reduced its emissions of VOCs through consolidation of facilities and additional controls on existing sources. It further asserted that this resulted in a net emissions reduction of ozone precursors and that "*A portion of the net emission reduction is being 'banked' as an emission credit available for use to offset future net emissions increases.*"[39]

The letter went on to cite its "historic" emissions calculations (note calculations, not actual measurements) for VOCs and NOx based upon average daily production. It argued that since "historic" emissions were being eliminated, they should be considered "*emissions credits*"..." by which emissions identified in the new permit application should be offset. However, all of this calculation was the product of emissions modeling exercises and legitimately open to question.

WYDEQ-AQD responded that due to high ozone detections by its air quality monitor closest to a population center in the area, it had issued previous guidance. This guidance presented options that the operators could invoke as a way to meet their requirement to "demonstrate" overall reductions in ozone precursor emissions. WYDEQ-AQD went on to agree with the EnCana assertions that it had satisfactorily demonstrated a reduction in VOC emissions at the 32 wells in question by the amount of 525.2 tpy while increasing NOx emissions by 76.9 tpy when compared to emissions from EnCana facilities in April 2008.

It further acknowledged that EnCana had proposed to use "excess" VOC credits to offset increases in NOx and that it was in agreement with EnCana's proposal. It invoked the Argonne box study as justification for arguing that VOC reduction needed to be emphasized over NOx reduction as a preferred strategy for reducing ozone. Therefore, to "... *recognize* ..." EnCana's "... *early actions to voluntarily reduce NOx emissions before April 2008...*," WYDEQ-AQD proposed to accept EnCana's request.[40] From that moment, all other operators jumped on the bandwagon because WYDEQ had implemented a game changer in favor of the operators.

BP America submitted a letter, with its permit applications to WYDEQ stating the company's interpretation of the Argonne study and argued that emphasizing reduction of VOCs while allowing NOx to increase would be more effective at reducing ozone.[41] Subsequently, WYDEQ accounting of VOCs and NOx allowed annual increases of NOx emissions, presumably due to these arguments.

However, high ozone levels persisted. According to EPA, its own data showed 13 days in February and March with levels above its standard of 75 ppb. One 8-hour reading was as high as 124 ppb.[42]

I decided to follow up on the ozone issue and set up my own electronic ozone monitor. I chose a Dasibi Model 1003 PC, manufactured in the 1970s and now long out of production, but still classified by EPA as an approved "equivalent method" of measuring ozone. To acquire this monitor, I applied for funding from the Joint Interagency Office (JIO).

Recall that this group was tasked with the job of seeking projects to fund with money donated by the operators for the purpose of addressing environmental impact problems from gas development. However, as explained in Chapter 6, WYDEQ, a JIO permanent reviewing member blocked approval of my request for funds. Undeterred, I proceeded on my own with the help of modest financial assistance from two environmental advocacy groups, the Greater Yellowstone Coalition and the Wilderness Society.

I was consequently able to assemble an ozone monitoring station at my home north of the Pinedale Anticline gas field. To be credible, I went to necessary lengths to insure that my data would meet EPA standards. I built my monitoring station in strict accordance with the specific Parts of the Federal Code of Regulations dictating exact measures stipulated by EPA to be followed in building, locating and operating such a monitor.

I hoped this would assure WYDEQ that my findings would be fully compliant with EPA requirements, but I was naive. Months later the WYDEQ director personally informed me he would never accept any of my measurements.

Eventually, Linda Baker backed by several citizens in the area asked me if I could put together an inexpensive program of volunteer ozone monitoring around the Green River Valley area. Many of these folks wanted to be participants and they were scattered all around the Pinedale area, so I was immediately presented with a sample set that was usefully large and widely distributed. The idea quickly firmed up thanks to the existence of passive ozone measuring devices available on the open market. The last variable would be the reliability of the participants in staying with the program for almost two months.

Linda's help with modest funding and the need for frugal solutions led me to decide upon an ozone measurement system being offered by a company in Glendale, California called Vistanomics, Inc. This is a measurement system based upon exposing small paper coupons called "Eco-Badges" coated with an ozone sensitive chemical that changes color. The color density indicates the level of ozone in parts per billion that can be deduced by comparing the color of the exposed coupon with a color chart supplied with the coupons. However, I soon learned that this technique was not accurate enough for our needs so I invested my own money to purchase an electro optical coupon reader called "Zikua" offered by the same company.[43]

This device is hand held and very simple to use. One has only to slip an exposed coupon into a slot where a light beam passes through it onto a sensor. Measurement circuitry then calculates the level of exposure and presents the result on a small LED display. I found this device would produce consistent readings within a few parts per billion but I, nevertheless, always took several readings and logged the average. Still being uncertain of the device's calibration, I developed one more step in my process.

The one truly reliable ozone-monitoring instrument in my possession was the Dasibi monitor described earlier. I was operating that according to EPA protocols one of which was a rigorous calibration process using a reliable ozone calibration source. I used that monitor to develop a method of calibrating the paper coupons in a manner that immediately characterized each coupon as I was measuring it.

I would expose a calibration coupon for a set period of time next to the air-sampling intake of the Dasibi. I then measured the coupon several times with the hand reader and calculated the average level. At the same time, I observed the reading given by the Dasibi. This allowed me to not only compute and apply a correction factor to the coupon reading, but also I was able to calculate the color decay curve. This proved necessary to compensate for sometimes-long periods between participants' exposing the coupons and my receiving them. During the period between exposure and reading, the color underwent a gradual bleaching so it was necessary to apply a variable correction factor tied to the elapsed time between exposure and reading.

Participants were given a set of coupons and an instruction sheet explaining how to expose them, information to record for each coupon's exposure circumstances, and finally, how to seal each coupon in aluminum foil to block further exposure to light. This last step was intended to reduce color bleaching as much as possible. Over the six-week period of the project, I collected scores of exposed coupons. In general, participants were diligent in their adherence to instruction but as always some proved inevitably unreliable, thus creating data gaps.

All of this produced useful and fascinating results that I applied to a spreadsheet. The spreadsheet slowly filled in as the weeks of measurement results accumulated and patterns began to emerge. I noticed higher readings on a regular basis at homes located in low-lying depressions and also homes situated at the base of hills where the slope was quite inclined. A bigger surprise was discovery of frequent high readings northward from the Jonah and Anticline gas fields in valley locations far from the gas fields and seemingly up wind. These findings seemed unlikely to be a result of winds but there was no other transport mechanism to invoke. Eventually, WYDEQ and my Dasibi data produced the answer.

As with the badge results, the results from the Dasibi were informative. On near calm days when a wind drift of a few miles per hour came my way from the gas fields and ozone readings were spiking there, a few hours later I experienced readings that spiked in the same manner. I plotted my Dasibi data from February 2008 through June 2008 and saw confirmation of this pattern.

An example was a notable ozone event on Feb 23, 2008. For the previous couple of days winds had been calm so ozone levels began to inch upward during this day. As a result, WYDEQ monitors at Jonah and Boulder showed significant increasing levels. Thru mid-day my monitor was showing levels in the mid-to-high 50s ppb. Around 1:00 P.M., a light wind started out of the south at the Jonah and Boulder monitor stations toward my location north of Pinedale. Within two hours, my monitor began a rapid climb in detected levels toward the low 90s ppb even though the WYDEQ-Daniel monitor southwest of me remained steady in the 70 ppb realm. This suggested the ozone event had a limited size and shape.

I proposed to Dave Finley that somehow, there seemed to be an ozone mass that remained intact and mobile under such conditions. He responded that recent findings from his own contractor's field investigations were revealing that indeed an ozone bubble was apparently being created that could move intact under the influence of a local light breeze. Ultimately, this finding helped drive nomination of Sublette County to be declared an EPA designated non-attainment area.

It wasn't until five years later in 2016 that I received a surprising revelation from Dr. Rob Field, a field researcher with whom I coordinated during this period. In an email exchange he commented that I had been the first person to discover what he called a "reverse drainage flow" of ozone from the Jonah/Anticline region that drifted northward in late afternoon. This phenomenon prompted a paper by Dr. Derek Montague, another U.W. professor with whom I coordinated and mentioned in an earlier chapter.[44] I had been unaware of any of this.

Over the years, WYDEQ expanded its number of monitoring stations to form a credible network and by 2015 included seventeen supplemental monitoring sites concentrated down the center of the Anticline. However, as I noted in Chapter 6, up through 2010 WYDEQ continued to seemingly collect data for the stated purpose of "understanding" the phenomenology at work rather than applying the network toward enforcement of mitigation actions.

When pressed in a 2011 public forum to describe actual actions being taken to regulate the operators, WYDEQ insisted that when an operator proposed to emit pollutants, he first had to obtain a permit.[45] The implication was that the mere act of obtaining permit paperwork constituted effective regulatory compliance. Minutes later another WYDEQ official declared that no permit applications had ever been denied, thus further implying that as far as they were concerned, emissions were being adequately controlled via paperwork.

• • • • •

Research papers have been published corroborating much of what I have related about the situation in Sublette County. Dr. Rob Field whom I mentioned earlier was the lead author of a paper that nailed down many issues and spotlighted other issues still harboring some unknowns.

He affirmed that emissions from the gas fields were responsible for driving episodes of ozone creation. His work confirmed episodes of high ozone presence I have cited from the EPA that he attributed to three primary sources. The first was combustion apparatus on well pads and truck traffic, the second was fugitive methane leaks from equipment, and the third was fugitive emissions leaks from condensate and produced water storage tanks on well pads. Furthermore, both types of fugitive emissions were the primary sources for non-methane hydrocarbons.[46]

His paper explained that as of 2011, modeling studies of winter ozone in the Sublette County environs have shown ozone creation to be critically dependent upon the presence of non-methane hydrocarbons. Although modeling techniques have improved, there remains a need for more definitive data on the distribution and specific types of ozone precursor emissions present, thus leaving certain questions unresolved.

Sampling activities in the field revealed expected influence upon emissions species concentrations due to local meteorological conditions as well as the types of on-pad operations underway. As an example, sampling close to active drill rigs revealed methane presence with ethane and toluene component values of 485 and 11.3 parts per billion by volume. On the other hand, sampling close to a well completion operation showed ethane and toluene values of 179 and 202 parts per billion by volume.[47]

Of particular interest to me were findings connected with the wastewater treatment facility in the New Fork River Valley. This facility is a large plant that processes produced water being transported primarily from well pads on the Anticline. When I questioned the plant's effectiveness in a County Commissioners meeting, I was heatedly told by an operator that all the hydrocarbons in this water were being removed and sent to market and the water left behind was clean. I was skeptical but lacked a means to validate the claims.

Dr. Field's sampling indicated that supposed advantages of cleaning up the produced water stream had a price attached. This sampling showed a significant increase in contributions of higher molecular weight species contributing to the presence of non-methane hydrocarbons. Also, elevated levels of aromatic emissions as well as toluene and a species of xylene were detected.[48] The stated caution was that because WYDEQ emissions inventory data did not include this facility, it was being overlooked as a possible major contributor to ozone precursors.

I look upon all of this as validation of my early doubts about modeling input data and accuracy of emissions inventories. I also take some satisfaction from the fact that I produced some original results despite shoestring funding and reliance upon simple equipment. Rob's statement to me that I was the first to detect ozone drift northward in the afternoon is particularly gratifying.

Emissions Statistical Legerdemain

As I have previously stated, beginning in 2005, WYDEQ invested its entire regulatory defense in the concept of an emissions inventory. This inventory was the product of reporting by operators included in their applications for permits to operate on each well pad. It is important to remember and I keep saying this: Operators' emissions numbers were not field data from actual instrumented measurements but were instead derived from computer calculations of estimated on-pad production processes. They were in fact modeled values.

Recall also that I had been assured by industry and BLM personnel that such inventory calculations were to be trusted because they relied upon industry standard models.[49] These assurances seemed dubious to me until actual empirical validation had been obtained. I was doubtful that adjustments had been properly included for combustion performance at altitudes of 7000 to 9000 feet above sea level where these gas fields are located. Nor was EPA's AP-42 a definitive guide because it addresses only combustion temperatures and omits altitude.

I further argued that models being used to produce emission inventories from production equipment such as storage tank batteries, dehydrator burners, and tank emission combustors were also highly dubious because of the lack of model validation for our altitude regime. EPA Region 8 officials agreed.

The 2008 Anticline EIS contained further inconsistencies that were easy to miss. Appendix F contained tables of "Actual Emissions Inventories" for Shell, Ultra, Questar, BP/Stone, and Yates who were five of the biggest operators on the Anticline. Three others were not included for reasons never explained. This appendix also included accountings of gas produced for a long list of wells owned by each operator.[50,51]

I selected several wells from each listed operator to verify their agreement with stated gas production figures recorded by the WOGCC. I did this because Jim Sewell assured me that gas production at the well is a reliable indicator of NOx created by production processes. It thus seemed worthwhile to confirm modeled data listed in the appendix was in agreement with records of WOGCC and by extension WYDEQ. The results were curious.

I randomly selected wells from four gas field administrative lease units named Mesa, Riverside, Warbonnet, and Rainbow. Ten wells in the Shell table revealed that six were under-reported and four were over-reported compared to WOGCC. What I mean here is that if the production numbers in Appendix F were lower than the numbers shown in the WOGCC record for those same wells, the result was classified as under-reporting. Conversely, higher numbers in Appendix F than in the WOGCC record were classified as over-reporting. The net difference in gas volume was an aggregate under-reporting by 11.918 million cubic feet (mmcf in gas production parlance).

Ten Questar well samples were checked and found to be in near perfect agreement with WOGCC records. The breakdown was as follows: one well was 1,500 cubic feet under-counted and the remaining nine wells were less than 1,000 below WOGCC records, usually as small as 100 cubic feet.

The Ultra table exhibited very poor agreement. Because of this finding, fourteen sample wells were checked against WOGCC data. Of the fourteen wells checked, 12 were under-counted by an aggregate volume of 18.059 mmcf.

Results from the BP/Stone table were even worse. This table contained only 32 wells and of eight samples checked, six were under-reported and two were over-reported yielding an aggregate under-reported volume of 24.585 mmcf. This prompted me to look at all 32 wells. Total under reporting for these 32 samples added up to 167.062 mmcf.

A likely criticism of the above narrative would be the small sample size described. My method was to select wells from the appendix list and check them for agreement with WOGCC. If agreement was found then that operator's comparison list was short. If disagreement was found, that operator's comparison list was longer. The goal was to ascertain trends and was limited by resources of time and labor to perform the comparisons. Also, I had no doubt that BLM would disregard this approach; so I limited my own expenditure of time on what would be a certain lost cause.

The implication from Appendix F was a high probability that total volume of produced gas in 2005 was seriously under counted and by extension, the resulting emissions were under estimated as well. Given that these appendices were referenced as the source data used in the air quality modeling exercise for the Anticline gas field, this was yet another reason to doubt the accuracy of air quality modeling.

One additional cautionary reminder must be added. Recall from the earlier section on the WOGCC that in 2012 while tabulating well production numbers, my assistant discovered that WOGCC had begun to post revised production numbers for hundreds of wells. These revisions were backdated as far as 2009 and significantly increased total produced gas volumes over what had been first posted. This also must be viewed as a source of emissions prediction under-estimation.

Appendix G of the Anticline EIS added even more suspicion to the emissions assertions it contained. Table G.8 listed "excluded sources" from the state permitted source inventory and the number of tons per year in emissions they represented. Reasons for exclusion included *"out of domain," "waiver <3.0 tpy,"* and *"< 1.0 tpy"*.[52,53] When all of the entries under 3 tpy were counted, the total was around 1,095 such sources. No clarification was given for how far below 3 tpy these sources fell; so as far as the reader could surmise, they could account for as little as a few hundred tpy or as many as 3200 tpy. If the higher figure was correct, a significant contribution had been overlooked for reasons of administrative book keeping.

The 2008 Final Supplemental EIS cited the use of a production decline factor to estimate out-year gas production declines and associated economic declines. This decline factor also was key to the related rate of NOx emissions decline. The EIS exhibited an "estimated average well production profile" which, as an example, showed well output at 12 percent of original volume five years after startup.[54]

The decline curve for each well is unique. Reporting to WYDEQ contained in permit applications was variable in that some operators presented decline equations in only some of their applications, and other operators did not offer clear quantification of that parameter at all. This information was completely eliminated from permit applications after 2005.

My own database of well decline factors did indeed support the general exponential nature of the rate of decline but my sampling suggested that the decline rate in the EIS could have been overstated by three to seven percent. The cumulative effect of this seemingly small difference could multiply air impacts considerably.

My database assembled from WYDEQ-AQD permit records and WOGCC records indicated further possible accuracy problems. By 2009, my data addressed 1,075 wells and chronicled their startup date, name, pad identifier, gas production, and emissions volumes for VOCs, HAPs, NOx, and CO. Here is what I found:

- Some early wells had been renamed by operators
- Some had been incorporated into multi-well pads
- Emissions for those individual wells had become incorporated into emissions reporting for the entire pad
- Reporting format imposed upon the operators by WYDEQ-AQD had changed several times since 2004
- Well emissions for old wells had declined as production declined
- Emissions were high for wells just going into production at maximum production
- All of this made accurate total emissions accounting highly suspect

After reviewing all of the materials discussed above as well as additional material, a fundamental trend seemed to emerge. The flow of information to

WYDEQ-AQD addressing emissions accounting was varied, shifting in standards, voluminous, ever changing, and most importantly, unverified. The administrative bookkeeping activities used to distill this flow of information was in need of assessment for accuracy and effectiveness.

If this information represented the foundation of the oft referenced "actual estimated emissions inventory," one had to seriously question the level of success by WYDEQ-AQD at maintaining a credible and accurate inventory figure. I wanted EPA to look into the following:

- What source materials were used to arrive at the final inventory figure?
- How rapidly was new source material incorporated into the inventory assessment?
- Was staff assigned to this task adequate to keep up with changes?
- What were the procedures used to reconcile past information with newer changes, i.e., changes in well names, consolidation of single wells into pad complexes, additions of new wells to pads, and resultant changes in pad emissions volumes?
- How were duplications of well data detected and compensated for?
- How well did this inventory agree with modeled estimates resulting from the computer codes being used by operators to report their estimated emissions for drilling and production?

My concerns did find their way to WYDEQ via EPA-Denver and the response contained vaguely worded assurances that the State was routinely assessing its methods to insure that the product was accurate. Not too long afterward at a public forum held by WYDEQ in Pinedale, its newest employee placed in charge of this topic stood up and recited the standard mantra common to all levels of government: "*We are working with industry to create an accurate emissions inventory.*" To this day I regard "we are working ..."as a blind intended to sound reassuring while avoiding any commitment to specifics.

Finally, recall that the Anticline EIS admitted to some serious errors on the part of BLM. At one point it addressed impacts to air quality from "*existing well field activities.*" The EIS made the point that since issuance of the Record of Decision (ROD) some five years earlier, "... *natural gas development in the PAPA (Pinedale Anticline Project Area) had occurred at a pace greater than that analyzed in the PAPA draft EIS.*" This was important because the ROD was BLM's formal document approving the project as described.

I restate here the warning I gave USFS director Kniffy Hamilton. That ROD "... *authorized the development of 700 producing wells or pads...*," and set a threshold of 376.59 tpy of NOx emissions from compression, and 693.5 tpy of NOx emissions from all sources in the field.[55,56,57]

BLM's EIS assumption of 700 producing wells and up to eight drilling rigs operating on the Anticline at any one time might be characterized as incompetent. As of five years later, there were approximately 457 producing wells and over 26 drilling rigs operating there. As a result, NOx emissions from all sources operating in the PAPA five years after approval of the ROD were estimated at 3,512.4 tpy which exceeded the 693.5 estimate by over a factor of five.[58,59,60,61]

I have previously noted that when the public cited this debacle as grounds for a slower rate of development being claimed as an option by BLM, the agency retreated behind an ambiguity imbedded in the phrase quoted above, "... *700 producing well or pads.*" It argued that the language made it unclear as to exactly how the emissions should be accounted, i.e., by individual wells or by pads. The difference was crucial because pads were expected to support from 12 to 18 wells, each thereby opening the door to as many as 12,000 wells.

• • • • •

A potentially very large source of emissions in the Jonah and Anticline EIS documents was compressor stations. As I related in the chapter about the BLM, I dug into the statistics of predicted emission volumes and compared them to WYDEQ-AQD permit approval statistics for the same emission species and found significant discrepancies for two stations in the Jonah. Specifically, the EIS greatly understated actual emissions versus those in the WYDEQ-AQD record of permitted emissions for the Jonah field Luman and Bird Canyon stations. Results for the stations in the Anticline were initially more in agreement.

However, the Anticline EIS stated that three compressor stations would be expanded through the year 2011 by 267,035 horsepower. Calculations of the NOx to be generated by this increase revealed that total tons per year could increase by an amount between 1790 tons per year (minimum) to 4344 tons per year (maximum) depending upon what emission factor was used in the calculation.[62]

At the close of 2011, I wrote to the local BLM office to demand a formal review of emissions predictions for gas compressor stations. Chapter 2 of the Anticline EIS documents contained commentary addressing these facilities and provided a table of emissions to be expected from them.[63] That table included specific emissions quantities in tons per year projected forward to year 2009, followed soon thereafter by additional BLM documents carrying those predictions into 2011. Year 2011 came and went with no formal review of these predictions, so I pressed for one.

I have described this affair in the chapter on the BLM but it warrants repeating here. Two months after I submitted my request, I received an uninformative three and one half-page reply. Only one paragraph addressed the specific point of my request. The remaining body of the letter included meandering comments about drill rig emissions that were of little clarity. BLM further denied that it had intended the Anticline EIS to be a declaration of not-to-be-exceeded environmental impact objectives.

In the Jonah and Anticline EIS documents, projected emissions were specifically identified as those to be expected from actual project development operations *including* unspecified but assured mitigation efforts. BLM acknowledged its responsibility for air quality protection by noting in its 2006 Anticline environmental impact statement that, *"BLM also has responsibility in regard to air quality."*[64] Also, BLM had committed to achieving *"air quality performance objectives"* in its earlier round of Jonah Field development comments.[65,66,67]

However, in its reply to me, BLM redefined the EIS content as being none of this. Instead, BLM asserted that its EIS objectives were only statements of "expectations."[68] This contradicted what many other federal and NGO stakeholders were led to believe. The outcome of all this was that BLM "predictions" which were flawed in the first place had now turned into mere "expectations" and the more rapid pace of development meant that we could "expect" the consequences to be far worse than even the "predictions" indicated.

• • • • •

Returning to well pad emissions, dependence of emissions inventory accuracy upon administrative bookkeeping became increasingly obvious as I reviewed permit application reporting submitted by operators. Review of submissions from Yates Petroleum, Forest Oil Corp., EnCana, and BP America was instructive. One of the smaller operators cited the requirement to "*demonstrate*" performance related to "*potential*" visibility impacts.

That operator went on to explain uncertainty as to what emissions it was obligated to report and most telling was a query as to who "… *performs [the] demonstration* …" that lessening of visibility impacts would happen. The operator also warned that knowledge of how to calculate visibility impacts was lacking within the company.

I was pretty sure this cited weakness applied to all operators and to WYDEQ-AQD as well as BLM and USFS. This seemed another confirmation of the need for field measurements to validate all the emissions estimation techniques being used by these players.

A report to the BLM from one of the largest operators presented a litany of semi-quantitative and narrative comments addressing accomplishments and intentions.[69] Under "Emissions Reductions," it acknowledged operators' obligation to demonstrate that impact levels would be even less than would result from the 80% emission reduction requirement described in the FEIS but noted that WYDEQ-AQD had yet to establish a Best Available Technology (BAT). This seemed to show that WYDEQ-AQD's ability to cope with issues of air quality management always fell short of operators' rate of gas field development.

The operators' reporting summarized wells drilled by slant vs. vertical methods, completion flaring events, and number of wells drilled. It then claimed that: *"Actual emissions from 2006 through 2008 and predicted impacts from emissions in 2009 may be assumed to be vastly lower than the predicted inventories ... and predicted impacts."* Operators further claimed, *"Given all of the above factors, actual emissions ... and expected impacts ... may be assumed to be significantly lower than the predictions in the FEIS."* This operator's perspective concluded with assurances that *"... impact levels from the JIDP will be less than the impact levels of the 80% emission reduction scenario described in the FEIS."*

I remained unconvinced and believed these claims required more empirical proof because of far reaching consequences effecting the Class I airsheds and our ozone problem.

Missing Gas

In recent years, EPA has been expressing concerns over venting of natural gas into the atmosphere from gas field operations. One of its concerns is the fact that methane is 25 times more efficient than carbon dioxide at trapping heat in the atmosphere. There is even speculation that if losses to the atmosphere are as high as some estimates, the result from a methane green house effect could offset the advantage of burning natural gas over other fossil fuels. One study done in the Denver-Julesburg Basin of Colorado estimated that around four percent of gas from operations there is going into the atmosphere excluding pipeline losses.[70]

I first became curious about what I termed "missing gas" in Sublette County because my activist friend Carmel Kail first mentioned to me that on the WOGCC web site there is reporting of actual gas production by well and also actual gas sold by well. She noted that the two figures were not the same. When she tried to obtain an explanation from the state, she was subjected to buck passing replies that led nowhere.

Her initial query to the WOGCC produced the reply that the missing gas was "probably" used on site but accounting rules imposed on operators did not require a breakdown of differences between sales and production. She was then referred to the State Department of Revenue & Taxation (DR&T).

A DR&T representative cited a state statute that allows venting conducted on-site under authority of the WOGCC to be exempt from taxation. The reply went on to note that while operators are still obligated to report vented gas volumes to DR&T for tax purposes, such reporting is confidential by reason of another state statute. The responder then opined that WYDEQ tracks gas venting in-state and she might go to that department.

When she queried WYDEQ, a staffer informed her that, *"I understand your concerns and we are looking at ways to improve venting/blowdown reporting. We anticipate that compliance with the conditions of the venting/blowdown permits will minimize emissions and increase the accuracy of reporting. Venting information submitted to the Department of Revenue is confidential and cannot be* accessed by WYDEQ."[71]

Thus, Carmel was successfully held at bay by a bureaucratic system of agencies that did not exhibit any internal coordination except perhaps to deny public access to such information. My assistant and I therefore proceeded to invest thousands of hours retrieving and compiling production and sales statistics for analysis from the internet. We confined our efforts to the Jonah and the Anticline fields and our source of data was the WOGCC web site. In addition to comparing production, I wanted to gain a perspective on the amount of gas burned on well pads by burners and combustors that are part of the production process.

For that, I scrutinized a sampling of operator permit applications. I specifically reviewed their work packages that described the number of process heaters and waste gas combustors present and fuel consumption statistics for them. At the start, I fully expected that when I reached the end of this particular exercise, I would see a close correlation between what was "missing" (the difference between production and sales) and what was burned. That did not happen.

I analyzed figures for five major operators, Ultra, EnCana, Questar, Shell, and BP America. For each operator, I collected statistics on all development unit leases they owned. I analyzed all the data from the WOGCC start date of 1991 up through December 2011. Statistics for all listed active wells were gathered and tabulated. Also, the number of wells underway but not yet online, were tabulated because this would show how much additional volume could be expected to develop over the next few years. Combined totals were 3,081 active wells with 1,414 still to come as of December 2011.

Total recorded production of gas from active wells was 6.043 trillion cubic feet (TCF) while total sales were 6.006 TCF. This equates to 37 billion cubic feet (BCF) not accounted for over the 20-year period analyzed. Stated another way, 0.61 percent of produced gas was unaccounted for. WOGCC stated on its web site that gas sales for all fields in Sublette County totaled 10.26 TCF. This would include fields in the Big Piney region to the southwest of the Jonah and Anticline. Applying the 0.61 percent figure to that volume results in 63 BCF having possibly gone missing. This is meaningful within the context of methane gas acting 24 times more effectively as a green house gas than CO_2.

Worthy of note was the variation between operators, ranging from 5.65 percent of produced gas unaccounted for in EnCana's Hacienda Unit to 0.29 percent for BP America's Antelope Unit. These variations suggested that some operators were practicing better field methods than others.

Burned Gas

The next step after adding up all the missing gas was to add up the gas burned on the well pads as part of the production process. Again my expectation was that the two figures would be similar but first, a review of the elements of production are in order.

Remember, the production process includes passing the gas through a dehydration facility to absorb water from it before sending it to market. This drying process uses glycol that absorbs the water; the glycol-water mixture is heated in a chamber called a reboiler fired by a gas burner to drive the water out so the glycol can be reused.

This dehydration cycle creates vapors that are disposed of by burning them in a stack called a VOC combustor. Also, condensate and produced water are generated in huge volumes that are stored in large on-site tanks. Gas dissolved in these liquids flashes to vapor into the void space at the top of the tanks; this gas is also burned in the VOC combustor stacks. Much later on, a few operators eliminated their storage tanks by sending the water to the central processing facility for recovery and marketing of the vented gases. I talked about that facility several pages earlier

Extracting the pertinent statistics from the operator work packages that accompanied their permit applications was a tedious and time-consuming process. Each operator has its own nomenclatures and nuances for chronicling the data. Also, each operator used slightly different values for the heating capacity of the gas used as fuel on site, and each operator characterized operations of its equipment such as combustor fuel consumption in language with differing degrees of clarity. Thus, it was necessary to simplify the task.

This was accomplished by sampling well combinations possessing production characteristics typical of the larger population addressed earlier regarding gas produced versus gas sold. To study all of the thousands of wells for the entire time period would literally have required at least one additional year of full-time work. Therefore, to make the task more manageable, I selected 2010 and a subset of wells from each production unit because by then the wells chosen had been continuously active for sufficient time to have an established production profile behavior.

I was surprised by what I found because I had originally expected similarity between the volumes of "missing" and "burned" gas. However, burned gas volumes exceed missing gas volumes by varying quantities. Also, there were big differences between development units and between operators.

Patterns that emerged showed: (1) VOC/BTEX combustors that burn storage tank vapors accounted for only a few percent of burned gas; (2) large numbers of dehydrator heaters accounted for most of the burned gas; (3) there was no clear relationship between missing gas and burned gas; (4) the total volume of burned and missing gas from just this subset of wells for year 2010 added up to 3.2 BCF which apparently went into the atmosphere as greenhouse gas and natural gas combustion pollutants. These findings are generally in agreement with the EPA belief that glycol dehydrators cause 80 percent of the industry's annual HAP and VOC emissions.[72]

These patterns have important implications. First, recalling that emissions inventories are not based on actual measurements of emissions but rather modeled emissions, (remember, they are referred to as *"modeled actual emissions"* or *"actual* emissions *estimates"* in the EIS documents), it seemed imperative to obtain *actual* emissions measurements for glycol re-boilers and VOC/BTEX combustors for all air pollutants of concern. These were listed in the 2006 draft Anticline EIS to be NOx, sulfur dioxide (SO_2), VOCs, CO, particulate matter, formaldehyde, and HAPs (benzene, toluene, xylene, ethylbenzene, n-hexane).[73]

All were discussed in the context of having been modeled. Even if flaws in modeling related to wind parameters had not been accepted as reason for reruns, dehydrator modeling should have been rerun incorporating actual measured emissions values. Predicted impacts could then have been based on more and better data. A related task should have been obtaining actual emissions data on the VOC combustors and re-boilers.

Overall, unaccounted-for gas and model-based company reporting methods on combustion-caused emissions reinforce previously identified reasons for being skeptical of the WYDEQ emissions inventory.

Wasted Years?

Terry Svalberg and I agree that BLM and WYDEQ engaged in a magician's form of diversion for the purpose of holding their opposition at bay. They kept us busy watching one hand waving about superficial efforts supposedly addressing our concerns in order to distract us from seeing the other hand giving operators everything they wanted.

During our many personal meetings in the years following our Jonah and Anticline involvements, Terry and I have often performed a sort of post mortem on our efforts. I often lamented that my efforts were ultimately a waste of a decade of my life. Terry always disagrees and assures me that I made a difference by forcing WYDEQ as well as the operators to pay more attention to their activities impacting air quality.

Epilogue

The history I have related only scratches the surface. As I wrote this book, dozens of incidents and conversations came flooding back into my memory. Many were too trivial to include here when contemplated individually but as a group conveyed greater patterns. Including that additional material would have pushed this book's page count to twice what it is and rendered it tedious. Perhaps that failure has happened anyway but I believe I have told a uniquely personalized broad-based story.

Other incidents such as my routine conversations with three friends and colleagues on staff at the University of Wyoming as well as a president of that university are important. Suffice it to say here that around 2006 I was told by my faculty friends about U.W. administrators' warnings not to be vocal about environment issues in Sublette County. Their fear was that it could jeopardize huge oil and gas industry donations to the new energy school.

When I heard this I marched right over to the office of the university president. To my surprise I was allowed to see him immediately. I begged him to maintain a balance between the energy school and the university's

Haub School of Environment and Natural Resources. He assured me that both would always "be at the table with equal status." To all appearances, that didn't happen.

Now years later I realize I followed a path that involved me in far more elements of the full story than most individuals are unlucky enough to experience and assemble into one narrative. These elements spanned the spectrum of inner sanctum politics, concealed agendas, powerful outside influence, corporate maneuver, agency battles over jurisdictions, and science so inconvenient that it had to be marginalized or dismissed.

I felt a sense of urgency to get this book onto the street as a warning to other communities facing gas and oil development. They need to be as forewarned as possible but I know in my heart of hearts these warnings will likely not make a big difference. As did Sublette County when similarly warned by a sister county, other communities will assure themselves they will do things differently.

Now a new threat emanates from Washington D.C. The 2016 presidential election has resulted in the appointment of agency Directors having a history of opposition to environment stewardship issues. That attitude has already manifested itself in the form of government agencies purging their websites of references to environmental issues and climate change. In the case of EPA and BLM websites, many of this book's endnote source internet links have disappeared.

Furthermore, the new director of EPA and many other new high-level federal appointees signal a determination to hand off environmental protection responsibility to the states. I believe I have shown that Wyoming is a good example of why that is a really bad idea.

This seems to be further validated by the WYDEQ, VOC/NOX cap and trade offset scheme. It was supposed to reduce NOx by a ratio of 1.1:1 per new tpy created and reduce VOCs by a ratio of 1.5:1 per new tpy as an ozone mitigation tactic. The outcome has been doubtful but because EPA failed to challenge the scheme WYDEQ appears to have successfully ignored the mandate that NOx emissions be reduced by 80 percent.

I mentioned at the end of the chapter on the EPA that Cindy Beeler expressed satisfaction over the accomplishments we achieved in addressing air quality issues. I am now concerned that all of that is being overturned with a stroke of a pen. In fact, this is happening almost weekly in the Trump White House.

I routinely tell my Sublette County story with the preamble: "I have seen the dark side of government, industry, news media, and environmental organizations." For me this all became revelational because as I have admitted throughout this book, I went into the fight in a state of naiveté. My personal climb toward enlightenment has been dark. I have concluded that one man cannot make a difference nor can a single person successfully fight city hall.

I recall a bit of wisdom I read in Stephen Schneider's book *Science as a Contact Sport*. He described his battles and scars resulting from his years of climate change advocacy and scientific research to back up his ideas. His work brought him into contact with notable personalities one of whom was Margaret Mead.

He related a conversation with her in which he lamented an experience that resulted in his job position being at risk because of his most recent unpopular public positions. She chastised him for moping about it and basically challenged him to man up. She then followed with sage advice that I try to paraphrase more directly in the context of the regulators and operators I have written about in this book:

> *That's exactly the way it's going to be…these people are being threatened…you're saying [to them] the way they define the [air quality issue] and the way they see [their operations and regulatory behavior] is not the only way and that's not going to float for many of them. Have a generational perspective. Don't just think about it from day to day; it will change slowly over time.*[1,2]

The last sentence matches well with a comment I heard from a former oil and gas executive who lived near Pinedale and added his voice to the local chorus against the way gas development was being pursued. He stated that oil and gas companies would have to go through a few more generation changes in top management before the business need for pro-active environmentalism would become part of their business model.

I hope these changes kick in sooner than generational time scales because we may be on a shorter string than we want to believe. This has been underscored yet again by recent events. In January, February, and March of 2017 winter average ozone levels in the Upper Green River Basin exceeded the federal recommended health risk standard of 70 ppb by 20 percent.[3] This was noted to have occurred despite operators' "voluntary efforts" to avoid such events.

True to form, the Petroleum Association of Wyoming behaved like an oil and gas version of the Flat Earth Society. A spokesman suggested that wood stoves and vehicle idling were to blame. The idea is silly. Both of those sources existed long before ozone appeared in Sublette County. It wasn't until the gas guys came on the scene that degraded air quality in general and ozone in particular became a new normal.

All of this has acquired new relevance because BLM released its draft environmental statement for the Normally Pressurized Lance Field (NPL) in July of 2017. It proposes 3500 new wells at a rate of 350 per year over ten years on a project area of 140,859 acres utilizing multi-well pads supporting "up to" 64 wells each.[4]

My reading of the EIS revived my frustrations and disbelief. It was as though BLM copied and pasted all of the same promising rhetoric from previous EIS documents onto clean sheets of paper. On top of this and even though a different modeling company was used, the same errors involving wind modeling of which I had been so critical were being repeated. Even worse, this time the wind rose cited by the modeler invoked winds averaged across a 10-year period between 2000 and 2010.[5] Finally, great emphasis was placed upon emissions inventories maintained by WYDEQ as constituting definitive information about the state of the atmosphere and its current emissions load.

I had intended to end my involvement with Sublette County gas field commenting on future projects but I could not let this one go without sending some kind of comment. Thus, I sent BLM-Pinedale a nine-page letter warning that the same mistakes were being made. I also provided my original series of wind graphs by direction into the Class I and Class II areas by season and cautioned that any idea of modeling on the basis of a decadal average was wrong.

In 2017, BLM announced intentions to consider drilling on the Hoback Rim in an eight square mile private area where it, nevertheless, holds the mineral rights. Its preferred plan is to ban drilling on top of the acreage and employ slant drilling from adjacent sites. At a minimum, this could damage any underground well water supplies as happened in Pavillion, Wyoming.[6,7,8,9,10,11]

True to form, Wyoming politicians can be inventive when energy development seems threatened. The state's Governor Mead urged BLM to keep open the option of drilling from an adjoining Wyoming School Trust parcel.[12] Also, newly elected U.S. Representative Liz Cheney is busy crafting legislation to pass more control over oil and gas development on federal lands to the states, something I have shown to be a bad idea in Wyoming.[13]

In 2018, BLM announced plans to lease 286,000 that stretch across a region south of the lower tip of the Wind River Mountain Range.[14] Public concerns are initially focused on interruption of a deer migration route but if actually developed, the area could also add to air pollution and lake acidification threats in that mountain vista during the winter season.

A different issue that keeps rearing up in a manner requiring update here is that of the original idea of trading NOX for VOCs. Recall that there was great confidence in this concept and its utility for beating down ozone emissions. In closing out this book project, I contacted Derek Montague to verify my memory regarding his own presentation in Pinedale under WYDEQ sponsorship. He had told the audience that there was counter intuitive evidence showing that removal of VOCs from emissions might be more effective at holding down ozone formation than limiting NOx.

His reply was bitter sweet. He stated that

> "... A couple of years ago we were leaning to O3 production being VOC limited, implying that VOC reductions could reduce O3. However recent work and changed circumstances have reduced the certainty of this conclusion so that the most knowledgeable person on these issues now here at UW, tells me that the situation is much less clear than was formerly believed. In fact his statement to me this morning is that all bets are off and no one now knows for sure whether VOC controls or NOx controls would nowadays be of greatest benefit in the UGRB."[15]

R. Perry Walker

It can now be argued with more certainty (as I did in the Air Quality Task Group) that Sublette County has become a test range in which a large improvised atmospheric experiment is underway. Not surprisingly, that experiment is not going well for air quality and the citizens exposed to it.

• • • • •

I now look back with some worry. Susan Caplan, my BLM contact at the state level was forced to seek a transfer due to political pressures from her immediate superiors and eventually retired. I dropped out of the game and left Wyoming because I could no longer willingly watch state and county politicians operate on the principle that the environment is here for the sole purpose of turning a buck. Now I have learned from Terry Svalberg that he retired at the end of 2017. Health issues and burnout are to blame and he is doubtful the USFS will fill his vacancy.

Thus, it looks like we three may be destined to become Sublette County's "Last Air Menders."

Terms and Abbreviations

APD – Application for Permit to Drill
AWOS – Automated Weather Observing System
BACT – Best Available Control Technology
BAT – Best Available Technology
BTEX – Benzene-Toluene-Ethylbenzene-Xylene
CAA – Federal Clean Air Act
CALPUFF – Modeling computer program for simulating air pollution dispersion
CO – Carbon Monoxide
COA – Conditions of Approval
DOT – Department of Transportation
Deciview – stepwise incremental variation in visibility equal to about 10% per step
EIA – Environmental Impact Assessment
FEIS – Final Environmental Impact Statement
FACA – Federal Advisory Committee Act
FONSI – Finding of No Significant Impact

HAP – Hazardous Air Pollutants
JIDP – Jonah Infill Development Project
JIO – Jonah Interagency Office
mcf – Thousand cubic feet
MDP – Master Development Plan
NADP – National Atmospheric Deposition Program
NAAQS – National Ambient Air Quality Standard
NEPA – National Environmental Policy Act
NOx – Nitrogen oxides
PAWG – Pinedale Anticline Working Group
PSD – Prevention of Significant Deterioration
O3 – Ozone
RMP – Resource Management Plan
ROD - Record of Decision
SO2 – Sulfur Dioxide
SEIS – Supplemental Environmental Impact Statement
SIP – Site Implementation Plan
tpy – Tons per year
TSP – Technical Support Document
VOC – Volatile Organic Compound
WAAQS – Wyoming Ambient Air Quality Standard
WAQSR – Wyoming Air Quality Standards Regulation
WOGCC – Wyoming Oil and Gas Conservation Commission
WRAP – Western Regional Air Partnership
WYDEQ – Wyoming Department of Environmental Quality
WYDEQ-AQD – Air Quality Division of WYDEQ

Appendices

Appendix 1

Agency Roles and Authorities Related to Air Quality[1]

EPA

The Environmental Protection Agency (EPA) administers the Federal Clean Air Act (CAA), (42 U.S.C. 7401 et seq.) to maintain the National Ambient Air Quality Standards (NAAQS) that protect human health and to preserve the rural air quality in the region by assuring the Prevention of Significant Deterioration Class I and Class II increments for SO_2, NO_2, and $PM10$, are not exceeded. EPA has delegated this CAA authority to the States of Montana and Wyoming.

Wyoming DEQ

Wyoming regulates pollutants emitted into the air through the Wyoming Environmental Quality Act (W.S. 35-11-101 et. seq.). Wyoming is also authorized by an approved State Implementation Plan (SIP) to administer all requirements of the Prevention of Significant Deterioration (PSD) permit program under the Clean Air Act. Additionally, the approved Wyoming SIP contains a number of programs which provide for the implementation, maintenance, and enforcement of the National Ambient Air Quality Standards, including a New Source Review program for minor source permitting

---

[1] U.S. Department of the Interior, Bureau of Land Management, Wyoming State Office, Buffalo, WY: U.S. Department of the Interior, Bureau of Land Management, Buffalo Field Office, "Powder River Basin Oil and Gas Project Final EIS and Proposed Plan Amendment," Vol. 3, (WY-070-02-065), January 2003.

which requires, among other things, application of Best Available Control Technology (BACT) for all new or modified sources regardless of size or source category. Included as well are authorities for the control of particulate emissions, including fugitive particulate emissions from haul roads, access roads, or general facility boundaries. Wyoming is also delegated responsibility to operate an approved ambient air quality monitoring network for the purpose of demonstrating compliance with the National and Wyoming Ambient Air Quality Standards.

Bureau of Land Management

NEPA requires that federal agencies consider mitigation of direct and cumulative impacts during their preparation of an EIS (BLM Land Use Planning Manual 1601). Under the CAA, federal agencies are to comply with State Implementation Plans regarding the control and abatement of air pollution. Prior to approval of RMPs or Amendments to RMPs, the State Director is to submit any known inconsistencies with SIPs to the Governor of that state. If the Governor of the State recommends changes in the proposed RMP or Amendment to meet SIP requirements, the State Director shall provide the public an opportunity to comment on those recommendations. (BLM Land Use Planning Manual at Section 1610.3-2.)

Forest Service

The Forest Service administers nine (9) wilderness areas (WAs) that could be affected by direct effects associated with the project: Bridger WA; Fitzpatrick WA; North Absaroka, Absaroka-Beartooth, and Washakie WAs, next to Yellowstone NP; Teton WA; U.L. Bend WA; Cloud Peak WA; and Popo Agie WA with mandatory Class I designation. As federal land mangers, the Forest Service could act in a consultative role to stipulate that the BLM modeling results, or any future EPA or State-administered PSD refined modeling results (if justified), triggers adverse impairment status. Should the Forest Service determine impairment of WAs, then BLM, the State, and/or EPA may need to mitigate this predicted adverse air quality effect.

National Park Service

Three areas administered by the National Park Service– Yellowstone National Park, Devils Tower National Monument, and Bighorn Canyon National Recreation Area– could be affected by direct effects associated with the project. As federal land mangers, the Park Service could act in a consultative role to stipulate that the BLM modeling results, or any future EPA or State-administered PSD refined modeling results (if justified), triggers adverse impairment status. Should the Park Service determine impairment of NPS-administered Class I areas, then BLM, the State, and/or EPA may need to mitigate this predicted adverse air quality effect.

Other notes about Wyoming Department of Environmental Quality (WDEQ) roles and authorities:

— WDEQ-Air Quality Division (WYDEQ) regulates stationary sources; WDEQ cannot regulate mobile sources (e.g. drill rigs, trucks, trains, etc.)

— WDEQ-WYDEQ's role in Bureau of Land Management (BLM) and United States Forest Service (USFS) National Environmental Policy Act (NEPA) actions is technical oversight. We review air quality analysis protocols, draft environmental impact statements (EISs) and final EISs for technical accuracy and consistency with other NEPA documents. We supply information on current ambient air and air quality related values (AQRV) conditions, WYDEQ permitted sources and current Best Available Control Technology (BACT) practices, and general information regarding our regulatory program.

— The direction to regulate specific pollutants comes mainly from the Environmental Protection Agency (EPA).

Appendix 2

FINAL DRAFT - AIR QUALITY MONITORING PLAN

SUBMITTED TO PINEDALE BLM FIELD OFFICE
THROUGH THE PINEDALE ANTICLINE WORKING GROUP

FEBRUARY 2005

• • • • •

AIR QUALITY MONITORING PLAN
PINEDALE ANTICLINE PROJECT AREA

Prepared by the Air Quality Task Group
Submitted to the Pinedale Anticline Working Group
February 18, 2005

DRAFT

1.0 Introduction

This document is the Monitoring Plan produced by the Air Quality Task Group of the Pinedale Anticline Working Group (PAWG). This Monitoring Plan is intended to initiate fulfillment of the goals and objectives of the Adaptive Environmental Management Process (Appendix C) contained in the July 2000 Bureau of Land Management Record of Decision for the Pinedale Anticline Oil and Gas Exploration and Development Project (ROD/EIS).

The ROD/EIS which authorized creation of the PAWG and its Task Groups to facilitate Adaptive Environmental Management applies only to the Pinedale Anticline Project Area (PAPA) and not to areas outside that project area. However, Appendix C of the ROD/EIS states the Monitoring Plan should contain information on existing monitoring that is within and adjacent to the Pinedale Anticline Project Area (PAPA) and the Jonah Field II Project Area (Jonah II). Therefore, this Monitoring Plan contains information on existing monitoring that is within and adjacent to the PAPA-Jonah II existing emissions sources, modeling exercises and mitigation measures within the PAPA-Jonah II, and recommendations on future monitoring within the PAPA-Jonah II. This document will be presented annually to the PAWG leadership who will use the information contained therein to advise the Bureau of Land Management (BLM) on the status of the air resource in the PAPA-Jonah II.

While the focus of the Air Quality Task Group will be related to emissions directly from the PAPA-Jonah II, we need to acknowledge that many of the emissions measured at these areas are actually from regional sources in SW Wyoming, other western states and even other countries. The Western Regional Air Partnership (WRAP) is currently working towards developing a source attribution assessment to evaluate the percentage of permitted emissions from these other sources that may impact each Class I Area in the western US. The task group shall consider using the information from the assessment (once it is available) to evaluate the significance of local emissions vs. regional emissions related to potential cumulative impacts to adjacent Class I Areas.

The February 18, 2005 Monitoring Plan should be considered an initial plan that contains the information the Air Quality Task Group was able to pull together in the relatively short timeframe (four months) permitted by the PAWG. The Report is not final, and should be considered as a work in progress which will be modified after the Air Task Group can perform a more comprehensive analysis of monitoring needs

1.1 Air Quality Task Group

The primary function of the Air Quality Task Group (task group) is to prepare and oversee the implementation of the Monitoring Plan, review the monitoring data annually, submit a monitoring report and make additional recommendations or changes to the Monitoring Plan as necessary.

The task group is made up of representatives of Federal, State and Sublette County governments, environmental protection groups, the public, and the oil and gas industry whose membership in the Task Group was approved by the PAWG. . Technical advisors have also participated in meetings and have included employees of the USDA Forest Service, the BLM and the Environmental Protection Agency and oil and gas operators. There have been five meetings of the task group. All meetings have been open to the public; during the meetings the task group reserved time for input or questions from anyone who is not a member of the task group. Meeting minutes were taken and are available at the Pinedale BLM office and online at the Pinedale BLM Office link from the BLM website at Future task group meeting will take place semi-annually, at a minimum, and may take place more often if needed.

A consensus process was used to develop the plan and will also be used during the plan implementation stage. The consensus guidelines can be found in Part C of the Appendix.

1.2 Background

The Pinedale Anticline Project Area (PAPA) is shown on Map 1 (use Figure 1, PAPA ROD) and covers approximately 197,345 acres of Federal, State, and private land, of which 80% is BLM managed. The PAPA extends from west of the town of Pinedale Wyoming approximately 25 miles to the south-southeast of the town of Pinedale and is generally bounded on the east by U.S. Highway 191. The Jonah II Project Area covers approximately 59,600 acres of Federal, State and Private land, of which 95% is BLM managed. The Jonah II sits at the southwestern end of the PAPA, and extends approximately 8 miles from north to south and approximately 15 miles from east to west.

The Pinedale Anticline ROD/EIS approved exploration and development of the PAPA for conventional oil and gas production. Drilling of wells and production and transportation of oil and gas necessitates equipment that emits air pollutants. These activities raise the question of whether air emissions from these projects will significantly affect the local or distant air resource and whether those effects can be measured, monitored and mitigated. Additionally, the proximity of the Jonah II necessitates examining the possible effects of the two project areas together, as specified in the ROD/EIS. The emissions in question are "criteria pollutants" (e.g. nitrogen dioxide, particulate matter, sulfur dioxides), volatile organic compounds (VOCs) and hazardous air pollutants (HAPs). The long term, air resource impacts in question are local visibility and local health impacts as well as air quality related values (e.g. visibility, acid deposition) in the Bridger-Teton Wilderness "Class I" area. The BLM has identified levels of impact that they consider "significant" when reviewing effects in potentially impacted areas, these criteria can be found in Appendix B.

There are several air quality monitoring stations and processes (atmospheric models) in place both in and distant from the project areas. An additional air quality monitoring station has been sited in the PAPA and will be operational later in 2005. Air quality models for southwest Wyoming have been in place since 1997, when the Southwest Wyoming Technical Air Forum (SWWYTAF) modeled sources in southwestern Wyoming, eastern Idaho, northeastern Utah and northwestern Colorado. SWWYTAF used the CALPUFF modeling package to evaluate the degree of degradation to air quality, visibility, and other AQRVs in the Bridger and Fitzpatrick Wilderness areas from anthropogenic and biogenic sources. The model and inventory were not maintained due to technical advances, however, the 1995 MM5 data developed for the project is still used extensively for regulatory and NEPA modeling purposes. These models have evaluated air quality on a cumulative basis for the entire southwestern region of Wyoming. (Map #2).

Air emission mitigation measures within the PAPA-Jonah II are extensive and include those mandated by BLM in RODs/DRs, the Wyoming Department of Environmental Quality and the Environmental Protection Agency as well as numerous voluntary measures implemented by oil and gas producers in the PAPA-Jonah II.

1.3 Annual Air Quality Monitoring Report Objectives

The Task Group is recommending (see section 6.4) the initiation of an Annual Air Quality Monitoring Report in order to characterize the annual air quality, effectiveness of ongoing mitigation and regulation, and to provide an coordinated means of managing the monitoring strategies in response to the field development levels and industry practices. Possible objectives of the annual report are:

- *Characterize current air quality status of the project area*
- *Summarize data collected during the reporting period*
- *Evaluate mitigation measures annually*
- *Capture recommendations from other air quality experts on monitors and networks*
- *Recommend long term monitoring strategies*
- *Recommend long term mitigation strategies*
- *Report of development activities (example: # rigs, # wells drilled, etc)*
- *Characterize new mitigation measures implemented (new regulations, gathering system installed, etc)*

2.0 Existing Air Quality Monitoring

Several air quality monitors are operating within the BLM Pinedale Field Office Area, including the PAPA-Jonah II, and in the nearby Bridger-Teton Forest. For the purposes of this monitoring plan the term "Air Quality Monitoring" will be defined as collecting data utilizing established protocols for the purposes of monitoring ambient air pollutant concentrations or air quality related value parameters. This can be either a site such as a lake or other sensitive receptor, or a fixed piece of equipment designed to measure a certain air quality related parameter. Table 1 shows the different air quality monitoring currently conducted in the area and the values and parameters which are measured. Data summaries from existing air quality monitors will be provided in the annual Monitoring Report as available.

2.1 Regulated Pollutant Monitoring

Regulated pollutants are those that have ambient air quality standards associated with them or are regulated through the Wyoming Air Quality Standards and Regulations. Presently there are three regulated pollutant monitors operating or planned in or adjacent to the PAPA-Jonah II.

The "Jonah" station is located approximately ten (10) miles west of Highway 191 on EnCana leasehold. The station began operation in October of 2004 and includes a nitrogen oxides (NO_x) monitor, an ozone monitor, a continuous PM_{10} monitor, a meteorology station. The station is currently funded by EnCana. EnCana will continue to fund the monitoring until November 2005 and then the Wyoming Department of Environmental Quality – Air Quality Division (WDEQ-WYDEQ) will take over the equipment and operation. The objective of this station is to monitor criteria pollutants downwind of the Jonah Field.

The "Boulder" station is located approximately four (4) miles southwest of the town of Boulder west of Highway 191 along Paradise Road. Equipment for the station has been installed and full operation is expected to begin in January of 2005. The station will have a NO_x monitor, and ozone monitor, a continuous PM_{10} monitor and a meteorology station. The station is being funded cooperatively by Shell Rocky Mountain Production and WDEQ-WYDEQ for three years. The objective of this station is to monitor criteria pollutants and optical parameters downwind and crosswind of the PAPA.

The "Daniel" station will be located approximately 5 miles south of the town of Daniel. The WDEQ-WYDEQ will be funding this station and a request for proposals has been issued. The station will have a NO_x monitor, and ozone monitor, a continuous PM_{10} monitor and a meteorology station. The objective of this station is to monitor criteria pollutants upwind of the PAPA-Jonah II oil and gas development.

2.2 Atmospheric Deposition/Concentration Monitoring

2.2.1 Long-term Lake Sampling

The Bridger-Teton/Shoshone Forests have 5 long-term lake monitoring sites that measure lake chemistry 3 times a year (spring, mid-summer and late fall). The lakes sampled include Hobbs, Black Joe and Deep lakes in the Bridger-Teton National Forest, and Ross and Lower Saddlebag lakes in the Shoshone National Forest. This sampling is done to provide baseline data for the lake chemistry and also to determine if there are any changes in the lake chemistry occurring.

This sampling involves collection of macroinvertibrates at the inlets and outlets of the lakes and zooplankton samples collected from the deepest part of the lake on the mid-summer trip.

In addition, Upper Frozen Lake, a very sensitive lake with an acid neutralizing capacity less than 25 ueq/l, is being sampled on an annual basis in the Bridger Wilderness. Because of the remoteness of this lake, and hazards associated with access to the lake it is only sampled during the mid-summer trip. It is believed that this very sensitive lake will be the first to show any changes from acid deposition in the wilderness.

Data for the long-term lakes is currently being analyzed for trends. Data for the long-term lakes sampled by the Forest Service are available at the Bridger-Teton National Forest, Pinedale Ranger District.

By the Rasping in My Lungs

2.2.2 Synoptic Sampling

Synoptic sampling is an ongoing process to help provide background data on lake chemistry across the forest outside of the project area, and to identify lakes which are very sensitive to any changes in chemistry due to atmospheric deposition. The synoptic sampling has mostly been part of larger scale operations such as the 1985 USGS/EPA Western Lakes Sampling, the 1983-87 Shoshone NF Sampling and the 1997 EPA Lakes Survey in the Wind River Mountains, though additional samples are collected when time allows. This data will be used to identify very sensitive lakes, and to help prioritize future long-term sampling needs.

Data for the synoptic lakes sampled by the Forest Service are available at the Pinedale Ranger District.
Graphics:1

2.2.3 National Atmospheric Deposition Program

The NADP program measures the chemical deposition of precipitation samples (rain and snow) at over 200 sites around the US, Virgin Islands and Porto Rico on a weekly basis. The Bridger-Teton and Shoshone Forests operate two sites Gypsum Creek site is located approximately 26 miles NNE of Pinedale, near where the Green River exits the Forest. The South Pass site is located approximately one-half mile north of Highway 28 at South Pass. An additional site (Pinedale), is located approximately 4 miles northeast of Pinedale, between Fremont and Half Moon Lake and is sampled by the BLM. The NADP protocols require sample collection every Tuesday.

Data and summary information for these NADP sites can be found at http://nadp.sws.uiuc.edu.

2.2.4 Clean Air Status and Trends Network

In 1986, EPA established the National Dry Deposition Network (NDDN) to obtain field data on rural deposition patterns and trends at different locations throughout the United States. The network consisted of 50 monitoring sites that derived dry deposition based on measured air pollutant concentrations and modeled dry deposition velocities estimated from meteorology, land use, and site characteristic data. In 1990, amendments to the Clean Air Act necessitated a long-term, national program to monitor the status and trends of air pollutant emissions, ambient air quality, and pollutant deposition. In response, EPA in cooperation with the National Oceanic Atmospheric Administration (NOAA), created the Clean Air Status and Trends Network (CASTNet) from NDDN.

CASTNet provides atmospheric data on the dry deposition component of total acid deposition, ground-level ozone and other forms of atmospheric pollution. CASTNet is considered the nation's primary source for atmospheric data to estimate dry acidic deposition and to provide data on rural ozone levels. Used in conjunction with other national monitoring networks, CASTNet can help determine the effectiveness of national emission control programs. Established in 1987, CASTNet now comprises over 70 monitoring stations across the United States. The longest data records are primarily at eastern sites. EPA's Office of Air and Radiation operates a majority of the monitoring stations; however, the National Park Service operates approximately 30 stations in cooperation with EPA. In Wyoming, there are 3 CASTNET sites one in Yellowstone National Park, one at Centennial and one at Pinedale. The Pinedale site is collocated with the Pinedale NADP site, approximately 4 miles northeast of Pinedale between Fremont and Half Moon Lakes.

Each CASTNET dry deposition station measures:

- Weekly average atmospheric concentrations of sulfate, nitrate, ammonium, sulfur dioxide, and nitric acid.
- Hourly concentrations of ambient ozone levels.
- Meteorological conditions required for calculating dry deposition rates.

Dry deposition rates are calculated using atmospheric concentrations, meteorological data, and information on land use, vegetation, and surface conditions. CASTNet complements the database compiled by NADP. Because of the interdependence of wet and dry deposition, NADP wet deposition data are collected at all CASTNet sites.

Together, these two long-term databases provide the necessary data to estimate trends and spatial patterns in total atmospheric deposition.

Data and summary information for the Pinedale CASTNET site can be found at: http://www.epa.gov/castnet/

2.2.5 Wyoming Air Resources Monitoring System

The **W**yoming **A**ir **R**esources **M**onitoring **S**ystem (WARMS) has measured concentrations of nitric acid, particulate nitrate, total nitrate, particulate ammonium, sulfur dioxide and sulfate in Wyoming since 1999. There are four WARMS stations in Wyoming: Buffalo, Sheridan, Newcastle and a station located near Fremont Lake, east of Pinedale.

WARMS data from the network start-up period from 1999 and 2000 may be unreliable. The years 1999 and 2000 was an experimental start-up period during which many changes were made to instruments, such as the pump and its regulation system. Therefore, the concentrations measured may be less reliable than those measured during the remaining period of record. However, there doesn't appear to be a large discrepancy between the concentrations measured during the start-up period and those measured after. BLM is considering performing a detailed QA study to determine the reliability of these early data.

Data from the WARMS network is available by contacting Susan Caplan at the Wyoming State BLM office.

2.2.6 Bulk Deposition Sampling

This program is administered by the USFS, and involves collecting precipitation (rain and snow) from 2 sites in the Bridger Wilderness (Hobbs and Black Joe Lakes) to determine the chemical composition of atmospheric deposition. These sites are sampled every 2 weeks in the summer (rain), and every 4 weeks in the winter (snow). Precipitation samples are sent to the U.S. Forest Service's Rocky Mtn. Research Station for chemical analysis and the Bridger-Teton N.F. Air Quality Specialist analyzes and summarizes the data and writes an annual report for the WY DEQ and industry. These sample sites are co-located with Long-term Lake Sampling sites to allow study of cause and effect of pollutants on aquatic systems.

Data for these Bulk Deposition sites and copies of the annual report are available at the USFS Pinedale Ranger District.

2.3 Visibility

2.3.1 Aerosol Monitoring

There is one "Interagency Monitoring of Protected Visual Environments" program (IMPROVE) site near Pinedale serviced by the USFS Pinedale Ranger District personnel. The Bridger aerosol sampling site (BRID) is located at the White Pine Ski Area approximately 12 miles northeast of Pinedale. This system collects airborne particles through a system of filters and pumps that run every third day. The filters are then removed (once a week) and sent to the University of California Davis, Crocker Nuclear Lab, for analysis. This information provides chemical analysis and size range of particulate matter present. These samples provide data on particle size and chemical makeup which is then used to reconstruct visibility conditions based on light extinction and scattering. IMPROVE protocols require samples to be changed every Tuesday.

Data from the IMPROVE network is available at:
http://vista.cira.colostate.edu/views/

2.3.2 Scene Monitoring

Scene monitoring existed in the Bridger Wilderness Area from 1986 to 1994. Scene Monitoring has been installed or is planned at the Jonah, Boulder and Daniel stations.

Archived photos from the Bridger Scene Monitor can be found at:

http://www.fsvisimages.com/

Scene monitors, which consist of digital cameras, are operating at the Jonah and Boulder stations. Although these sites have not been operating long enough to have a long-term archive, near real-time digital images can be viewed at www.wyvisnet.com along with images taken at 9:00 a.m., 12:00 p.m. and 3:00 p.m. for the previous two weeks.

2.3.3 Optical Monitoring

A nephelometer is operating at the Boulder site. The nephelometer provides accurate measurement of light scattered by aerosols and gases in the ambient air and is used to calculate visibility parameters (visual range). Visual range is calculated at the Boulder site and can be viewed at the Boulder Live Sites link at www.wyvisnet.com .

The IMPROVE monitoring site also includes a transmissometer station which consists of a light source and a receiver placed about 5.08 KM apart. The receptor measures how much of the light reaches the receptor vs. how much is scattered or absorbed by particles in the air, and is used to calculate the Standard Visual Range. The transmitter is located above Fremont Lake near the White Pine Ski area, and the receiver is collocated with the Pinedale NADP site.

3.0 Existing Emission Inventories and Tracking

Several inventory and tracking mechanisms exist within and adjacent to the PAPA-Jonah II. Emission inventories, in general, contain pollutant emission rates from sources within a specified area collected on a certain date (i.e. a snapshot in time). The emissions inventories described in this section contain emission rates (usually in tons per year) of regulated pollutants from several types of sources (e.g. point sources, mobile sources, biogenic sources). Emission tracking is similar to an emission inventory but the results are continually updated. An emission tracking report can track changes in emissions from different reporting periods.

3.1 Background

Many sources of air pollutants exist locally or emissions are transported in from a distance to act separately or combine, as determined by meteorology and may impact air quality in the Pinedale area and the nearby Wind River Mountains. A short description of emission sources in the Pinedale Anticline and Jonah Fields is outlined below. Also presented are descriptions of other local emission sources, regional emission sources, and out of state emission sources.

3.1.1   Well Site Construction:
   Removal of vegetation cover with heavy construction equipment in order to build roadways and drilling sites
   NOx and PM from internal combustion
   PM (dust) from construction activities and removal of vegetation cover

3.1.2   Well Drilling:
   Drilling gas wells and associated activities including movement of drilling rigs from location to location
   NOx from diesel combustion
   PM from combustion (PM2.5) and from movement and servicing of rigs

3.1.3   Well Completions:
   Fracturing the formation to produce gas, flaring and associated activities
   Hydrocarbons (VOC, HAP) and PM if flaring is incomplete
   NOx from fracking trucks and pumps

3.1.4   Production:
   Separation of liquids, dehydration of gas, and storage of liquids in tanks, and associated servicing
   VOCs and HAPs, amounts dependent on emission control device
   NOx, CO from heaters
   PM from road travel during daily operator visits

3.1.5   Gas Transmission:
   Compression of gas in order to move the gas down the pipeline with some dehydration and hydrocarbon liquid storage and other maintenance
   NOx and CO from natural gas fired engines / turbines, VOCs and HAPs from pipeline and flowline depressurization

3.1.6 Other Pinedale Anticline and Jonah II Field Sources:
Includes trucking of liquids from well sites to water disposal facilities and condensate sale facilities, and water evaporation facilities
NOx and PM from trucking
VOCs and HAPs (hydrocarbons/methanol) from water evaporation facilities

3.1.7 Other Local Sources
Natural (wild fires, windblown dust)
Anthropogenic (controlled burns, tourist traffic, recreation, dirt roadways, wood / coal burning fireplaces, gas stations, misc. small industry, agricultural burning, agricultural tillage)

3.1.8 SW Wyoming Sources
Natural (wild fires, windblown dust)
Anthropogenic (chemical (trona) industry, mining, power plants, I-80 corridor and other highways, oil and gas operations, household firewood burning, agricultural burning, and railroad)

3.1.9 Regional Sources:
see Draft WRAP " Attribution of Haze Report (Phase I)" 12/04 for attribution of emissions from other states to the Bridger Teton National Forest.

3.2 Amended Letter of Agreement for NO_x Tracking

The WDEQ-WYDEQ and the Bureau of Land Management (BLM) signed a *Letter of Agreement for Tracking Nitrogen Oxide Emissions* (Letter of Agreement) within the BLM Rock Springs District in June of 1997. The Letter of Agreement was a result of provisions within the Moxa Arch and Fontenelle Environmental Impact Statements to track changes in total NO_x emissions only within the airshed of the BLM Rock Springs Field District. In April of 2000, the WDEQ-WYDEQ and the BLM signed an *Amended Letter of Agreement for Tracking Nitrogen Oxide Emissions* within the Rock Springs, Pinedale and Kemmerer Field

Office Areas. The Amended Letter of Agreement called for the issuance of annual reports, beginning December 1, 2000, to track changes in emissions from existing sources and new sources for the period beginning January 1, 1996.

The Amended Letter of Agreement specifies that the WDEQ-WYDEQ will be responsible for tracking all new, abandoned and/or modified NO_x emission sources located within the BLM Rock Springs, Pinedale and Kemmerer Field Office Areas including emission rate in tons/year. This data includes permits and waivers the WDEQ-WYDEQ New Source Review Section has issued from January 1, 1996 to October 31 of each reporting year.

The Amended Letter of Agreement specifies that the BLM will maintain a database of new, abandoned and/or modified sources based upon the information provided by WDEQ-WYDEQ, including those on private, State and Federal lands. The BLM should also maintain a record of active drilling rigs and document plugged, abandoned and reclaimed oil and gas wells. The BLM also agreed to discuss the Pinedale Anticline and Jonah II Project Areas separately in the annual report, including on-the-ground calculated potential NO_x emissions. The BLM then prepares a report to be distributed to the WDEQ-WYDEQ, USEPA Region VIII, USDA-Forest Service Regions 2 and 4 and any other interested individual or party upon request.

The WDEQ-WYDEQ submitted their most recent report to BLM on December 6, 2004, which tracked potential emissions in the specified area from January 1, 1996 to October 31, 2004. The BLM submitted their last annual report on December 22, 2000. At the January 5, 2005 meeting of the PAWG, the PAWG decided to send a letter to BLM urging them to fulfill their annual obligations under the agreement.

3.3 Industry Emission Monitoring

Emissions are calculated and monitored by individual companies as required by State permitting and reporting requirements. Gathering system companies track fuel gas use and perform stack tests to determine emissions from gas compression facility engines. Compressor station emissions are modeled during the permit process to determine any air quality impacts. All major source compression facilities and most minor engine sources report emissions on an annual basis to WDEQ.

Producing companies track gas and liquid production from their individual wells. These volumes form the basis to calculate potential to emit emission levels for permitting requirements and determine when emission control devices are required to be installed at the gas processing facilities. Some operators track their gas well facility emissions in order to manage their emission control device equipment and other companies track emissions internally on a field wide basis, including greenhouse gas emissions.

3.4 WDEQ-WYDEQ Statewide Emission Inventory

The WDEQ-WYDEQ is in the process of completing a statewide emission inventory for the year 2002. The inventory is expected to be complete in mid-2005. Pollutants inventoried include particulate matter less than 2.5 m diameter ($PM_{2.5}$), particulate matter less than 10 m diameter (PM_{10}), sulfur dioxide (SO_2), nitrogen oxides (NO_x), volatile organic compounds (VOCs), carbon monoxide (CO), lead (Pb), hydrogen sulfide (H_2S), hazardous air pollutants (HAPs), and estimations of ammonia (NH_3). Inventory data are being collected by source category. Categories include stationary point sources, area sources, mobile sources, fire, and biogenic/geogenic sources.

In addition, WDEQ –WYDEQ has recently requested an inventory of NOx/VOC/HAP actual emissions for 2004 activity from companies who operate in the Pinedale Anticline and Jonah.

3.5 Pinedale Resource Management Plan Revision Inventory

Pinedale RMP: emission inventory for BLM sources within the Pinedale field office area. Estimated emissions from BLM sources within the Pinedale FO area will be presented in the Pinedale RMP.

Air Pollutant	Alternative1	Alternative 2	Alternative 3	Alternative 4
NO_x	15,036	15,925	8,344	13,709
SO_2	62	68	14	53
PM_{10}	2723	2947	1415	2467
$PM_{2.5}$	2346	2468	1405	2158

Emissions were calculated for the following activities: coal bed methane (CBM) development, coal mining, lands and realty actions, livestock grazing, off-highway vehicle (OHV) use, resource roads, saleable mineral development, vegetation management (including fire), and conventional natural gas development. Activities related to cultural resources, paleontology, recreation, transportation and access, noxious weed control, wild horses, and wildlife and fish are assumed to be minor sources of air emissions. Information provided by the Pinedale Field Office was used to generate BLM activities.

4.0 Modeling

Federal, State, Tribal and local agencies have responsibilities under the Clean Air Act to protect and enhance air quality by various air quality management tools. These tools include air quality monitoring, emissions tracking and air quality modeling. These atmospheric dispersion models estimate air quality by mathematically relating air pollutant emissions with atmospheric processes and terrain. Air quality models are required to estimate air quality that would result from the construction of proposed emission sources.

Modeling is the method of choice specified in the Clean Air Act for estimating potential concentrations from proposed PSD sources.

5.0 Existing Mitigation Measures

There are extensive mitigation measures being utilized in the project areas. These include BLM best management practices, measures that apply only to the PAPA as well as a variety of measures voluntarily executed by operatiors.

5.1 BLM Best Management Practices (Reference PAPA ROD Appendix A, Section A-2)

- Compliance with all applicable air quality laws, regulations and standards.
- Analyses of risks involved with the development of sour gas pipelines and treatment facilities. *Note:There is no sour gas production in the project areas.*

- Emission of fugitive dust shall be limited by all persons handling, transporting, or storing any material to prevent unnecessary amounts of particulate matter from becoming airborne to the extent that ambient air standards are exceeded.
- Necessary air quality permits to construct, test and operate facilities will be obtained from the WDEQ-WYDEQ.
- Operators will comply with all applicable local, state, tribal, and federal air quality laws, statutes, regulations, standards and implementation plans.
- Operators may be required to cooperate in the implementation of a supplemental coordinated air quality monitoring program or emissions control program.
- No open burning of garbage or refuse will be allowed at the well sites or other facilities.
- To avoid the incremental risk of exposure to carcinogenic toxins from producing wells, no well will be located closer than 1,320 feet from a dwelling or residence.

5.2 PAPA specific mitigation measures
(Reference PAPA July 2000 Record of Decision, Appendix A, Section A-3)

- To avoid the incremental risk of exposure to carcinogenic toxins from compressor facilities, any compressor facility located closer than four miles to a dwelling or residence will require additional NEPA analysis prior to the final selection of the site and authorization to construct.

5.3 Jonah II Specific Mitigation Measures

- Roads and well pads that prove to be susceptible to wind erosion will be appropriately surfaced or have dust inhibitors applied to reduce fugitive dust.

- Operators will establish and enforce speed limits to reduce fugitive dust concerns as well as for human health and safety reasons.
- Jonah II Project Area will be tracked as a subset of the current tracking agreement (See Section 3.2).

5.4 PAPA specific mitigation measures within the Questar leashold (Reference Questar Winter Drilling Proposal FONSI/DR)

These mitigation measures only apply to operations conducted by Questar on BLM lands.

- Drilling operations and the number of rigs utilized are greatly limited from November 15 to April 30 each year.
- The number of drilling sites (pads) utilized during the project life is restricted
- Construction of a condensate pipeline will eliminate condensate tanks and greatly reduce condensate hauling by large trucks
- Construction of a water pipeline will eliminate water tanks and greatly reduce water handling by large trucks
- Flareless completions will be utilized
- Tier II compliant drilling rig engines or alternative fuel engines will be utilized
- A WDEQ-WYDEQ inspector will be funded for 5 years
- Implement additional air quality mitigation measures such as: use of selective or non-selective catalytic reduction on compressors; increased diameter of sales pipelines; increased water or magnesium chloride applications or other treatments on all surface disturbances including resource roads and pads.
- Evaluate the necessity of installing vapor recovery systems on pump stations, storage, and processing equipment located at the Pinedale Compressor Station.

5.5 Voluntary Industry Mitigation Measures

- Some operators outside of the PAPA utilize flareless or reduced emission completions.
- Some operators are utilizing well completion procedures which greatly reduce the length of time wells have the potential to flare.
- Some operators have installed compressors which have actual, tested emissions much lower than those anticipated in the original environmental assessments and permitted under WDEQ regulations.
- Some operators install control devices on all of their production equipment (dehydration units, condensate tanks) at start-up of the facility instead of within the time frames allowed by the WDEQ regulations which result in lower VOC and HAP emissions.
- Some operators utilize accelerated or enhanced reclamation techniques that help reduced wind born dust sources.
- Some operators utilize water, including reclaimed water, as a means to control road dust and during location / road construction.
- Several operators utilize newer US EPA Tier 1 compliant diesel/electric drilling engines resulting in lower NOx and particulate emissions.
- Some operators participate in the US EPA STARS program which promotes and documents voluntary reduction of green house gas emissions.
- Most operators operate emission source related equipment (i.e.: Kimray glycol pumps) well below the permitted levels.
- One operator is utilizing crew busses to reduce vehicle traffic.
- One operator, in cooperation with local law enforcement, is attempting to restrict vehicle traffic on the north end of The Mesa Road.

5.6 Wyoming Air Quality Standards and Regulations Mitigation Requirements

Any source constructed in the Pinedale Anticline area is subject to Wyoming's Air Quality Standards and Regulations and any resulting guidance documents prepared by the WDEQ-WYDEQ. Chapter 6 of these regulations establishes permitting requirements for all sources constructed and/or operating in the State of Wyoming. This chapter includes Section 2 which covers general air quality permitting requirements for construction, modification, as well as minor source permits to operate. Section 2 also requires that any proposed facility utilize what the State of Wyoming determines to be best available control technology (BACT) during the permitting process.

Oil and gas production facilities in the Pinedale Anticline area must follow guidance specifically prepared for this industry, as well as even more restrictive requirements associated with guidance developed specifically for development in the Pinedale Anticline and Jonah Fields. This guidance sets specific emission control technology requirements on equipment designed for natural gas dehydration and condensate storage.

Compressor stations constructed and operated in the Pinedale area are also required to meet permitting and best available control technology requirements.

Completion flares are regulated by several different regulatory agencies, the WDEQ-WYDEQ; Wyoming Oil and Gas Conservation Commission; and the Bureau of Land Management. Though currently regulated, industry and the WDEQ-WYDEQ are working together on guidance to set and implement new best available control technology standards that will require that each operator obtain a flaring permit and will result in further reduce emissions from flaring operations in the Jonah and Pinedale Anticline Fields and facilitate in the tracking of flaring events.

WDEQ-WYDEQ does not presently have the regulatory authority to permit mobile or nonroad sources (i.e.: drilling rigs). The US EPA does have standards for new nonroad engines manufactured between 1996 to 2015, depending on horsepower, which progressively reduces the NOx, hydrocarbon, and PM through improved engine design over the phase-in period. Nonroad diesel fuel is also regulated by the US EPA, starting in June 2007 when the sulfur content is reduced from 5000 ppm (maximum) to 500 ppm (low sulfur diesel) and again in June 2010 to15 ppm (ultra-low sulfur diesel).

The following sections (5.3.1-5.3.4) describe control requirements mandated by WDEQ-WYDEQ for oil and gas production and transportation operation in the PAPA-Jonah II fields.

5.6.1 Wellsite Hydrocarbon Liquid Storage Tank and Pressurized Vessels

For single well site facilities, projected VOC flashing emissions of 30 tons per year (TPY) or greater must be controlled within 90-days of the First Date of Production. Projected emissions are based on initial production.

For PAD well site facilities or single well site facilities converted to PAD facilities, all flashing emissions must be controlled upon the First Date of Production for the PAD.

Flashing emission control devices or systems must reduce the mass content of VOCs in the vapors routed to the control by at least 98% by weight. Typical controls for flashing emissions are smokeless combustion chambers. Less common are vapor recovery units which required a source of electricity or natural gas-fired compressor unit to operate.

5.6.2 Wellsite Dehydration Units

For single well site facilities projected potential emissions of 15 TPY VOC/5 TPY total HAPs or greater, but less than major source levels, must be reduced to less than the 15 TPY VOC /5 TPY HAPs thresholds through limited operational practices within 40-days of the First Date of Production or must be controlled within 90-days of the First Date of Production. Projected potential emissions greater than major source levels must be controlled within 90-days of the First Date of Production. Projected potential emissions are based on initial production and operational parameters of the dehydration equipment.

For PAD well site facilities or single well site facilities converted to PAD facilities, all dehydration unit emissions, regardless of potential, must be controlled upon the First Date of Production for the PAD facility.

Dehydration unit emission control systems must reduce the mass content of VOCs and HAPs in the vapors routed to the system by at least 90% by weight for VOCs and at least 95% by weight for HAPs. The most common control method for dehydration unit emissions is a condenser/combustor system.

At PAD facilities all pilot flames associated with combustion devices must be recorded and monitored in order for continual operation to be demonstrated. The recording and monitoring equipment must be operational upon start up of the combustion device.

5.6.3 Compressor Stations and Wellsite Compression

Compressor Stations and wellsite compression are subject to Wyoming Air Quality Standards and Regulations (WAQSR) Chapter 6 Section 2 permitting requirements when constructing or modifying a compressor station/engine in the PAPA-Jonah II. Depending on the amount of proposed compression, the station may also be subject to WAQSR Chapter 6, Section 3 (Title V Permitting) and National Emission Standards for Hazardous Air Pollutants for Reciprocating Internal Combustions Engines (RICE MACT) (40CFR Part 63 subpart ZZZZ).

Typically, larger compressor engines are outfitted with either and oxidation catalyst or non-selective catalytic reduction to control either NO_x and/or CO and formaldehyde emissions. The type of control and emission rate is determined on a case-by-case basis during the permitting BACT review.

Typically, a large compressor station will have a dehydration unit onsite. Controls on reboiler still vent and flash tank vent emissions can consist of a vapor recovery device (BTEX controller) or an enclosed combustion device. Both devices must be designed to reduce the mass of VOCs and HAPs vented to the device by 95%.

6.0 Air Quality Monitoring Funding

This section will discuss the current funding for ongoing and proposed air quality related monitoring. Status of current funding for existing and proposed air quality monitoring is presented in the table found in Table 2 . Table 2 also estimates expected shortfalls in funding of the monitoring shown for the next 5 years.

6.1 National Atmospheric Deposition Program (NADP)

This consists of 3 NADP sample sites, at Pinedale, Gypsum Creek and South Pass. In the past, the Gypsum Creek sampling has been financed by Exxon-Mobil, the South Pass site has been financed by Simplot Phosphates. As of 12/31/04, WY DEQ removed the conditions from the permits for ExxonMobil and Simplot Phosphates which required them to fund these 2 sites leaving these monitoring sites unfunded. **The FS with assistance from WY DEQ was able to come up with funding to keep these sites in operation until 9/30/05, but need assistance to secure a long-term funding source to maintain these sites.** The Pinedale site is financed by the BLM, and will continue to be funded by them in the future.

6.2 Bulk Deposition Sampling

This consists of two sites in the Bridger Wilderness where precipitation samples (rain and snow) are collected and analyzed for chemistry to determine elements being deposited in the wilderness. In the past the bulk deposition sampling program has been funded by ExxonMobil and Simplot Phosphates as part of their permits to operate as granted by the WY DEQ. The permit conditions for these companies were modified, so they will not be required to fund monitoring at these sites after 12/31/04. **The Forest Service has established short term funding to keep this monitoring operational through 9/30/05, but they need assistance find a long-term funding source to maintain these monitoring sites.**

6.3 Long-term Lake Monitoring

The USFS Bridger-Teton and Shoshone NFs have historically funded the long-term lake monitoring program for the Wind River Range. The Forest Service will continue to fund this monitoring in the future.

6.4 Interagency Monitoring of Protected Visual Environments (IMPROVE)

The IMPROVE monitoring consists of two types of monitors, aerosol, and optical (transmissometer). In the past, the USFS has funded the operation of these monitors from the Washington Office (WO). However, due to declining budgets across the country, the WO has made the decision to no longer fund the optical monitoring past 9/30/05. They will continue to fund the aerosol portion of the program into the future. **The FS needs assistance to acquire long-term funding to maintain the transmissometer instrument at Pinedale.**

6.5 Wyoming Air Resources Monitoring System (WARMS)

The BLM has purchased equipment and maintained the WARMS site in Pinedale. It is expected that they will continue to fund and maintain this site.

6.6 Clean Air Status and Trends Network (CASTNet)

The Pinedale CASTNet site has historically been financed by the EPA. It is expected that they will continue to fund and maintain this site in the future.

6.7 Ambient Air Monitors

This includes the ambient air monitors which are, or will be installed in the Jonah field, the Boulder area and the Daniel area. These will be discussed separately since each has a different funding source.

The monitor in the Jonah Field was purchased by Encana. Encana will provide funding for the first year of operation, and then it will be the obligation of the State to provide funding. As a part of the agreement between the State and Encana, the monitor will operate in the original location for 3 years, at which time it can be re-located. **Beyond the 3 year period, no funding has been identified.**

The Boulder monitor was jointly purchased by WY DEQ and Shell. **They will share the cost of operation for the 1st 3 years, with no funding for operation identified beyond that point.**

The Daniel monitoring site is not yet operational. The WY DEQ has committed to the purchase of equipment for this site, and the costs for operation for 3 years. **No funding has been identified beyond the first 3 years.**

7.0 Future Work/Data Gaps

The following is a listing of work the Air Quality Working Group see as important to be done to adequately assess the effectiveness of current monitoring and mitigation for the Jonah-Pinedale area, and to provide adequate feedback to the PAWG.

- Assess the current conditions of air quality (NOx, PM10 and PM2.5)
- Review NOx tracking report, and provide a quantifiable assessment .
- Assess the effectiveness of current monitors. Is there a need for more and different types (HAPs, VOC, PM2.5, SO2), different locations?
- Research monitors that are available (need technical input and advice).
- Analyze existing data to determine if there are "significant" trends appearing (specifically long-term lake, IMPROVE and NADP data).
- Examine commonalities and overlap with other working groups.
- Develop a format and outline for the annual air quality report.
- Develop suggestions for equitable contributions of funding from operators, government entities and others for air quality monitoring and monitoring equipment. (develop a formula)

By the Rasping in My Lungs

8.0 Monitoring/Mitigation Recommendations

8.1 USFS AQRV Monitoring continuation

The 1977 Amended Clean Air Act establishes several Mandatory Class 1 Airsheds throughout the country, and required Federal Land Managers to protect them from degradation. Air Quality Related Values (AQRV's) were established for each of these Class 1 areas, and include things which are affected by air pollution, like visibility, and the health and diversity of aquatic and terrestrial ecosystems. The U.S. Forest Service has monitored Air Quality Related Values,(AQRV's), in the Bridger, Fitzpatrick and Popo Agie Wilderness Areas since 1984. This AQRV monitoring was initiated to establish a baseline of data against which future trends and impacts could be measured. With the greatly increased pace of energy development in the Upper Green River Valley this data, and it's continued collection, will be of enormous value in tracking changes to the Class 1 Air Resource, and in the validation of modeling efforts. **We recommend that the PAWG request BLM to provide financial and administrative support for ongoing U.S. Forest Service AQRV monitoring.**

8.2 Bridger Scene Monitoring reinitiation

Scene Monitoring was established in 1986 to measure visibility changes and trends in the Bridger Wilderness Area. A camera was located on top of Fortification Mountain and took pictures of the Mount Bonneville Area three times a day from September 22, 1986 through April 17, 1994. Approximately 7,500 35mm color slides were reviewed to develop a historical photographic archive. In 1994 it was determined that a sufficient archive of images had been produced, and due to budgetary and other constraints the site was dismantled and decommissioned. Since 1994 there has been considerable new development in the area west of the Bridger Wilderness. This new development has greatly increased the amount of both particulate and NOx, as well as other emissions which affect visibility. Although the new monitoring sites at Jonah, Boulder, and Daniel will have combinations of Web Cams and Nephelometers, they will not represent visibility conditions in the Bridger Wilderness. The availability of historical data and the relatively low cost of operating the Bridger

Camera site make it a good option for inclusion in the recommended Air Quality Monitoring Plan. Recent concerns about Visibility Impairment to the Bridger Wilderness warrant the reinitiation of Scene Monitoring at the Fortification Mountain Site. **We recommend that the PAWG encourage BLM to support the U.S. Forest Service in reestablishing and maintaining Scene monitoring at the Bridger Camera Site.**

8.3 BLM fulfillment of Amended Letter of Agreement for NO_x Tracking

At the January 5, 2005 PAWG meeting the Air Quality Working Group asked the PAWG to write a letter to the BLM requesting that they reinitiate the NOx Tracking as agreed to in the Amended Letter of Agreement for NOx Tracking. The Air Quality Working Group felt this information is vital for us to be able to move forward to determine effectiveness of mitigation, and to suggest gaps in monitoring. The PAWG has sent this letter to the BLM, but we have not received a reply to that letter. BLM Staff have indicated that they have been told by Management to conduct the NOx tracking, though no timelines or schedules for completion of this project are available. **We recommend that the PAWG continue to press the BLM for a timely completion of the NOx Tracking so the Air Quality Work Group can continue to move forward and evaluate current conditions.**

8.4 Initiation of an Annual Air Quality Annual Report submitted to PAWG

The Air Quality Working Group would like to present to the PAWG an annual report which will summarize the most recent quality assured data from air quality related monitoring. Though a format has not yet been established, we expect at a minimum that the report would include the most recent data, and analysis of that data with historical data (trends). We also expect this report to identify any gaps in the monitoring and recommendations on additional needs for air quality monitoring and data analysis. **We recommend that the PAWG establish a timeline for completion of these reports, so work of the various work groups can be coordinated. Please determine if these reports will be due on an annual or fiscal (preferred) basis, as** well as recommendations on what else the PAWG would like to see in the annual report.

8.5 PM$_{2.5}$ and SO2 monitoring

Existing air quality monitoring lacks data regarding potential impacts to public health related to the combustion of hydrocarbon based fuels. **We recommend a PM$_{2.5}$ monitor be setup in Pinedale to monitor the air quality.** *Although the natural gas does not contain hydrogen sulfide, no monitoring data exists in field for SO$_2$.*

8.6 Initiation of an Activity Tracking Program by BLM

The Air Quality Workgroup would like to request that the PAWG request that the BLM provide and accurate up to date (monthly) summary of activities occurring within the gas fields. This should include the number of drill rigs operation, amount of compression being used, number of wells flared, approximate traffic counts etc. The Air Quality Work group feels this level of Activity Tracking is essential if the PAWG is going to try to integrate work from the various work groups. Without a report as described above, work groups will be using different numbers for their analysis, which will potentially result in inconsistent conclusions. The group feels this level of standardization is needed to integrate and support finding of the various work groups. BLM coordination with the work groups should insure all needed items for this Activity Tracking Program will be included. **We recommend that the PAWG ask the BLM to coordinate with the work groups to develop an Activity Tracking program for the Jonah and Pinedale Anticline gas fields.**

Appendix A

Agency Roles and Authorities Related to Air Quality

EPA

The Environmental Protection Agency (EPA) administers the Federal Clean Air Act (CAA), (42 U.S.C. 7401 et seq.) to maintain the National Ambient Air Quality Standards (NAAQS) that protect human health and to preserve the rural air quality in the region by assuring the Prevention of Significant Deterioration Class I and Class II increments for SO_2, NO_2, and $PM10$, are not exceeded. EPA has delegated this CAA authority to the States of Montana and Wyoming.

Wyoming DEQ

Wyoming regulates pollutants emitted into the air through the Wyoming Environmental Quality Act (W.S. 35-11-101 et. seq.). Wyoming is also authorized by an approved State Implementation Plan (SIP) to administer all requirements of the Prevention of Significant Deterioration (PSD) permit program under the Clean Air Act. Additionally, the approved Wyoming SIP contains a number of programs which provide for the implementation, maintenance, and enforcement of the National Ambient Air Quality Standards, including a New Source Review program for minor source permitting which requires, among other things, application of Best Available Control Technology (BACT) for all new or modified sources regardless of size or source category. Included as well are authorities for the control of particulate emissions, including fugitive particulate emissions from haul roads, access roads, or general facility boundaries. Wyoming is also delegated responsibility to operate an approved ambient air quality monitoring network for the purpose of demonstrating compliance with the National and Wyoming Ambient Air Quality Standards.

Bureau of Land Management

NEPA .requires that federal agencies consider mitigation of direct and cumulative impacts during their preparation of an EIS (BLM Land Use Planning Manual 1601). Under the CAA, federal agencies are to comply with State Implementation Plans regarding the control and abatement of air pollution. Prior to approval of RMPs or Amendments to RMPs, the State Director is to submit any known inconsistencies with SIPs to the Governor of that state. If the Governor of the State recommends changes in the proposed RMP or Amendment to meet SIP requirements, the State Director shall provide the public an opportunity to comment on those recommendations. (BLM Land Use Planning Manual at Section 1610.3-2.)

Forest Service

The Forest Service administers nine (9) wilderness areas (WAs) that could be affected by direct effects associated with the project: Bridger WA; Fitzpatrick WA; North Absaroka, Absaroka-Beartooth, and Washakie WAs, next to Yellowstone NP; Teton WA; U.L. Bend WA; Cloud Peak WA; and Popo Agie WA with mandatory Class I designation. As federal land mangers, the Forest Service could act in a consultative role to stipulate that the BLM modeling results, or any future EPA or State-administered PSD refined modeling results (if justified), triggers adverse impairment status. Should the Forest Service determine impairment of WAs, then BLM, the State, and/or EPA may need to mitigate this predicted adverse air quality effect.

National Park Service

Three areas administered by the National Park Service– Yellowstone National Park, Devils Tower National Monument, and Bighorn Canyon National

Recreation Area– could be affected by direct effects associated with the project. As federal land mangers, the Park Service could act in a consultative role to stipulate that the BLM modeling results, or any future EPA or State-administered PSD refined modeling results (if justified), triggers adverse impairment status. Should the Park Service determine impairment of NPS-administered Class I areas, then BLM, the State, and/or EPA may need to mitigate this predicted adverse air quality effect.

Appendix B

BLM Significance Criteria

- Potential impacts are considered significant if:
- Potential total near-field concentrations are greater than WAAQS or NAAQS;
- Potential total far-field concentrations are greater than WAAQS or NAAQS;
- Potential cumulative near-field concentrations are greater than PSD Class II increments;
- Potential cumulative far-field concentrations in Parks and Wilderness Areas in the region are greater than PSD Class I increments;
- Potential decrease in visibility in Parks and Wilderness Areas in the regions are greater than FLAG threshold;
- Potential decrease in ANC in sensitive lakes in the region are greater than levels of acceptable change (LAC); or
- Potential cumulative deposition total loadings are greater than USFS levels of acceptable change.

R. Perry Walker

Appendix C

Consensus Decision Making

What is consensus?

Consensus is a process for group decision-making. It is a method by which an entire group of people can come to an agreement. The input and ideas of all participants are gathered and synthesized to arrive at a final decision acceptable to all. Through consensus, we are not only working to achieve better solutions, but also to promote the growth of community and trust.

Consensus vs. voting

Voting is a means by which we choose one alternative from several. Consensus, on the other hand, is a process of synthesizing many diverse elements together.

Voting is a win or lose model, in which people are more often concerned with the numbers it takes to "win" than with the issue itself. Voting does not take into account individual feelings or needs. In essence, it is a quantitative, rather than qualitative, method of decision-making.

With consensus people can and should work through differences and reach a mutually satisfactory position. It is possible for one person's insights or strongly held beliefs to sway the whole group. No ideas are lost, each member's input is valued as part of the solution.

A group committed to consensus may utilize other forms of decision making (individual, compromise, majority rules) when appropriate; however, a group that has adopted a consensus model wil use that process for any item that brings up a lot of emotions, is something that concerns people's ethics, politics, morals or other areas where there is much investment.

What does consensus mean?

Consensus does not mean that everyone thinks that the decision made is necessarily the best one possible, or even that they are sure it will work. What it does mean is that in coming to that decision, no one felt that her/his position on the matter was misunderstood or that it wasn't given a proper hearing. Hopefully, everyone will think it is the best decision; this often happens because, when it works, collective intelligence does come up with better solutions than could individuals.

Consensus takes more time and member skill, but uses lots of resources before a decision is made, creates commitment to the decision and often facilitates creative decision. It gives everyone some experience with new processes of interaction and conflict resolution, which is basic but important skill-building. For consensus to be a positive experience, it is best if the group has 1) common values, 2) some skill in group process and conflict resolution, or a commitment to let these be facilitated, 3) commitment and responsibility to the group by its members and 4) sufficient time for everyone to participate in the process.

Forming the consensus proposals

During discussion a proposal for resolution is put forward. It is amended and modified through more discussion, or withdrawing if it seems to be a dead end. During this discussion period it is important to articulate differences clearly. It is the responsibility of those who are having trouble with a proposal to put forth alternative suggestions.

The fundamental right of consensus is for all people to be able to express themselves in their own words and of their own will. The fundamental responsibility of consensus is to assure others of their right to speak and be heard. Coercion and trade-offs are replaced with creative alternatives, and compromise with synthesis.

When a proposal seems to be well understood by everyone, and there are no new changes asked for, the facilitator(s) can ask if there are any objections or reservations to it. If there are no objections, there can be a call for consensus. If there are still no objections, then after a moment of silence you have your decision. Once consensus does appear to have been reached, it really helps to have someone repeat the decision to the group so everyone is clear on what has been decided.

Difficulties in reaching consensus

If a decision has been reached, or is on the verge of being reached that you cannot support, there are several ways to express your objection:

Non-Blocking Methods

>Non-Support ("I don't see the need for this, but I'll go along.")
>Reservations ("I think this may be a mistake but I can live with it.")
>Standing aside ("I personally can't do this, but I won't stop others from doing it.")
>Withdrawing from the group. Obviously, if many people express non-support or reservations or stand aside or leave the group, it may not be a viable decision even if no one directly blocks it. This is what is known as a "lukewarm" consensus and it is just as desirable as a lukewarm beer or a lukewarm bath. Non-blocking, but drastic.

Blocking

("I cannot support this or allow the group to support this. It is immoral." If a final decision vilates someone's fundamental moral values they are obligated to block consensus.)

If consensus is blocked and no new consensus can be reached, the group stays with whatever the previous decision was on the subject, or does nothing if that is applicable. Major philosophical or moral questions that will come up with each affinity group will have to be worked through as soon as the group forms.

Roles in a consensus meeting

There are several roles which, if filled, can help consensus decision making run smoothly. The facilitator(s) aids the group in defining decisions that need to be made, helps them through the stages of reaching an agreement, keeps the meeting moving, focuses discussion to the point at hand; makes sure everyone has the opportunity to participate, and formulates and tests to see if consensus has been reached. Facilitators help to direct the process of the meeting, not its content. They never make decisions for the group. If a facilitator feels too emotionally involved in an issue or discussion and cannot remain neutral in behavior, if not in attitude, the s/he should ask someone to take over the task of facilitation for that agenda item.

A vibes-watcher is someone besides the facilitator who watches and comments on individual and group feelings and patterns or participation. Vibes-watchers need to be especially tuned in to the sexism of group dynamics.

A recorder can take notes on the meeting, especially of decisions made and means of implementation and a time-keeper keeps things going on schedule so that each agenda item can be covered in the time allotted for it (if discussion runs over the time for an item, the group may or may not decide to contract for more time to finish up).

Even though individuals take on these roles, all participants in a meeting should be aware of and involved in the issues, process, and feelings of the group, and should share their individual expertise in helping the group run smoothly and reach a decision. This is especially true when it comes to finding compromise agreements to seemingly contradictory positions.

Appendix 3

Recommendations from the Air Quality Task Group to the PAWG, August 9, 2005

Below is a list of detailed recommendations from the Air Quality Task Group of the PAWG. We request that these recommendations be moved forward to the BLM in a timely manner. These recommendations were formulated as a result of Air Quality Task Group meetings on June 16th and July 28th, as well as an Interagency Roundtable meeting between the BLM, DEQ, EPA and FS in Cheyenne on July 21, 2005 in which Federal Land Managers and Regulators discussed the existing and planned air quality monitoring for SW WY as well as highlighting gaps in the current monitoring.

The first four recommendations relate to needs of the Air Quality Task Group for information and financing, while the remaining four are more administrative in nature.

1. Once again, we request the BLM to complete the NOx Tracking Report for 2005. The Task group feels we need this important piece of information to determine if current conditions are within those modeled in the Pinedale Anticline EIS. The Task Group requested this information in January, and we were told that the BLM was working on finalizing the report. To date we have not seen a completed report.

2. Because of high ozone events that were monitored by ambient air monitors at Jonah and Boulder monitoring stations in the first quarter of 2005, we request the BLM to initiate an annual VOC Tracking Report,

similar to the NOx Tracking Report cited above. VOC's are a pre-cursor to ozone development, so knowledge of the relative availability of VOC's in the air will allow us to better understand the process of ozone formation in this area. The State DEQ currently provides information on permitted VOC's with their portion of the NOx Tracking Report to the BLM. The BLM would just have to make small additions to cover non-permitted sources such as traffic and drill rig emissions to account for all VOC sources in the area to complete the report.

3. We request that un-obligated BLM Monitoring Funds ($50,000), be used toward an evaluation of air quality monitoring architecture analysis by independent scientists. This request was developed as a result of discussions at the July 21, 2005 meeting in Cheyenne where Federal Land Managers and Regulators discussed monitoring in the Pinedale area. It was felt by the group that there would be some value of an independent analysis of existing, and proposed monitoring, analysis of data gaps, as well as integrating under one Quality Assurance Plan all of the existing monitoring protocols. The group felt this would add credibility to the overall monitoring programs. Also, this project would include the development of a long-term exit strategy for the monitoring. The AQTG thought this proposal was warranted. This contract might be accomplished through existing BLM agreements, or possibly through coordination with the State DEQ. (Draft Statement of work is available.)

4. We request that the PAWG facilitate an industry meeting at the Ruckleshaus Institute (UW) through WYO-DEQ, and BLM for the purpose of developing open communication between operators and regulators that will develop new best management practices. This recommendation goes beyond just air quality, and may encompass all of the task groups. Many of the natural gas operators in the Pinedale Anticline (and Jonah too) area are being very innovative in finding ways to reduce emissions and increase efficiencies. While adjacent operators may be aware of what is being done, we have been told there is not a mechanism to openly discuss these things. We feel this may be a great mechanism

for open dialogue between operators, regulators, Federal Land Managers and the general public.

5. We request a more formalized response mechanism for BLM to address PAWG and TG recommendations in a clear, concise and timely manner. While the processes for making recommendations, from the TG's to the PAWG and from the PAWG to the BLM are relatively clear, it seems that there is no mechanism in place to communicate back from the BLM to the PAWG and then the TG's. For example, all of the Task Groups submitted recommendations for funding in February and April. However, no notification of how the money would be allocated was delivered to the Task Groups. This information became known to the TG's only when it was printed in the newspaper in July. It seems that there is a need for better communication.

6. We request presence of BLM decision maker at our meetings to better understand our complex subject matter, meeting dynamics, and to provide some consistent messages and updates to the BLM management. The AQTG feels that we are doing a great job in our discussions and developing our recommendations, however we feel that this information is not getting to BLM managers in a timely manner. We feel the BLM management could directly benefit from a better understanding of the complex and far reaching problems related to AQ they are facing.

7. We recommend the creation of a unified PAWG (Pinedale Area Working Group) for the BLM Pinedale Area Office. New NEPA analysis for the Jonah Infill Project proposes a group similar to the PAWG and Task Groups to oversee adaptive management in that project area. The AQTG feels the result is a fragmentation of the skill base currently focused on the PAPA. This is in part because there will be duplication of administrative processes (more meetings, more reports....). It seems to be much more logical to develop one AQTG for the Area Office with the charge of looking at the entire air quality picture across the Field Office. Incorporation of this type of adaptive management group in the Revision of the Resource Management Plan would be appropriate.

8. The Air Quality Task Group would like to stress the urgency of securing funds for existing Forest Service monitoring sites (Bulk Deposition, NADP and Transmissometer) for FY 2006 which starts October 1, 2005. If this monitoring is to continue, agreements will need to be initiated in early September to be in place on October 1. We request that the PAWG convey this urgent message to the BLM.

Appendix 4

Ozone History

Ozone history from WYDEQ Boulder monitoring site from 2005 through 2012. Horizontal dotted line indicates EPA maximum recommended at that time. Accessed July 17, 2016. http://www.epa.gov/outdoor-air-quality-data/air-data-concentration-plot

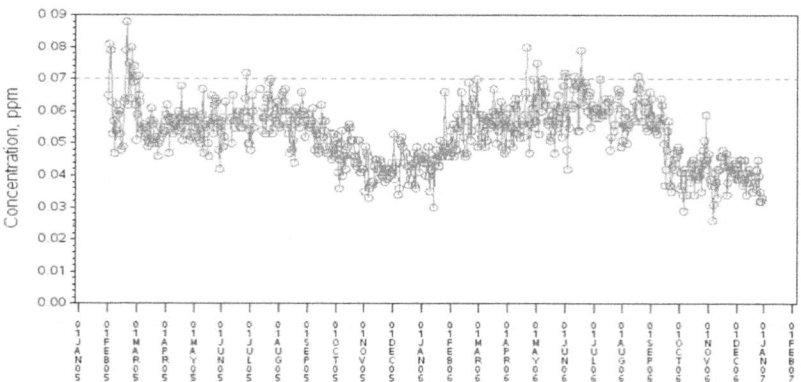

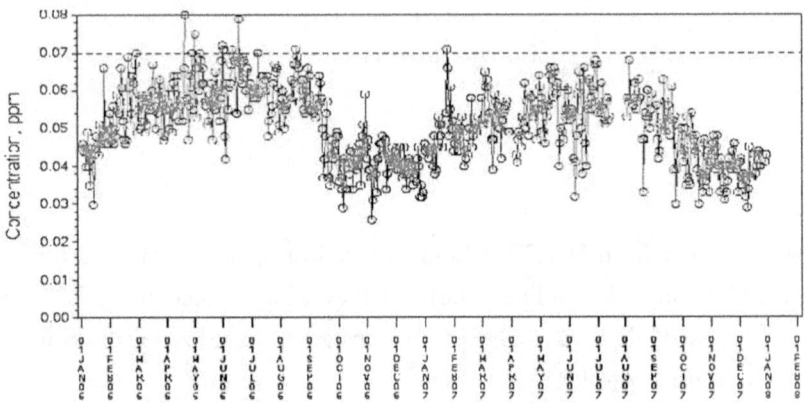

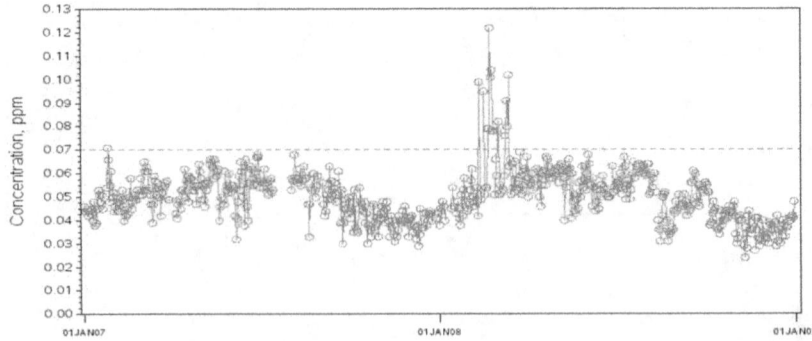

By the Rasping in My Lungs

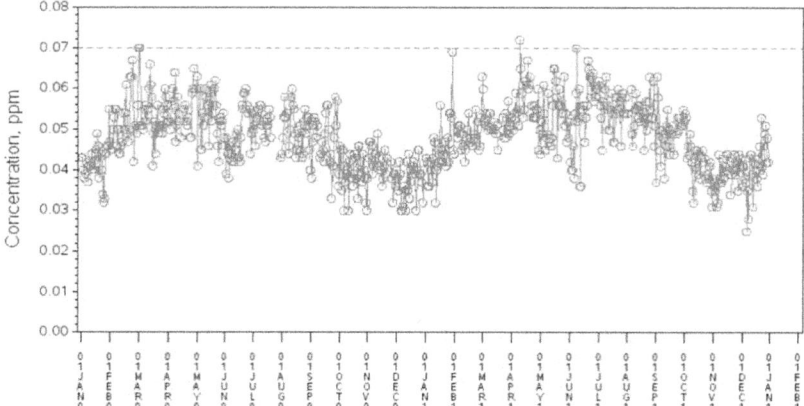

Daily Max 8-hour Ozone Concentrations from 01/01/09 to 12/31/10
Parameter: Ozone (Applicable standard is .070 ppm)
CBSA:
County: Sublette
State: Wyoming
AQS Site ID: 56-035-0099, poc 1

Source: U.S. EPA AirData <https://www.epa.gov/air-data>

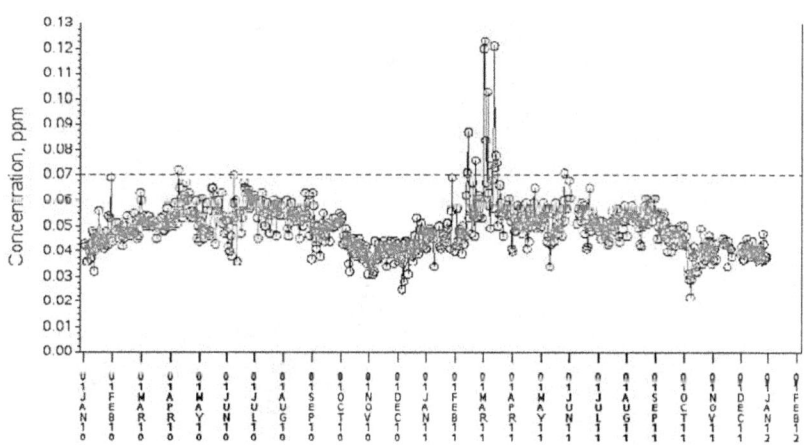

Daily Max 8-hour Ozone Concentrations from 01/01/10 to 12/31/11
Parameter: Ozone (Applicable standard is .070 ppm)
CBSA:
County: Sublette
State: Wyoming
AQS Site ID: 56-035-0099, poc 1

Source: U.S. EPA AirData <https://www.epa.gov/air-data>

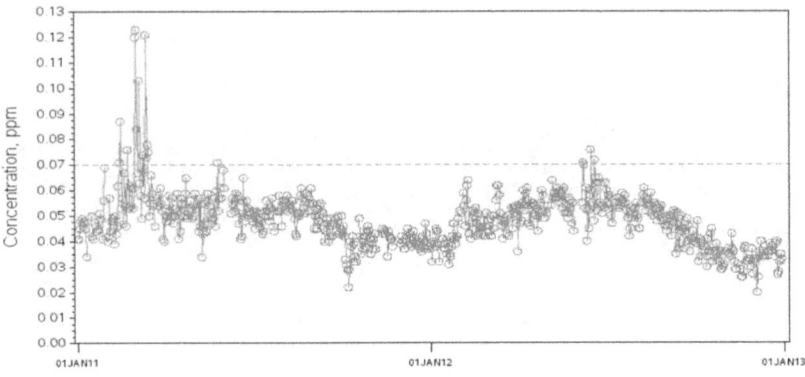

Corresponding EPA calculated ozone health threat index for each year shown in previous graphs (same EPA source website).

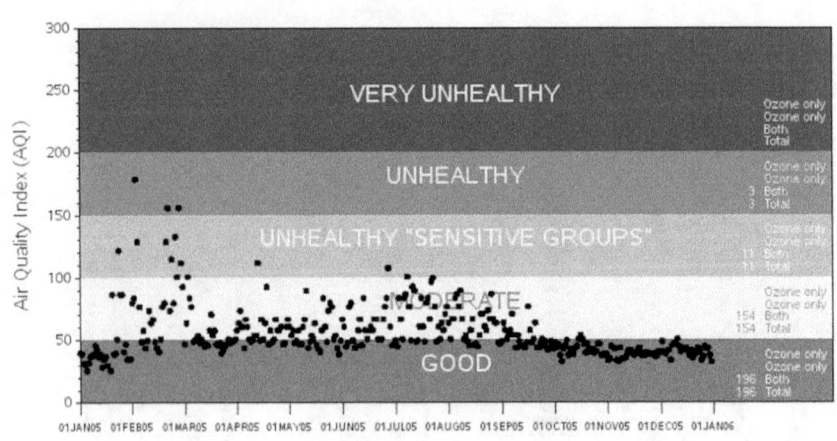

Daily Ozone and Ozone AQI Values in 2006
Sublette County, WY

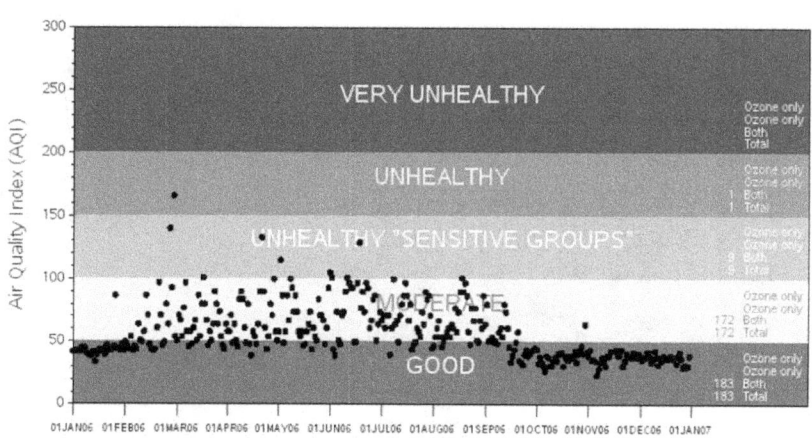

Daily Ozone and Ozone AQI Values in 2007
Sublette County, WY

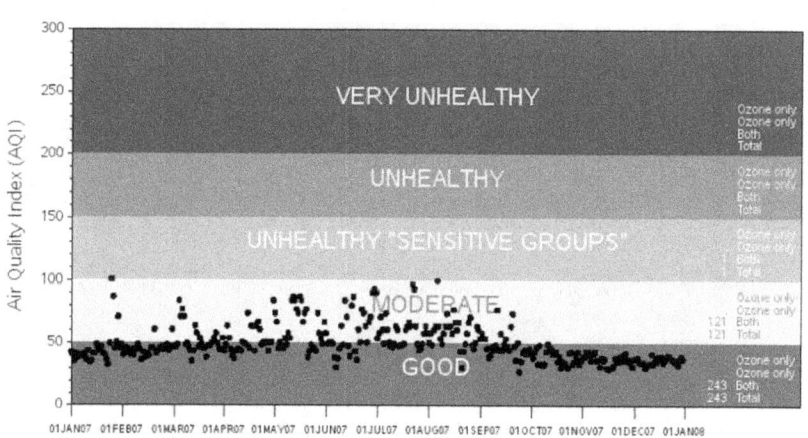

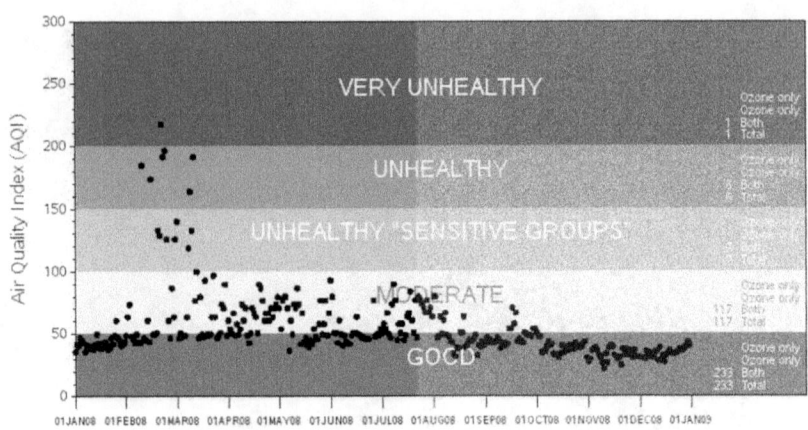

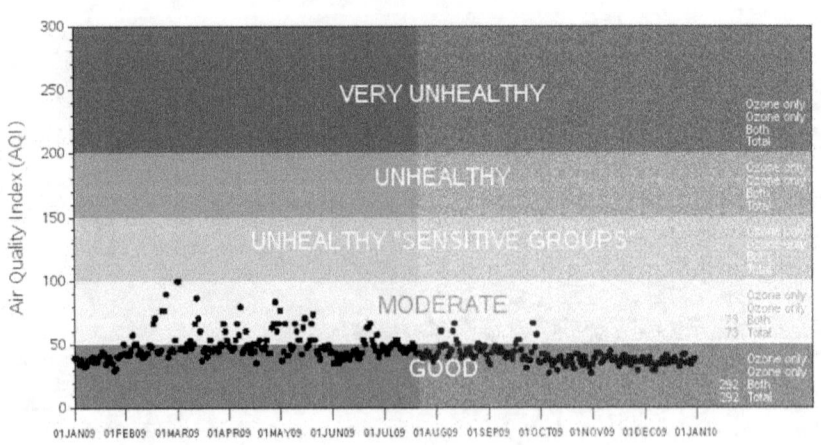

By the Rasping in My Lungs

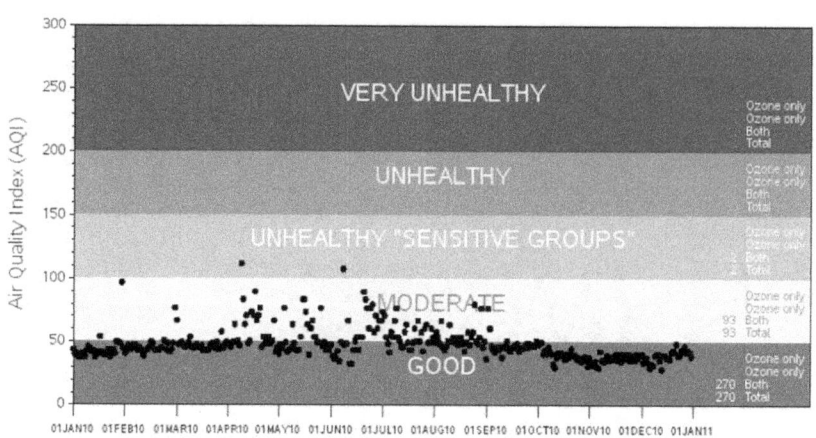

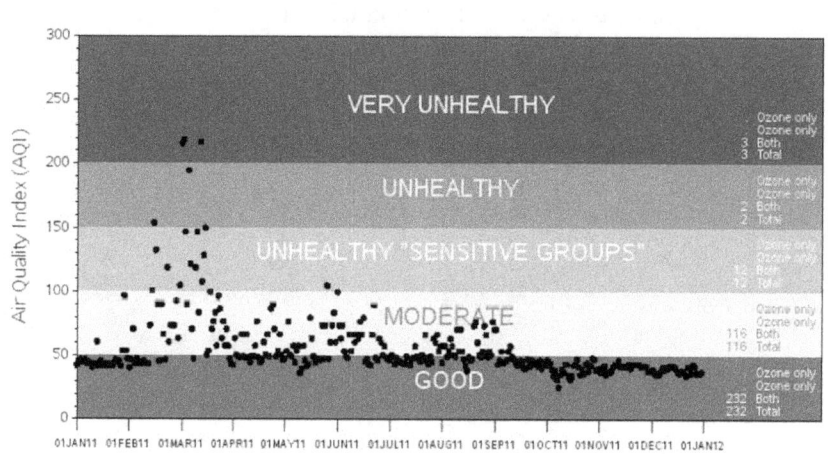

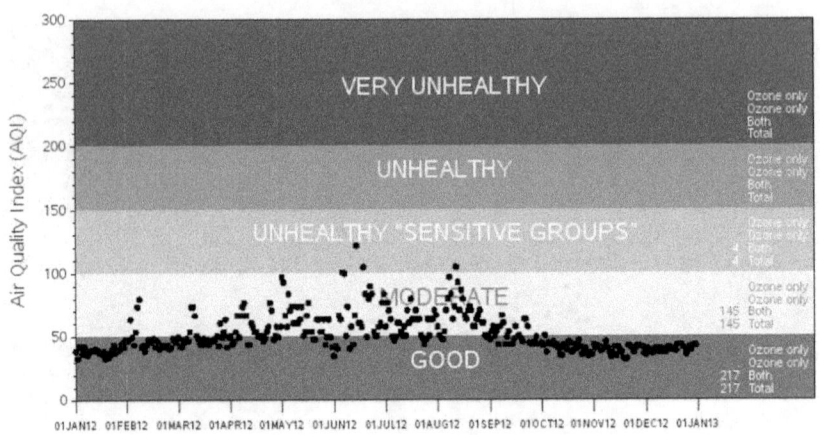

Summary of ozone history recorded by Jonah and Anticline air quality monitors.

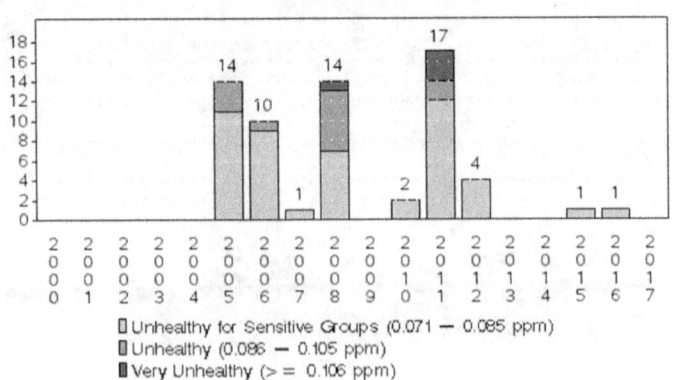

Appendix 5

Example API Well Emissions Tables

AP-4081 Permit Application Analysis – June 2006

EMISSIONS SUMMARY

Rainbow 5-31 (based on 6.0 BPD condensate, 1.3 MMCFD gas[1], 619 psig separator pressure)				
SOURCE	EMISSIONS (TPY) [2]			
	VOC	HAP	NOx	CO
500-bbl condensate tank				
UNCONTROLLED tank flash and S/W/B vapors	32.8	1.5		
CONTROLLED tank flash and S/W/B vapors Smokeless combustion device	0.7	insignificant	0.2	0.1
one 4.0 MMCFD dehydration unit w/ Kimray Model4015PV pump at maximum rate of 0.67 gpm				
POTENTIAL reboiler still vent emissions	40.1	24.9		
CONTROLLED reboiler still vent emissions reboiler overheads condenser/smokeless combustion device	1.6	1.0	0.2	0.1
one 0.500 MMBtu/hr separator heater one 0.750 MMBtu/hr reboiler heater			0.5	0.1
pneumatic equipment	11.1	1.4		
Total Uncontrolled Facility Emissions	83.9	27.8	0.5	0.1
Total Controlled Facility Emissions	13.4	2.4	0.9	0.3

HAP	Potential	Controlled
n-hexane	0.54	0.03
2,2,4-trimethylpentane	0.07	insig
benzene	4.74	0.22
toluene	12.05	0.50
e-benzene	0.54	0.02
xylenes	7.00	0.21
Total	24.9	1.0

[1] daily average production reported by [operator]
[2] rounded to the nearest 0.1 ton

Example API emissions summary table before elimination of toluene and benzene and implementation of VOC/NOX cap trade offset scheme.

AP-8838 Permit Application Analysis – June 2009

Riverside 4 PAD
175 BPD total condensate and 10 MMCFD total gas[1]
six wells: Riverside 6-4D, 5-4D, 14-4D, 11-4D, 3-9DX, and 4-9D

SOURCE	EMISSIONS (TPY)[2]			
	VOC	HAP	NOx	CO
Dehydration Unit				
UNCONTROLLED	80.7	51.8		
CONTROLLED	0.6	0.3	0.2	0.1
Pneumatic Heat Trace Pumps				
UNCONTROLLED	9.1	0.5		
CONTROLLED	0.2	Insig	0.4	0.1
Process Heaters Process Fugitives	4.0	0.2	1.3	1.1
Total Uncontrolled Facility Emissions	93.8	52.5	1.3	1.1
Total Controlled Facility Emissions	4.8	0.5	1.9	1.3

Offset Requirements

Emissions/Production	VOC (TPY)	NOx (TPY)
Equipment Design Emissions (10.0 MMCFD gas)	4.8	1.9
April 1 Emissions (1.6 MMCFD gas)	13.7	1.6
Difference	-8.9	+0.3
Offset Required	none	0.3*1.1=+0.3

[1] production rates reported by applicant, condensate routed to liquids gathering system via pipeline
[2] rounded to the nearest 0.1 ton

Example API emissions summary table after implementation of WYDEQ VOC/NOX cap and trade offset scheme. The offset directive instructed operators to reduce NOx by a ratio of 1.1:1 to achieve 1.1tpy reduction per new tpy generated; VOC reduction was by a ratio of 1.5:1 to achieve 1.5tpy reduction per new tpy. This was WYDEQ's approach to prioritizing VOC reductions over NOx in an attempt to lower ozone levels.

Appendix 6

Fracking Additives Exempt from Public Disclosure by WOGCC

Haliburton

19N Non-emulsifier (isopropanol, methanol, quaternary ammonium compound)
AS-9 Function undeclared (No information provided)
AQF-2 Foaming Agent (alkyl sufonate)
AQF-4 Foaming Agent (inner salt of alkyl amines)
BC-200 UC Crosslinker (sodium carbonate, propylene carponate, silica sand, alcohols, hexanol, 3 redactions)
BioVert NWB Viscosifier (polylactide resin)
BioVert CF Diverter/Fluid Loss Additive (fully redacted)
CAT-3 Activator (EDTA/ copper chelate family, all redacted)
Clayfix 2 Clay Control (alkylated quaternary chloride)
Cla-Web Clay Stabilizer (hydroxyalkylalkyammonium chloride family, all redacted)
Cla-Web II Clay stabilizer (Sodium chloride, redacted 7 times)
Cla-Web III Clay Stabilizer(2 quaterary amine salts, redacted 6 times))
CL-22 UC Crosslinker (potassium formate, 1 redaction)
CL-22 UCw Crosslinker (modified alkane, propylene carbonate, silica sand, propylene glycol, ethylhexanol, methanol, 4 redactions)
CL-41 Crosslinker (inorganic salt, lactic acid)
FDP-M1090-13 (methanol,quaternary ammonium chloride, homopolymer, 1 redaction)

FDP-S867A-07 Function undeclared (potassium chloride, resin)
FDP-S867B-07 Function undeclared (No information provided)
FDP-S948A-09 Function undeclared (Epoxy Resin)
FDP-S948B-09 Function undeclared (tetraethylenpentamine, polyoxypropolene diamine, polyamine polymer)
FDP-S948C-09 Function undeclared (methanol, phosphate ester)
FDP-S1021-11 Diverter/Fluid Loss Additive (fully redacted)
FDP-S1078-12 Friction Reducer (distillate, alcohols, ammonium chloride, 3 redactions)
FDP-S1176-15 Friction Reducer (petroleum distillate, sodium chloride, ammonium chloride, 4 redactions)
FDP-S1233-16 Surfactant (terpenes, terpanoids, isopropanol, hydrochloric acid, 6 redactions)
FDP-W905-08 Conformance Control (fully redacted)
FE-3A Function undeclared (No information provided)
FR-66 Friction Reducer (distillates, alcohols, sodium chloride, sorbitan, polyacrylamide copolymer)
FR-76 Friction Reducer (petroleum distillates, 2 redactions)
GBW-30 Breaker (hemicellulose enzyme, 1 redaction)
HAI-404M Corrosion Inhibitor (1-(Benzylquinolinium chloride, polyethyloxylated fatty amine salt, methanol, isopropanol, 6 redactions)
HC-2 Surfactant (sodium chloride, 1, reaction)
HPT-1 Conformance Control ("no hazardous substances," all redacted)
HPT-2 Surfactant ("no hazardous substances," redacted)
LCA-1 Solvent (paraffinic solvent)
LoSurf-300D Surfactant (naptha, naphthalene, trimethylbenzene, complex hydrocarbon, 1 redaction)
LGC-VI UC Gelling Agent (hydrotreated naptha, guar gum derivative)
Morflo II Surfactant (sulfonated hydrocarbons, 1 redaction)
MO-86 M Gelling Agent (ferric sulfate, amine, metal salt, 4 redactions)
MO-IV Breaker (fully redacted)
MUSOL A Solvent (glycol, monobutyl ether, oxylated alcohol)
OilPerm A Surfactant (trimethylbenzene, ethanol, naptha, 1 redaction)
OilPerm B Surfactant (terpenes, isopropanol, 4 redactions)

OilPerm FMM-1 Surfactant (terpenes, isopropanol, napthalene, 4 redactions)

OilPerm FMM-6 Demulsifier (ethanol, petroleum distillates, glycol, cyclohexanol, urea, pentanol, glycerine, 14 redactions)

Optiflo-HTE Breaker (silica sand, C.I. Pigment Red 5, walnut hulls, 3 redactions)

PermVis VFR-10 Friction Reducer (distillates, alcohols, ammonium chloride, 3 redactions)

RockOn MX 5-3447 Surfactant (methyl alcohol, sulfonate family, 5 redactions)

SandWedge OS Conductivity Enhancer (distillates, alcohols, 1 redaction)

SandWedge WF Conductivity Enhancer (naptha, napthalene, methanol, 2 redactions)

Scalecheck HTM Scale Inhibitor (organic acid salt, 3 redactions)

SuperFlo 2000 Surfactant (Terpenes, methanol, ammonia compound, distillates, citronella, 4 redactions)

SuperSet W Activator (methanol, ethoxylated nonylphenol)

WLC-5 Function undeclared (No information)

WLC-6 Fluid Loss Additive (redacted)

WG-39 Gelling Agent (polysaccharide)

WG_18 Gelling Agent (sodium bicarbonate, silica sand, 1 redaction)

WS-36M Emulsifier (ethoxylated fatty acids, alkyl aryl sulfonate, 3 redactions)

Schlumberger

B145 Friction Reducer

XE207 Friction Reducer

XE253 Proppant Transport Additive (synthetic organic polymer, esterfied organic polymer)

XE243 Diverting Agent (synthetic organic polymer, salt of aliphatic acid)

Sources and Endnotes

Chapter 1

1. U.S. Federal Fish and Wildlife Service, *Federally Listed, Proposed and Candidate Species, Wildlife Species: Greater Sage-grouse*. Accessed February 12, 2013. https://www.fws.gov/wyominges/Pages/Species/Species_Endangered.html`

2. H. Sawyer, R. M. Nielson, F. Lindzey, L. L. McDonald, "Winter Habitat Selection of Mule Deer Before and During Development of a Natural Gas Field," *The Journal of Wildlife Management*, 70(2) (2006): 396-403

3. D. Glick, *"End of the Road?"* Science and Nature, Smithsonian. Accessed April 8, 2013. http://www.smithsonianmag.com/sciencenature/pronghorn.html?c=y&page=1

4. Brian, Maffly, *"Deadly Roadkill in Wyoming Kills 21 Pronghorn,"* New West Environment, Feb 2, 2007. Accessed January 2013. www.newwest.net/index.php/topic/article/wyomingrecords_deadliest_roadkill_21_pronghorn_dead/C38/L38/

5. Becky Roher, "New Report Ties Roads to Fragmented Habitat," *Associated Press*, July 31, 2005.

6. Joe Baird, "Neighbor's plight is Utah's cautionary tale," *The Salt Lake City Tribune*, July 31, 2005. Accessed March 10, 2016. http://archive.sltrib.com/article.php?id=2903061&itype=NGPSID

7. A teacher in the Pinedale high school told me about these kids. She said many were living with their parents in campers and trailers that were poorly heated. This could be dangerous because fall temperatures at our high altitude would dive after sundown. Many of these "petro braceros" as I called, them had driven here from southern warmer climes and often had no clue what they were getting into.

8. Ibid, "Neighbor's plight is Utah's cautionary tale."

9. Ibid.
10. Ibid.
11. Whitney Royster, "Study: Drilling displaces grouse," *Casper Star-Tribune*, July 14, 2005.
12. Ibid.
13. Ibid.
14. Ibid.
15. Darryl Fears, "Decision not to list sage grouse as endangered is called life saver by some, death knell by others," *The Washington Post*, September 22, 2015. Accessed March 12, 2016. https://www.washingtonpost.com/news/energy-environment/wp/2015/09/22/fewer-than-500000-sage-grouse-are-left-the-obama-administration-says-they-dont-merit-federal-protection/
16. Alisa Opar,"Unprecedented Conservation Efforts Keep Greater Sage-Grouse off Endangered Species List," *Audubon, National Audubon Society*, September 22, 2015. Accessed March 12, 2016. https://www.audubon.org/news/ unprecedented- conservation-efforts-keep-greater-sage-grouse-endangered-species
17. Dawn Ballou, Personal email exchange, March 12, 2016.
18. "Anthrax Outbreak in Cattle," *Pinedale Roundup*, Pinedale Library Historical Archive, Sept 13, 1956.
19. Ron Aiken, Deer declines trigger mitigation, anger, *Pinedale Roundup*, Feb. 28, 2011.
20. Ibid.
21. "Finding the Missing Piece in the Climate Change Puzzle," *LLNL Science & Technology Review*, April 2003, pp. 4-12.
22. Dawn Ballou, "Pinedale area featured on Nightline TV Program," *Pinedale Online!*, July 25, 2004. Accessed Jan. 14 2016. http://www.pinedaleonline.com/ news/2004/07/Pinedaleareafeatured.htm
23. Wyoming Department of Environmental Quality, Air Quality Division, Letter to Operators: *"Interim Policy on Demonstration of Compliance with WAQSR Chapter 6, Section 2(c)(ii) for Sources in Sublette County,"* July 21, 2008. WYDEQ Operator Permit Archive, Cheyenne, WY.
24. U.S. Environmental Protection Agency, *What is NEPA?* EPA Region 8 website. Accessed April 3, 2014. http://www2.epa.gov/region8/national-environmental-policy-act.
25. M. Manguso, Residents React to DEQ Permit Hearing. *Pinedale Roundup* March 22, 2013.
26. This project never actually happened because EnCana sold its Jonah holdings before it ever got started.

27. Nicholas Groom, "Analysis: Thrifty truckers wary of pricey natural gas vehicles," *Reuters*, Mar. 22, 2013. Accessed January 24, 2016. http://www.reuters.com/article/us-trucks-naturalgasidUSBRE92L07620130322

28. Jim Motavalli, "Cheap Natural Gas Sells Cleaner Trucks, but not the Big Rigs Yet," *CBS Money Watch*, June 14, 2011. Accessed Jan. 24, 2016. http://www.cbsnews.com/news/cheap-natural-gas-sells-cleaner-trucks-but-not-the-big-rigs-yet/

29. Jeff St. John, "Pickens Wants Natural Gas-Fueled Big Rigs," *Greentech Media*, Jan. 21, 2009. Accessed Jan 24, 2016. https://www.greentechmedia.com/articles/read/pickens-wants-natural-gas-fueled-big-rigs-5568

30. Rhiannon Meyers, "Natural gas industry lobbyists ask feds to put LNG projects on fast track," *Fuelfix*, April 16, 2015. Accessed Jan. 12, 2016. http://fuelfix.com/blog/2015/04/16/natural-gas-industry-lobbyists-ask-feds-to-put-lng-projects-on-fast-track/#28323101=0

31. Eric Lipton, Energy Firms in Secretive Alliance with Attorneys General," *New York Times*, Dec. 6, 2014. Accessed Jan 24, 2016. http://www.nytimes.com/2014/12/07/us/politics/energy-firms-in-secretive-alliance-with-attorneys-general.html

32. Dawn Ballou, "WYDEQ Pre-Ozone Season public meeting," *Pinedale Online!*, Nov. 19, 2015.

Chapter 2

1. Kim McGuire, "EPA backs effort to cut drilling haze," *The Denver Post*, Oct. 12, 2005. Assessed on Dec 22, 2015. http://www.denverpost.com/news/ci_3107210/undefined

2. Whitney Royster, "BLM eyes tighter air restrictions," *The Casper Star Tribune*, Oct. 13, 2005.

3. Different sources cite different acreages ranging between 23,000 and 30,000.

4. Author's, letter to Director EPA Region 8 regarding EPA support for reduced rate of development, Oct 19, 2005.

5. Hall Sawyer, Ryan M. Neilson, Fred Lindzey, and Lyman L. McDonald, "Winter Habitat Selection of Mule Deer Before and During Development of a Natural Gas Field." *Journal of Wildlife Management 70(2)*, April 2006. 396-403

6. USEPA, Region 8, "Draft Environmental Impact Statement for the Pinedale Anticline Oil and Gas Exploration and Development Project, Sublette County, Wyoming CEQ#20060512," Letter from the Director EPA Region 8 to the State Director, Wyoming State Office of BLM, April 6, 2007.

7. National Oceanic and Atmospheric Administration, "New study explains wintertime ozone pollution in Utah oil and gas fields," media contacts: Monica Allen (NOAA) and

Katy Human (CIRES), Oct. 1, 2014. Accessed Dec. 22, 2015. http://www.noaanews.noaa.gov/stories2014/20141001utahwinterozonestudy.html

8. Author's petition letter to state politicians, "A Petition from the Citizens of Sublette County, Wyoming," April 8, 2007.
9. This phrase became a standard reference for us around Pinedale and referred to the brown tinted layer that could be seen hovering over the Jonah field and beyond.
10. United States Environmental Protection Agency Region 8, "Ref: 8P-AR," September 12, 2007.
11. United States Environmental Protection Agency Region 8, "Revised Draft Supplemental Environmental Impact Statement for the Pinedale Anticline Oil and Gas Exploration and Project Sublette County, Wyoming CEQ #20070542," Ref: EPR-N, February 14, 2008.
12. Ibid, 1.
13. Ibid, 2.
14. See Chapter 10, endnote 5 for explanation of the EPA tier system of emissions controls.
15. United States Environmental Protection Agency Region 8, "Revised Draft Supplemental Environmental Impact Statement for the Pinedale Anticline Oil and Gas Exploration and Project Sublette County, Wyoming CEQ #20070542," Ref: EPR-N, February 14, 2008. 2
16. Ibid.
17. Ibid, 3.
18. Ibid.
19. Ibid.
20. Ibid.
21. The ozone standard was in fact reduced to 70 ppb not long after.
22. Open Government Initiative, "Council on Environmental Quality: Open Government," Accessed Dec 30, 2015. https://www.whitehouse.gov/administration/eop/ceq/open
23. United States Environmental Protection Agency Region 8, "Revised Draft Supplemental Environmental Impact Statement for the Pinedale Anticline Oil and Gas Exploration and Project Sublette County, Wyoming CEQ #20070542," Ref: EPR-N, February 14, 2008. 5
24. Jo Becker and Barton Gellman, "Leaving no Tracks," *The Washington Post*, June 27, 2007. A01. Accessed December 30, 2015. http://voices.washingtonpost.com/cheney/chapters/leaving_no_tracks/

25. Ibid.
26. Ibid.
27. Eric Peterson, "Braving the political winds," *High Country News*, *Web Exclusive*, Aug. 20, 2008. Accessed Dec 31, 2015. http://l.hcn.org/articles/braving-the-political-winds
28. Joyel Dhieux, Personal email to author: "WYDEQ/AQD Emissions Inventory Credibility," U.S. EPA Region 8, Division of NEPA Compliance and Review, April 23, 2009.
29. U.S. Environmental Protection Agency, "Oil and Natural Gas Air Pollution Standards," Announcement of Public Hearings, Nov. 5, 2015. Accessed January 1, 2016. http://www3.epa.gov/airquality/oilandgas/
30. Kathy Fackelmann, "Milken Institute School of Public Health Names Glenn Paulson as Visiting Professor of Environmental and Occupational Health," Milken Institute School of Public Health, George Washington University, June 2, 2014. Accessed Jan 2, 2016. http://publichealth.gwu.edu/content/milken-institute-school-public-health-names-glenn-paulson-visiting-professor-environmental
31. EnCana Oil & Gas (USA) Inc. Letter to WYDEQ-AQD: "Demonstration of Emission Reductions, Emissions Offset, and Emission Credit for Sublette County Emission Sources," Dec 23, 2008. WYDEQ Operator Permit Archive, Cheyenne, WY.
32. Wyoming Department of Environmental Quality, Air Quality Division, DEQ Notice of Public Hearing: Public Hearing on EnCana 32 Well proposal, Jan 5, 2009, WYDEQ Operator Permit Archive, Cheyenne, WY.7-8
33. U.S. Environmental Protection Agency Region 8, "Ref: 8P-AR," Letter to the author, July 14, 2009.
34. Cara started out with us under the name Keslar but later became Casten as a result of marriage. Still later, she reverted to Keslar. To minimize confusion as to who I talk about, I stick with Keslar.
35. The Federal Register, "Oil and Natural Gas Sector: Emission Standards for New and Modified Sources," Environmental Protection Agency, EPA-HQ-OAR- 2010-0505; FRL-9929-75-OAR, RIN 2060-AS30, Aug. 18, 2015.
36. "EPA announces public hearings for the proposed oil and gas rules," Official email notice, Cindy Beeler, Energy Advisor, US EPA Region 8, August 27, 2015.
37. U. S. Environmental Protection Agency, "Summary of Requirements for Processes and Equipment at Natural Gas Well Sites," EPA's Air Rules for the Oil & Gas Industry. Accessed Jan 9, 2015. http://www3.epa.gov/airquality/oilandgas/pdfs/oil_well_site_summ_081815.pdf
38. Ibid, "Oil and Natural Gas Sector: Emission Standards for New and Modified Sources." 198-217.

39. Ibid, 237-251.

40. Ibid, EPA's Air Rules for the Oil & Gas Industry. 1

Chapter 3

1. T.A. Larsen, *History of Wyoming, Chapter 3 - "The Coming of the Union Pacific,"* (University of Nebraska Press, 1965), 17-38.

2. Wikipedia contributors, "Teapot Dome scandal," *Wikipedia, The Free Encyclopedia*, Accessed February 25, 2016. https://en.wikipedia.org/w/index.php?title=Teapot_Dome_scandal&oldid=704834872

3. Ibid.

4. Ibid.

5. U.S. Department of Energy, "Energy Department Sells Historic Teapot Dome Oilfield," January 30, 2016. Accessed February 25, 2016. http://energy.gov/articles/energy-department-sells-historic-teapot-dome-oilfield

6. Wikipedia contributors, "Mineral Leasing Act of 1920," *Wikipedia, The Free Encyclopedia*, Accessed February 25, 2016. https://en.wikipedia.org/w/index.php?title=Mineral_Leasing_Act_of_1920&%20%20%250D%20% 20%20oldid=704291665

7. *The Wilderness Society*, "New Information Documents Bush Administration's Land-Management Shift, Secret Policy Changes Made Oil and Gas Development the Dominant Use of Federal Public Lands, *Abuse of Trust*, May 26, 2004. Accessed Feb 7, 2016. http://www.resilience.org/stories/2004-05-23/new-information-documents-bush-administrations-land-management-shift

8. Ibid.

9. Ibid.

10. U.S. Department of The Interior, Office of the Secretary of The Interior, "New BLM Initiative to Enhance Environmental Protection During Oil and Gas Activity on Public Lands," Press News Release, June 22, 2004.

11. Ibid.

12. Ibid.

13. Alan C. Miller, Tom Hamburger, and Julie Cart, "A Changing Landscape," *Los Angeles Times*, August 25, 2004. Accessed February 14, 2016. http://articles.latimes.com/2004/aug/25/nation/na-bog25

14. Ibid.

15. Ibid.

16. Catherine Chuang, "Finding the Missing Piece in the Climate Change Puzzle," *Science*

& *Technology Review*, Lawrence Livermore National Laboratory, UCRL-52000-03-04, April 16, 2003. Accessed Feb 26, 2016. https://str.llnl.gov/str/April03/Chuang.html

17. Author's letter to Shane DeForest, Manager, Pinedale BLM Field Office, "*Subject: 2011 Compressor Station Emissions Review*," Dec. 6, 2011.

18. U.S. Department of the Interior, Bureau of Land Management, *Record of Decision Requirements and Conditions for the Pinedale Anticline Oil and Gas Exploration and Development Project and the Jonah Infill Drilling Project*, Reply to 3160(WYD01), Jan. 30, 2012.

19. Wikipedia contributors, "Kathleen Clarke (Bureau of Land Management)," *Wikipedia, The Free Encyclopedia*, Accessed February 28, 2016. https://en.wikipedia.org/w/index.php?title=Kathleen_Clarke_(Bureau_of_Land_Management)&oldid=636963256

20. Author's op-ed, "Mixed Messages, "*Casper Star Tribune*, Sept. 5, 2005.

21. John Heilprin, "Feds Cite Lapses in Oil, Gas Drilling," *Associated Press*, July 20, 2005.

22. "BLM on pace to set record for permits," *Associated Press*, July 28, 2004. Accessed Feb 10, 2016. http://billingsgazette.com/news/state-and-regional/wyoming/blm-on-pace-to-set-record-for-permits/article_802be8c9-002e-5494-98ea 6ebdee9f34e4.html

23. My own APD archive suggests that we experienced close to that if not more so I wondered how extensively the report had dug into permit statistics in our county.

24. John Heilprin, "Feds Cite Lapses in Oil, Gas Drilling," *Associated Press*, July 20, 2005.

25. Dave Alberswerth and Dave Slater, "BLM issued Record Number of Drilling Permits in 2004, But Most Went Undrilled," *The Wilderness Society*, December 16, 2004.

26. Associated Press, "BLM defends oil, gas inspections," *Billings Gazette*, July 23, 2005.

27. Associated Press, "BLM on pace to set record for permits," *Billings Gazette*, July 28, 2004. Accessed Feb 10, 2016. http://billingsgazette.com/news/state-and-regional/wyoming/blm-on-pace-to-set-record-for-permits/article_802be8c9-002e-5494-98ea-6ebdee9f34e4.html

28. Juliet Eilperin, "Interior May Delay Oil and Gas Projects; BLM Officers Can Defer Leases on U.S. Lands While Resource Plans Developed, "*Washington Post*, August 18, 2004. A17. Accessed February 14, 2016. http://www.washingtonpost.com/wp-dyn/articles/A9536-2004Aug17.html

29. Ibid.

30. Robert Gehrke, "Oil industry providing workers for BLM office," *The Salt Lake Tribune*, July 9, 2005.

31. Ibid.

32. Whitney Royster, "Some fear comments ignored," *Casper Star-Tribune*, March 27, 2005.
33. Ibid.
34. Prill Mecham, BLM Pinedale Field Manager, "BLM Committed to PAWG's Success," *Casper Star Tribune*, Sept. 5, 2005.
35. U.S. Department of The Interior, Bureau of Land Management, *Draft Jonah Infill Development Project Analysis Environmental Impact Statement (JIDPA)*, Appendix D., 2005. D-2.
36. Dawn Ballou, "BLM seeks nominations for the Pinedale Anticline Working Group (PAWG)," *Pinedale Online!*, Oct. 29, 2010.
37. Kelpie Wilson, "Good Riddance to Gale Norton," *truthout.org*, March 22 2006. Accessed November 15, 2016. http://www.alternet.org
38. Dan Berman, "Former Secretary Gale Norton won't face charges," *Politico*, December 11, 2010. Accessed May 28, 2015. http://www.politico.com/news/stories/1210/46257.html
39. Wikipedia contributors, "Gale Norton," *Wikipedia, The Free Encyclopedia*, Accessed December 16, 2016. https://en.wikipedia.org/w/index.php?title=Gale_Norton&oldid=761134331
40. Dawn Ballou, "PAWG Votes to Disband," *Pinedale Online!*, October 29, 2012.
41. Ibid.
42. Ibid.
43. Ibid.
44. Ibid.
45. Ibid.
46. Ibid.
47. Ibid.

Chapter 4

1. USDA-Forest Service, "Bridger and Fitzpatrick Wilderness Air Quality Related Values Action/Monitoring Plan," File Code 2120, September 8, 1984.
2. Ibid, 3.
3. Ibid, 4.
4. U.S Department of Energy Office of Fossil Energy Washington, D.C 20585, "Rocky Mountain States Natural Gas Resource Potential and Prerequisites to Expand Production," DOE/FE-0460, Sept 2003.

5. Ibid.

6. Ibid.

7. Author's letter to Regional Forester Harv Forsgren, Subject:"Certification of Impairment," USFS Region 2, 324 25th St., Ogden, UT 84401, March 23, 2009.

8. United States Forest Service, Intermountain Region, United States Department of Agriculture, Forsgren Response to Walker Letter Requesting Visibility Impairment Certification, File Code 2580, April 22, 2009.

9. Ibid.

10. Ibid.

11. Ibid.

12. Department of Agriculture, U.S. Forest Service, Written Comments on Revised Pinedale Anticline Project Area Oil and Gas Development Project (PAPA) Supplemental Environmental Impact Statement (SEIS) Draft Document, File Code 2580, February 7, 2008.

13. Ibid.

14. Dawn Ballou, "BTNF offers Wyoming Range gas lease parcels," *Pinedale Online!*, April 26, 2005.

15. Ibid.

16. U.S. Forest Service Bridger-Teton National Forest, Big Piney Ranger District, "Oil and Gas Leasing on Portions of the Wyoming Range in the Bridger-Teton National Forest, Draft Supplemental Environmental Impact Statement, Sublette County, Wyoming," Volume 1, "Purpose and need for Action, Leasing and Analysis History of the Project Area", Chapter 1, April 2016. 9-11.

17. Ibid, 9-12.

18. Dawn Ballou, "Hoback Gas Well Project, Notice of Staking Just the First Step of the Process," *Pinedale Online!*, September 23, 2006.

19. United States Government Federal Register, Department of Agriculture, Forest Service, Bridger-Teton National Forest Big Piney Ranger District, Wyoming, Eagle Prospect, "Notice of intent to Prepare an Environmental Impact Statement," Vol. 71, No. 7, January 11, 2006.

20. Alkali metals is a term that refers to a column in the periodic table consisting of lithium, sodium, potassium, rubidium, cesium, and francium.

21. Dawn Ballou, "Hoback gas well public meeting notes," *Pinedale Online!*, February 2, 2006.

22. Dawn Ballou, "3 Exploratory gas wells planned near Bondurant, EIS public comment deadline February 13, 2006," *Pinedale Online!*, January 23, 2006.

23. United States Government Federal Register, Department of Agriculture, Forest Service, Bridger-Teton National Forest Big Piney Ranger District, Wyoming, Supplemental Analysis to Consider Potential Field Development (Master Development Plan) Subsequent to Proposed Exploratory Drilling by Plains exploration and Production Company (PXP) Within the South Rim Unit on the Big Piney Ranger District, Supplemental Notice of Intent to Prepare an Environmental Impact Statement, Vol. 72, No. 236, Notices, December 10, 2007.

24. U.S. Department of The Interior, Bureau of Land Management, Draft Supplemental Environmental Impact Statement for the Pinedale Anticline Oil and Gas Exploration and Development Project Sublette County, Wyoming, Vol. 1, 2.3 Existing Development Within the PAPA, 2.3.3 Drilling Rigs, December 2006. 2-11.

25. Ibid.

26. Ibid.

27. U.S. Department of The Interior, Bureau of Land Management, Revised Draft Supplemental Environmental Impact Statement for the Pinedale Anticline Oil and Gas Exploration and Development Project Sublette County, Wyoming, Chapter 2, Public Participation, Existing Development and Alternatives, 2.4 Alternatives, 2.4.3 Alternative D, December 2007. 2-43.

28. Draft Supplemental Environmental Impact Statement for the Pinedale Anticline Oil and Gas Exploration and Development Project Sublette County, Wyoming, 4.9 Air Quality, 4.9.4 Cumulative Impacts, December 2006. 4-70

29. I learned of this meeting from Terry. He had been the USFS representative in the meeting and told me he had challenged the appropriateness of making the decision out of view of the public but was ignored. Attendees included WYDEQ, USFS, EPA-Denver, and BLM.

30. U.S. Department of The Interior, Bureau of Land Management, (PAPA 2006b), Draft Supplemental Environmental Impact Statement for the Pinedale Anticline Oil and Gas Exploration and Development Project, Sublette County, Wyoming, Chapter 3 Affected Environment, 3.11 Air Quality, 3.11.2 Impacts to Air Quality from Existing Wellfield Activities. Washington D.C.: Government Printing Office, December 2006; 3-62,3-63.

31. Ibid, 4.7 Visual Resources, 4.7.1 Scoping Issues, 4.7.2 Impacts Considered in the PAPA DEIS, 4-50; 4.7.3 Alternative Impacts, 4.7.3.1 Summary of Impacts Common to All Alternatives, 4-51.

32. Author's comment letter to Big Piney District Ranger Greg Clark, "Scoping Notice-Eagle prospect/Noble Basin," January 16, 2008.

33. Koshmri, Mike, "Crowds Herald Noble Deal," *Jackson Hole News and Guide*, October 10, 2012.

34. U.S. Department of Agriculture, U. S. Forest Service, News Release, *Final Supplemental Environmental Impact Statement Released for Oil & Gas Leasing in the Wyoming Range*, Federal Register, December 16, 2016.

35. Christine Peterson, "Forest Service says no leasing on 40,000 acres in the Wyoming Range," *Billings Gazette*, December 16, 2016. Accessed March 4, 2017. http://billingsgazette.com/news/state-and-regional/%20%20%20%20%20%20wyoming/forest-service-says-no-leasing-on-acres-in-the-wyoming/article_%20%20%20%20%209cdb8ca0-c9f5-5532-b85d-6be4ab910a3c.html

36. Tal Kopan, "Could a President Trump reverse Obama's regulations on 'Day One'?", *CNN Politics*, September 28, 2016. Accessed March 4, 2017. http://www.cnn.com/2016/09/28/politics/trump-executive-action-obama/index.html

37. Emmarie Huetteman, "How Republicans Will Try to Roll Back Obama Regulations," The New York Times, January 30, 2017. Accessed March 4, 2017. https://www.nytimes.com/2017/01/30/us/politics/congressional-review-act-obama-regulations.html?_r=0

38. Alex Chadwick and Elizabeth Arnold, "Bush Administration Lifts Wilderness Road Ban," *NPR-Environment*, May 6, 2005. Accessed July 22, 2016. http://www.npr.org/templates/story/story.php?storyId=4633374

39. AllGov/Controversies, "Roadless Rule Controversy," Copyright 2015. Accessed March 4, 2017. http://www.allgov.com/departments/department-of-agriculture/united-%20%20%20%20%20states-forest-service?agencyid=7277

Chapter 5

1. FACA or Federal Advisory Committee Act enacted in 1972 to insure objective accessible advice is available to the public. The Act sets up the process for establishing, operating, overseeing, and ending such advisory bodies. Accessed March 9, 2017. https://www.gsa.gov/portal/content/104514

2. Charles Helm and Mario Morelli, "Stanley and the Obedience Experiment: Authority, Legitimacy, and Human Action," Political Theory, Sage Publications, Inc., Vol. 7, No. 3, Aug. 1979. 321-345.

3. Pinedale Anticline Working Group Air Quality Task Group (AQTG), Meeting Minutes, Nov. 3, 2004.

4. The phrase nuances is meant to refer to the large latitude that seemed to exist when it came to exercising emission mitigation actions through regulatory enforcement. If WYDEQ did not want to burden industry with mitigation requirements, such latitude would be invoked, i.e., "that is not a criteria pollutants so we have no authority to regulate it."

5. It took me several years to finally understand this statistical game that was imbedded in the Clean Air Act.
6. In particular, I had become much disliked by the directors of WYDEQ and AQD because of my criticisms in state and local news media of their dismal involvement to date in protecting the county's formerly pristine air quality.
7. Pinedale Anticline Working Group Air Quality Task Group (AQTG), Meeting Minutes, November 30, 2004.
8. U.S. Environmental Protection Agency, "*National Ambient Air Quality Standards (NAAQS).*" Accessed July 19, 2015. http://www.epa.gov/airprogm/oar/criteria.html
9. U.S. Environmental Protection Agency, "*Prevention of Significant Deterioration (PSD) Basic Information.*" Accessed July 19, 2015. http://www.epa.gov/NSR/psd.html
10. U.S. Environmental Protection Agency, "*What is the National Environmental Policy Act?*," Accessed July 29, 2015. http://www2.epa.gov/nepa/what-national-environmental-policy-act
11. California Air Pollution Control Officers Association Best Available Control Technology Clearing House, CAPCOA BACT Clearinghouse Resource Manual, Glossary of Some Air Pollution Terms. Accessed July 19, 2015. http://www.arb.ca.gov/bact/docs/definitions.htm
12. U.S. Environmental Protection Agency, "*List of Mandatory Class I Federal Areas.*" Accessed July 19, 2015. http://www.epa.gov/visibility/class1.html
13. Linsey DeBell, "Introduction to the IMPROVE program's new interactive web-based data validation tools," Accessed March 10, 2017. https://www3.epa.gov/ttn/amtic/files/ambient/monitorstrat/debell.pdf
14. "Deciview, A Standard Visibility Index, Interagency Monitoring of Protected Visual Environments," *IMPROVE Newsletter*, Vol. 2, No. 1, April 1993. Accessed March 10, 2017. http://www.shodor.org/os411/courses/411c/module07/unit03/page07.html
15. The reader should pay attention to these recommendations as the meeting chronicle continues because they evolved into a toned down, sanitized, final incarnation largely for reasons of politically correct interagency courtesey.
16. Email to AQTG members from Co-Chair Cara Keslar, WYDEQ-AQD Air Quality Engineer, Cheyenne, WY, Dec. 14, 2004.
17. Cara Keslar email to AQTG Members, "*USFS Decision Participation in our TG,*" Dec. 23, 2004.
18. Author's email to AQTG Members, "*USFS Decision Participation in our TG,*" Dec. 23, 2004.
19. Email from Mike Golas to author, December 22, 2004.

20. Pinedale Anticline Working Group Air Quality Task Group (AQTG), Meeting Minutes, Jan. 4, 2005.

21. U.S. Department of The Interior, Bureau of Land Management, **Draft** Supplemental Environmental Impact Statement for the Pinedale Anticline Oil and Gas Exploration and Development Project Sublette County, Wyoming, Vol. 1, 2.3 Existing Development Within the PAPA, 2.3.3 Drilling Rigs, Dec. 2006. 2-11.

22. Pinedale Anticline Working Group Air Quality Task Group (AQTG), Meeting Minutes, Jan. 25, 2005.

23. A few years later, that would be one of the points of justification by BLM which it would use to assert that we had exceeded our authority and used to explain our dissolution as a task group.

24. Air Quality Task Group Prioritization of Air Quality Recommendations for the PAWG March 18, 2005. Ultimate outcome for this list was as follows:
All priority 1A: items continue to be funded by USFS.
Priority 1B: Shell provided funding to purchase, install and operate the camera for 5 years. USFS has been funding since.
Priority 2A: Analysis was done in house (2010) by USFS.
Priority 2B: Never happened
For those unfamiliar with the function of a transmissometer, it is an instrument that measures visible haze.

25. Pinedale Anticline Working Group Air Quality Task Group (AQTG), Meeting Minutes, April 7, 2005.

26. These designations come from a somewhat complicated set of EPA criteria. Simply stated here, Tier 1 engines across a spectrum of horsepower emit 6.9 grams per horsepower-hour of NOx. Tier 2 engines drop that NOx emission rate to 5.2 grams for 50-100Hp engines and 4.5 grams per horsepower-hour for 100-750Hp engines. The reader can obtain more detail by Googling EPA with the tier search word. *(Information extracted from "USAEPA Emission Standards for Tier 1-3 engines)* Accessed Jan. 22, 2017. http://www.ourair.org/wp-content/uploads/epatierstnds.pdf.

27. Pinedale Anticline Working Group Air Quality Task Group (AQTG), Meeting Minutes, June 16, 2005.

28. In retrospect years later, this is a bitter comment because all of our efforts to do so were rebuffed. If nothing else, this illustrates how the managers of the field offices impart their personal characteristics into the conduct of business by their respective field offices.

29. Pinedale Anticline Working Group Air Quality Task Group (AQTG), Meeting Minutes, July 16, 2005.

30. Pinedale Anticline Working Group Air Quality Task Group (AQTG), Meeting Minutes, July 28, 2005.

31. My reaction was, "Ah yes, dodge and deny. Who can prove otherwise?"
32. In two years I would find myself responsible for more EPA resistance toward BLM by Delwiche's ultimate superior, the Director of Region 8.
33. Unfortunately, as with most of our ideas, this one never happened.
34. This was in recognition that gas exploration was beginning to boom in that region.
35. This was an acknowledgement that more up wind and down wind measurements were needed to establish the true impact of the Jonah and Anticline fields on local air quality.
36. Future measurement data confirmed these findings and in so doing revealed a previously underappreciated phenomenon in the atmospheric science community. That phenomenon was an amplification of ultraviolet radiation's effect in catalyzing ozone precursors by reason of its reflecting off ground snow-cover back into the sky, thereby performing a double pass through the air column.
37. Mike's point here would become more relevant with regard to a "toxics air study" the Sublette County Board of Commissioners would soon authorize. They finally had to react to growing citizen anger over gas field air pollution and elected to do so by spending over $1 million for an outsourced study that looked primarily for the "all others." This is discussed in the section about the commissioners.
38. Pinedale Anticline Working Group Air Quality Task Group Concerns and Recommendations Drafted for Submission to the PAWG and BLM, Aug. 9, 2005.
39. Pinedale Anticline Working Group Air Quality Task Group (AQTG) Meeting Minutes, Oct. 18, 2005.
40. I was never able to discover what EC-1 was. In the EIS released to the public the options were identified as Alternatives A through G and BLM Preferred Alternative.
41. U.S. Department of the Interior, Bureau of Land Management, "Record of Decision, Jonah Infill Drilling Project Sublette County, Wyoming," Appendix A, Jonah Infill Drilling Project Administrative Requirements, Conditions of Approval, and Mitigation, Air Quality, Item 2c, March 14, 2006. A-3.
42. This was an unsubstantiated rumor that seemed to emanate from BLM but lacked specific source attribution.
43. Here again, Shell, through its representative, ran successful inter- ference which made the requirement for consensus become the ultimate blocking influence.
44. This was a strategy aimed at reducing fugitive methane leaks which EPA recognized as a major source.
45. Pinedale Anticline Working Group Air Quality Task Group Updates to Concerns and Recommendations submitted to the PAWG, Oct 25, 2005.
46. Pinedale Anticline Working Group Air Quality Task Group (AQTG) Meeting Minutes, Dec. 1, 2005.

47. Pinedale Anticline Working Group Air Quality Task Group (AQTG), Meeting Minutes, Jan. 11, 2006.

48. Pinedale Anticline Working Group Air Quality Task Group (AQTG), Meeting Minutes, Feb. 16, 2006.

49. The section on WYDEQ deference to EnCana's NOx credits idea illustrates this practice.

50. Pinedale Anticline Working Group Air Quality Task Group (AQTG), Meeting Minutes, Jan. 25, 2007.

51. Pinedale Anticline Working Group Air Quality Task Group (AQTG), Meeting Minutes, April 7, 2009.

52. Pinedale Anticline Working Group Air Quality Task Group (AQTG), Meeting Minutes, April 30, 2009.

53. Pinedale Anticline Working Group Air Quality Task Group (AQTG), Meeting Minutes, June 4, 2009.

Chapter 6

1. Author's letter, "Subject: New Source Applications," Personal comment letter re. BP America Corona 14 Pad facility, To Dan Olson, Administrator of WYDEQ-AQD, March 4, 2004.

2. See www.oilandgas/investor.com, Sept 2002 issue, 3.

3. Benjamin Storrow, "Ultra Petroleum, Wyoming's largest natural gas company, on edge of bankruptcy," *Casper Star Tribune*, Feb. 18, 2016.

4. Tiffany Kary and Steven Hurch, "Ultra Petroleum Files for Bankruptcy, Citing $3.9 Billion Debt," Bloomberg, May 1, 2016.

5. Perhaps cautionary about hubris at the corporate executive level, it seems a dark irony that 12 years later Ultra would buy Shell's Anticline holdings for $925 million, report a $3.2 billion loss two more years after that and ultimately file for Chapter 11 bankruptcy because gas prices tanked.

6. Wyoming Department of Environmental Quality, "Comments on Permit Application AP-1514 BP America Corona 14 PAD," June 7, 2004.

7. Wyoming Dept. of Environmental Quality, Division of Air Quality Permit AP-4081 Application Analysis, June 12, 2006.

8. Wyoming Dept. of Environmental Quality, Division of Air Quality, Permit CT-4633 Application Analysis for EnCana Oil and Gas (USA) Permit Application AP-2177, February 7, 2008.

9. Ibid.

10. Wyoming Dept. of Environmental Quality, Division of Air Quality, Permit Application Analysis for Newfield Exploration Permit Application CT-7074, February 5 and February 25, 2008.
11. Wyoming Department of Environmental Quality, "Air Quality Permit CT-7074," Letter of Reply from WYDEQ–AQD Director Dave Finley, February 25, 2008.
12. This reference to stationary sources was a sore spot for us in the Air Quality Task Group for quite some time. The CAA drew distinctions and exempted sources deemed mobile from important restrictions. For a long time WYDEQ held back on regulating drill rigs because they were argued to be non-stationary sources.
13. Wyoming Department of Environmental Quality/Air Quality Division Permit Application Analyses, WYDEQ Operator Permit Archive, Cheyenne, WY, 2004- 2011.
14. U.S. Department of The Interior, Bureau of Land Management, "Draft Supplemental Environmental Impact Statement for the Pinedale Anticline Oil and Gas Exploration and Development Project, Sublette County, Wyoming, Chapter 3 – Affected Environment, Section 3.11 - Air Quality, Sub-Section 3.11.2 – Impacts to Air Quality from Existing Wellfield Activities." Washington: Government Printing Office, December 2006. 3-61 through 3-63.
15. U.S. Department of The Interior, Bureau of Land Management, "Revised Draft Supplemental Environmental Impact Statement for the Pinedale Anticline Oil and Gas Exploration and Development Project, Sublette County, Wyoming, Chapter 3-Affected Environment, Section 3.11.2 Impacts to Air Quality from Existing Wellfield Activities." Washington: Government Printing Office, December 2007. 11,3-71.
16. U.S. Department of The Interior, Bureau of Land Management, "Final Supplemental Environmental Impact Statement for the Pinedale Anticline Oil and Gas Exploration and Development Project, Sublette County Wyoming, Volume 1, Chapter 3 - Affected Environment." Washington: Government Printing Office, June 2008. 3-5,3-74,75.
17. U.S. Department of The Interior, Bureau of Land Management, "Revised Draft Supplemental Environmental Impact Statement for the Pinedale Anticline Oil and Gas Exploration and Development Sublette County, Wyoming, Project, Chapter 4 –Environmental Consequences, Section 4.9 – Air Quality, Section 4.9.3.1 – Summary of Impacts Common to All Alternatives. Washington: Government Printing Office, December 2007. 4-81 – 4-87.
18. Ibid.
19. U.S. Department of The Interior, Bureau of Land Management, "Draft Air Quality Technical Support Document for the Jonah Infill Drilling Project Environmental Impact Statement dated 2004, Section 4.0 Mid-Field and Far Field Analysis, Sub Section 4.2 Project Alternative Modeling Scenarios." TRC Environmental Corp., November 2004. Washington: Government Printing Office, February 2005. 49.

20. Wyoming Department of Environmental Quality, Air Quality Division, Letter to Operators: "Interim Policy on Demonstration of Compliance with WAQSR Chapter 6, Section 2(c)(ii) for Sources in Sublette County," July 21, 2008. WYDEQ Operator Permit Archive, Cheyenne, WY.

21. Edwards, Peter M. et al., Oct. 16, 2014, High winter ozone pollution from carbonyl photolysis in an oil and gas basin, *Nature Research Letter*: Vol. 514. 351-354.

22. USEPA, EPA compilation of reporting on Sublette County gas field ozone history, Air Data. Accessed February 20, 2016. http://www.epa.gov/airdata

23. Mead Gruver, "Wyo. Ozone season ends, 13 days topped EPA limit," *Associated Press, Bloomberg Businessweek*, April 14, 2011.

24. Joy Ufford, "EPA moves on CURED ozone request," *Sublette Examiner*, Dec 26, 2011.

25. Ron Aiken, "More can be done about ozone," *Sublette Examiner*, Apr 4, 2011.

26. Cat Urbigkit, "Wyo Gas Field Operators react to increased ozone," *Casper Star-Tribune*, March 5, 2011.

27. Dawn Ballou, "Ozone and SCRs," *Pinedale online*, March 24, 2011.

28. Joy Ufford, "EPA moves on CURED ozone request," *Sublette Examiner*, Dec 26, 2011.

29. U.S. Environmental Protection Agency, National Risk Management Research Center, Office of Research and Development. Greenhouse Gas Technology Center, Southern Research Institute. ETV Joint Verification Statement: "*Emissions Control of Criteria pollutants, Hazardous Pollutants, and Greenhouse Gases*," *Application: Natural Gas Dehydration-Quantum Leap Dehydrator; Statement of Evaluation Testing conducted under the Environmental Technology Verification (ETV) Program*, 2003; S-2.

30. Dayton, Kelsey, "New Ozone Task Force Not Enough, Some in Pinedale Say," *Casper Star-Tribune*, January 18, 2012.

31. Benjamin Storrow, "Progress Report: More Work is Needed on Ozone Near Pinedale," *Casper Star-Tribune*, March 20, 2014.

32. WYDEQ, Memorandum to Wyoming Oil & Gas Production Facility Operators, Subject: Status of Proposed Revisions to O&G Permitting Guidance, April 20, 2007, 7.

33. Ibid, 2-3.

34. Ibid, 3.

35. Ibid, 4.

36. Ibid, 6.

37. Ibid, 6.

38. Ibid, 7.

R. Perry Walker

39. Dave Finley, Director of WYWYDEQ-AQD, personal email reply to my comments re. WYDEQ, Memorandum to Wyoming Oil & Gas Production Facility Operators, Subject: Status of Proposed Revisions to O&G Permitting Guidance, June 6, 2007.

40. Her entire opposition was focused on the 4-cylinder compact auto engine used to drive the QLD pump equipment. She invoked AP-42 and its generalized approach to assessing NOx emissions as her justification but the inventors put her objections to rest with subsequent field testing. Nevertheless, neither she nor her boss would yield the point. As a result, EnCana abandoned its plan to buy 50 units out of reluctance to adopt what AQD would not. I was furious and this poisoned all further relationship I had with AQD.

41. "US NOAA Study finds high ozone levels in Jonah/Pinedale gas field," Platts Gas Daily, January 23. 2009.

42. Whitney Royster, "EPA: Weather didn't cause ozone spikes," Casper Star-Tribune, August 2, 2006.

43. Wyoming Department of Environmental Quality, "Exceptional Event Demonstration Package for the Environmental Protection Agency, South Pass Wyoming Ozone Standard Exceedances," May 23-26, 2007.

44. Ibid, 5-6.

45. U.S. Dept. of the Interior Bureau of Land Management Wyoming State Office Cheyenne, Wyoming, Pinedale Field Office, Pinedale, Wyoming and Rock Springs Field Office Rock Springs, Wyoming, FINAL ENVIRONMENTAL IMPACT STATEMENT JONAH INFILL DRILLING PROJECT, SUBLETTE COUNTY, WYOMING, (Volume 2 of 2)-Appendix, Attachment F-1, United States Department of the Interior Bureau of Land Management State of Wyoming DRAFT CHARTER, Jan., 2006. 312-314.

46. Ibid.

47. Ibid.

48. WYDEQ-AQD, Letter of reply from Dave Finley, Administrator, Air Quality Division to Mike Steiwig, Chairman Jonah Infill Office, Pinedale WY, August 8, 2007.

49. Author's letter of challenge to Mike Steiwig, Chairman Jonah Infill Office, September 4, 2007.

Chapter 7

1. These figures were routinely quoted in emissions tables that accompanied every permit application sumitted by these operators to WYDEQ.

2. These statements were submitted to the Commissioners in my letter of July 23, 2006.

3. Tami H. Funk, Draft Final Report STI-908003.08-3423-DFR, Sonoma Technology Inc., 1455 N. McDowell Blvd., Suite D, Petaluma, CA 94954-6503, August 15, 2008. http://www.tetonscience.org/data/contentfiles/file/downloads/pdf/CRC/CR-CAirQualityOil/Funk%20and%20Hafner.%202008.%20Air%20Monitoring%20Network%20Assessment%20for%20Southwest%20Wyoming%20NXPowerLite).pdf

4. I discuss my work on wind behavior across the Jonah and Anticline in detail in The Science Chapter later.

5. Author's written reply to County Commissioner John Linn's request for assistance in interpreting *"Final Briefing on Southwest Wyoming Air Monitoring Network Assessment"* by Sonoma Technology. Inc., November 1, 2008.

6. Garfield County experienced much the same gas development scenario as was being experienced by Sublette County. Citizens there were frequently featured in national news coverage investigating their claims of health impacts from wells that were as close to homes as 500 meters.

7. U.S. Depart. Of Health and Human Services, Public Health Service, Agency for Toxic Substances and Disease Registry, Division of Health Assessment and Consultation, Atlanta, Georgia 30333, Health Consultation, Garfield County, "Public Health Implications of Ambient Air Exposures to Volatile Organic Compounds as Measured in Rural, Urban, and Oil & Gas Development Areas," Garfield County, Colorado, March 13, 2008. Last Accessed Oct. 9, 2016. http://www.atsdr.cdc.gov/hac/pha/Garfield_County_HC_3-13-08/Garfield_County_HC_3-13-08.pdf

8. Sublette County Commissioners, WYDEQ, and Wyo. DEH, "Request for Proposal Ambient Air Monitoring Provisions Sublette County Human Health Risk Assessment Air Toxics Inhalation Project," Deadline for submittal Oct 27, 2008.

9. **C**hronic **O**bstructive **P**ulmonary **D**isease

10. Cat Urbigkit, "Residents question Sublette County health risk assessment results," Casper Star-Tribune, April 2, 2011.

11. Ibid.

12. Ibid.

13. Greg Gordon, "Enough is Enough," Opinion Essay, High Country News, A Web Exclusive, June 28, 2004. Last accessed Oct 6, 2016. http://www.hcn.org/wotr/14849

14. Joan Barron, "Governor Taps Planning Chief," Casper Star-Tribune Capital Bureau, March 29, 2005.

15. The Governor stated in his press release announcing her promotion that *"Mary is fair-minded, she is capable and, more than that, she truly always has the best interests of the state at heart. Over the course of her career, she has worked through some extremely contentious issues in land-use planning and has earned the respect and praise of those involved.*

She will make a fine state planning coordinator, and the state is lucky to have her." I absolutely agree with that assessment.

16. I am very uncertain of this date. At the time I never imagined I would be referring to it a decade later so I kept no notes. Furthermore and almost amazingly, I can find no reference whatsoever in the news media that describes the trip. I believe it was in 2005 but Terry Svalberg believes 2009. I doubt this, however. What I am certain of was that it was in the fall......September or October.

17. Dustin Bleizefffer, Star-Tribune energy reporter, "Freudenthal protests oil, gas sale," Casper Star-Tribune, June 9, 2004.

18. News Release issued by Questar Corp., "Questar Receives Wyoming Governor's Conditional Endorsement of Pinedale Winter-Drilling Plan," PRNewswire, Aug. 3, 2004. Accessed Sept. 26, 2016. http://www.prnewswire.com/news-releases/questar-receives-wyoming-governors-conditional-endorsement-of-pinedale-winter-drilling-plan-71515782.html

19. Whitney Royster, Environmental Reporter, "Governor: Don't cut protest period," Casper Star-Tribune, May 4, 2005.

20. Ibid.

21. A search on the IOGCC turns up several web sites that are quite flashy and effusive about environmentally sound extraction of oil and gas. However, Wikipedia states that it started out as a confederation of six oil and gas states that grew to 48 domestic and international members. As of 2010, seven members were industry executives and lobbyists. See https://en.wikipedia.org/wiki/Interstate_Oil_and_Gas_Compact_Commission

22. "EnCana recognized by Governor - Receives stewardship award," News release from the office of Governor Freudenthal, *Pinedale Online!*, October 16, 2006. Accessed Oct. 14, 2016. http://www.pinedaleonline.com/news/2006/10/EnCanarecognizedbyGo.htm

23. Ibid.

24. "Gov. blasts BLM plan," Casper Star-Tribune, No attribution, May 3, 2008. Accessed Sept. 26, 2016. http://trib.com/news/state-and-regional/gov-blasts-blm-plan/article_b5bfa9e1-3610-5d7d-8fd5-e30d69f24d01.html

25. Cory Hatch, "Gov. criticizes Forest," Jackson Hole News and Guide, May 24, 2008.

26. Emilene Ostlind, "BLM stays course in Wyoming gas patch despite mule deer decline," March 30, 2011. Accessed September 28, 2016. http://www.hcn.org/issues/43.5/blm-stays-course-in-wyoming-gaspatch-despite-mule-deer-decline

27. U.S. Department of the Interior, Bureau of Land Management. *Final Supplemental Environmental Impact Statement for the Pinedale Anticline Oil and Gas Exploration and Development Project, Sublette County, Wyoming, Volume 2, Appendix 11, Alternative D, Mitigation.* Washington: Government Printing Office, June 2008. 11-1,2,3

By the Rasping in My Lungs

28. Ibid, 11-2

29. Cory Hatch, "Wells versus wildlife: Drill plan draws flak," Jackson Hole News & Guide, July 2, 2008, updated Oct 14, 2013.

30. Joint Appropriations Committee (JAC) Budget Hearings, 01/11/2006, Transcript of recorded proceedings, index numbers 27, 1:43, 2:14. Accessed September 23, 2016. https://legisweb.state.wy.us/budget/2006/jac0111.htm

31. Dustin Bleizeffer, Casper Star-Tribune energy reporter, "Fracking debate heats up," June 11, 1009. Last accessed Oct 12, 2016, http://trib.com/news/state-and-regional/fracking-debate-heats-up/article_2d9bbf0b-543d-5fe6-b791-8b1b0dd95828.html

32. Bebout, Cooper, Jennings, Lockhart, and Stubson, "Resolution – Environmental Protection Agency Regulations," Senate Joint Resolution No. SJ0006, 11LSO-0442, 2011.

33. Thomas Mann and Norman Orstein, "It's Even Worse Than it Looks: How the American Constitutional System Collided with the New Politics of Extremism," Published by Basic Books, ISBN 978-0465031337, May 1, 2012.

34. Laura Hancock, "Wyoming sues EPA over regional haze," *Casper Star- Tribune*, March 29, 2014.

35. Todd, Leah, "Wyoming First State to Block New Science Standards," *Casper Star-Tribune*, March 14, 2014.

36. Senator Wallop wrote an op-ed in the Casper Star which was published on Sunday, Dec. 18, 2005 but I didn't keep the article nor could I find it via the Casper Star online portals. I do have in my records the rebuttal I submitted and which the Star printed.

37. Derived from Energy Information Agency web site analytical data.

38. Energy and Natural Resources Law, "Wyoming Oil and Gas Conservation New Rules," July 19, 2016. Accessed September 18, 2016. https://www.wyomingbar.org/wp-content/uploads/7-19-2016_ENR_Webinar.pdf

39. Dustin Bleizeffer, Star-Tribune energy reporter, "Wyoming energy Industry fights fracking rules," Casper Star-Tribune, March 32, 2010.

40. "New Fracking Rules Approved," Wyoming Outdoor Council, posted June 8, 2010. Accessed Sept. 22, 2016. https://wyomingoutdoorcouncil.org/2010/06/08/new-fracking-rules-approved/

41. Appendix 5 contains the information I managed to pull together.

42. WOGCC URL address shows an illegible example. This report is for API No. 3525370 under the title "03/09/2010 Notice of Intent Other." Frustratingly, this seems the norm. An example of a selectively legible report, API no. 3528094, can be found at http://wogcc.state.wy.us/Sundryapi.cfm under the title "03/15/2011 Notice of intent Other." Accessed Sept. 22, 2016.

43. See Appendix 6 for the list of fracking fluid additives addressed by these requests for exemption from public disclosure.
44. Amy Joi O'Donoghue, "Herbert, Stewart meet with Trump for Monday's repeal of controversial planning rule," *Ladd Egan, KSL TV, Salt Lake city, UT*, March 27, 2017. Accessed March 28, 2017. http://www.ksl.com/?sid=43657105&nid=148&title=herbert-stewart-meet-with-trump-for-repeal-of-controversial-planning-rule
45. Charlie Passut, "Bill to Scrap BLM's 'Planning 2.0' Rule Heads to Trump's Desk," *NGI Natural Gas Daily*, March 7, 2017. Accessed March 28, 2017. http://www.naturalgasintel.com/articles/109669-bill-to-scrap-blms-planning-20-rule-heads-to-trumps-desk
46. Jeff Gearino, "A green victory: Wyoming Range gains protection from further energy development," Casper Star Tribune, Southwest Wyoming Bureau, Dec. 29, 2009. Accessed Sept. 30, 2016. http://trib.com/news/state-and-regional/a-green-victory/article_51646df7-3ba1-5875-97cc-a1620b413965.html
47. U.S. Department of the Interior, Bureau of Land Management News Release, Wyoming State Office, "U.S. Forest Service and BLM Officials Celebrate Wyoming Range Legacy Act," Release Date: 08/24/09. Accessed Sept 30, 2016, http://www.blm.gov/wy/st/en/info/news_room/2009/august/24legacy-act.html

Chapter 8

1. Laurie D. Goodman, Former EIS consultant to Ultra Petroleum, e-mail letter to author, April 23, 2017.
2. PSD means Prevention of Significant Deterioration. It originates with the Federal Clean Air Act and over the years, EPA has turned it into a bit of esoterica crafted more for the use of regulatory cognoscenti than for the novice citizen. I reconciled the concept in my own mind by thinking of it as the "space" below a regulatory not-to-exceed limit for a pollutant, in our case NOx. An example definition comes from an EPA memo in 1988: *"A PSD source will not be considered to cause or contribute to a predicted NAAQS or increment violation if the source's estimated air quality impact is insignificant (i.e., at or below defined de minimis levels).* Accessed February 3, 2017. https://www3.epa.gov/ttn/naaqs/aqmguide/collection/ nsr/reaffirm.pdf
3. Laurie D. Goodman, Former EIS consultant to Ultra Petroleum, e-mail message to author, April 23, 2017.
4. Ultra Petroleum Corp., 363 North Sam Houston Parkway East, Suite 1200, Houston TX 77060, May 25, 2006. *Notice of Annual Meeting of Shareholders to be held on June 29, 2006.* 21
5. Ibid, 22,23

6. Ibid, 23
7. Ibid, 23
8. Mike Golas drew this diagram on the chalk board to drive home his concern:

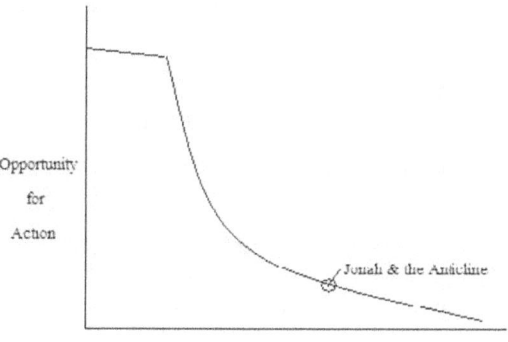

9. U.S. Environmental Protection Agency, National Risk Management Research Center, Office of Research and Development. Greenhouse Gas Technology Center, Southern Research Institute. ETV Joint Verification Statement: *"Emissions Control of Criteria pollutants, Hazardous Pollutants, and Greenhouse Gases,"* Application: *Natural Gas Dehydration-Quantum Leap Dehydrator; Statement of Evaluation Testing conducted under the Environmental Technology Verification (ETV) Program*, 2003;S-2.

10. Mead Gruver, Winter Ozone Problem Returns to Western Wyoming Gas Fields," *Associated Press, U.S. News,* March 20, 2017. Accessed April 2, 2017. https://www.us-news.com/news/best-states/wyoming/articles/2017-03-20/winter-ozone-problem-returns-to-western-wyoming-gas-fields

11. EnCana Oil & Gas (USA) Inc., *"EnCana to Sell its Jonah Field Operations in Wyoming to an Affiliate of TPG Capital for $1.8 billion,"* EnCana News and Stories, March 31, 2014. Last accessed April 12, 2014. http://www.EnCana.com/news-stories/news-releases/details.html?release=836442

12. Dawn Ballou, "Shell sells Pinedale gas Field assets to Ultra Petroleum, *Pinedale Online!,* August 14, 2014. Accessed February 15, 2017. http://www.pinedaleonline.com/news/2014/08/ShellsellsPinedalega.htm

13. Tiffany Kary and Steven Church, "Ultra Petroleum Files for Bankruptcy, Citing $3.9 Billion in Debt," *Bloomberg,* May 1, 2016. Accessed Jan 20, 2017. https://www.bloomberg.com/news/articles/2016-05-01/ultrapetroleum-files-for-bankruptcy-citing-3-9-billion-debt

R. Perry Walker

Chapter 9

1. Wyoming Outdoor Council, Jackson Hole Conservation Alliance, The Wilderness Society, Biodiversity Conservation Alliance, *Prime Hunting and Fishing Areas in Bridger-Teton Forest to be Opened for Oil and Gas Development*, Joint News Release, June 17, 2004.

2. Whitney Royster, Environmental Reporter, "Groups appeal leasing on winter range," Casper Star-Tribune, Thursday, May 19, 2005.

3. Bruce Pendery, "Excluding the public," Perspective, Casper Star, Casper, Wyoming, Sunday, July 31, 2005.

4. Harold Magistrale, "An Introduction to the Information Quality Act and its Application to Environmental Regulation," The Urban Lawyer, Vol. 38, No. 3, Summer 2006. 561-584 Accessed Sep 6, 2016 http://www.thecre.com/pdf/20120723_IQA%20History.pdf

5. George W. Bush's first term as President of the United States, Science and Technology, Wikipedia, July 2, 2016. Accessed Sept. 6, 2016. https://en.wikipedia.org/wiki/George_W._Bush%27s_first_term_as_President_of_the_United_States

6. This was the only time John Corra apparently felt compelled to respond to me because he called me personally to state "You have gotten our attention."

7. Laurie D. Goodman, Former EIS consultant to Ultra Petroleum, e-mail message to author, April 23, 2017.

8. Dawn Ballou, "Environmental groups offers guide to Pinedale Anticline," *Pinedale Online!*, May 6, 2009.

9. Leslie Waggener, Interview with Mary Lynn Worl, *Wyoming's Energy Boom, 1995-2010*, University of Wyoming American Heritage Center's Oral History Program, Accession #11749, October 11, 2010.

10. This board member added justification for that focus by commenting that in his view, GYC had been spending too much effort on protecting wolves. That stunned me because indeed, GYC had seemed fixated on advancing the cause of that critter so his statement had the ring of heresy.

11. Gary Gerhardt, "Rocky Mountain News, "Interior defends drilling policies- Bush administration rebuffs biologists' plea to slow oil, gas leases," Rocky Mountain News, February 26, 2005.

12. John Krist, "LNG Part 2: A Valley Transformed," *Ventura County Star*, August 25, 2005. Accessed January 24, 2017. http://www.venturacountystar.com/vcs/county_news/article/0,1375,VCS_226_4038190,00.html

13. Ray Ring, "Stargazer aims his scopes at gas industry," *High Country News*, May 1, 2006. Accessed January 24, 2017. Accessed January 24, 2017. http://www.hcn.org/issues/321/16273

Chapter 10

1. Author's written public comments to the Pinedale BLM Field Office addressing the Jonah Infill EIS, May 8, 2003.

2. This discussion is contained in chapter 1 of the many versions of the Anticline EIS identified by BLM as PAPA 2006a, through PAPA 2006g, PAPA 2007a through PAPA 2007e, and PAPA 2008a through PAPA 2008g.

3. William C. Malm, *Introduction to Visibility*, Colorado State University, Air Resources Div., National Park Service, CIRA-NPS Visibility Program, Section 6.2.1.

4. Barbara J. Finlayson-Pitts and James N. Pitts Jr., *Atmospheric Chemistry: Fundamentals and Experimental Techniques*. (New York: John Wiley & Sons, Inc. 1979). 14.

5. EPA established a complex set of standards for non-road diesel engines that specifies upper limits on air pollutant emissions according to horsepower ratings of such engines. The standards were designed to become effective over time with each higher tier rating replacing the previous until over the years of implementation, only the cleanest engines would be available to industry. The standards became effective for Tier 1 as of 1996 which set a limit for NOx emissions from engines above 750 horsepower at 6.9 grams per horsepower-hour run time. Requirements progressively tightened through Tier 4 which limits engines above 750 horsepower to 2.6 grams per horsepower-hour of run time. Compliance for Tier 4 was 2015. For detailed information see EPA website: https://www.dieselnet.com//standards/us/nonroad.php#tier3.

6. Arnie Heller, "Finding the Missing Piece in the Climate Change Puzzle," *LLNL Science & Technology Review*, Univ. of California for the U.S. Dept. of Energy, Lawrence Livermore National Laboratory, April 2003. 4-12.

7. U.S. Department of The Interior, Bureau of Land Management (PAPA 2007a), "Revised Draft Supplemental Environmental Impact Statement for the Pinedale Anticline Oil and Gas Exploration and Development Project, Sublette County, Wyoming," Chapter 3 - Affected Environment, Section 3.3 – Climate, Table 3.3.2., December 2007. 3-5. See also Jonah 2005a, same subjects.

8. U.S. Department of The Interior, Bureau of Land Management (Jonah 2005a) Draft Environmental Impact Statement, Jonah Infill Drilling Project, Sublette County, Wyoming, Volume 1, Chapter 3 – Affected Environment, Section 3.1 Physical Resources, Sub Section 3.1.2 - Air Quality, Tables 3- 4,3-5,3-6, (footnote 1), February 2005. 3-6. See also the same subjects of Jonah 2006c, PAPA 2007a, 2007d, 2007e, PAPA 2008e, 2008f.

9. U.S. Department of The Interior, Bureau of Land Management (Jonah 2006a), "Final Air Quality Technical Support Document for the Jonah Infill Drilling Project Environmental Impact Statement," Volume 1, Appendix A – Air Quality Assessment Protocol, second Appendix A – EPA Ozone Screening Methodology, Section 3.0 -

Screening Tables, third Appendix A – Development of Screening Tables. TRC Environmental Corp. September 2003, Revised October 2003. January 2006. 6.

10. Ibid, 13-15

11. Ibid, 18

12. Ibid, 24

13. U.S. Department of The Interior, Bureau of Land Management (Jonah 2005a) "Draft Environmental Impact Statement, Jonah Infill Drilling Project, Sublette County, Wyoming," Volume 1, Chapter 3 – Affected Environment, Section 3.1 Physical Resources, Sub Section 3.1.2 Air Quality, Tables 3- 4,3-5,3-6, (footnote 1), February 2005; 3-6. See also same subjects Jonah 2006b, PAPA 2007a, 2007d, 2007e, 2008e, and 2008f.

14. S.J. Campbell, R. Wanek, W. Coulston, "Ozone injury in the West Coast Forests: 6 Years of Monitoring," U. S. Department of Agriculture, U.S. Forest Service, General Technical Report PNW-GTR-722, June 2007.

15. M. Story, J. Shea, T. Svalberg, M. Hektner, and G. Ingersoll, "Greater Yellowstone Area Air Quality Assessment Update," *Proceedings of Annual Conference of the Greater Yellowstone Area Clean Air Partnership (GYACAP)*, 2005; 161. Accessed January 12, 2013. http://www.fs.fed.us/air/documents/GYA_airquality_assessment-YellSci_Conf.pdf

16. As I explained in earlier chapters, this was reported to me by Terry Svalberg who was present and who challenged the plan, but was ruled out of order.

17. See Appendix 7 for an example. All others are on file at the University of Wyoming American Heritage Center.

18. Note: all wind rose figure I have included herein are reversed from the standard convention in that all wind bars show the direction TOWARD which the winds are blowing instead of FROM which they are blowing. I have done this because my audiences have found the method to be more intuitive to understand.

19. See Appendices 7 for an example. Many more are on file at the University of Wyoming American Heritage Center.

20. Federal Clean Air Act, 42 U.S.C. 7474, Section 165 part 8(d)(2)(C)(i).

21. U.S. Department of The Interior, Bureau of Land Management (Jonah 2006a), "Final Air Quality Technical Support Document for the Jonah Infill Drilling Project Environmental Impact Statement," Volume 1, Appendix A – Air Quality Assessment Protocol, second Appendix A – EPA Ozone Screening Methodology, Section 3.0 - Screening Tables, third Appendix A – Development of Screening Tables. TRC Environmental Corp. September 2003, Revised October 2003. January 2006. 13-14.

22. Ibid, 14

23. Ibid, 15

24. Ibid, 18

25. Ibid, 24

26. U.S. Department of The Interior, Bureau of Land Management (Jonah 2006b), "Final Air Quality Technical Support Document for the Jonah Infill Drilling Project Environmental Impact Statement," Volume 1, Section 3.0 Near-Field Modeling Analyses, Sub-Section 3.2 – Meteorological Data. TRC Environmental Corp. January 2006. 20.

27. Ibid, Figure 3.1, p. 21

28. See Appendix 7 for an example of Big Piney AWOS (KBPI) data collected by the author from the University of Utah MesoWest website for this information. Many years are on file at the University of Wyoming American Heritage Center. Accessed at: http://mesowest.utah.edu/cgi-bin/droman/meso_base_dyn.cgi?stn=KBPI

29. U.S. Department of The Interior, Bureau of Land Management (Jonah 2006a), "Final Air Quality Technical Support Document for the Jonah Infill Drilling Project Environmental Impact Statement," Volume 1, Section 4.0 Mid-Field and Far-Field Analysis, TRC Environmental Corp. Jan 2006. 64

30. Ibid, Tables F.8.1 through F.8.17 and F.10.17 through F.10.20. F-69

31. Barbara J. Finlayson-Pitts and James N. Pitts Jr., *Atmospheric Chemistry: Fundamentals and Experimental Techniques*. (New York: John Wiley & Sons, Inc. 1979), 14.

32. U.S. Department of The Interior, Bureau of Land Management (Jonah 2005a) "Draft Environmental Impact Statement, Jonah Infill Drilling Project, Sublette County, Wyoming," DES-05-05, BLM/WY/PL-05/009+1310,(Volume 1 of 2), Chapter 4 – Environmental Consequences and Mitigation Measures, Section 4.1.2.11 Cumulative Impacts, February 2005. 4-25,26.

33. R.P. Walker, *"Natural Gas Flare Emission Monitoring Using a Miniature Fiber Optic Spectrometer,"* Coalbed Natural Gas Conference: I - Research, Monitoring, and Applications, Wyoming State Geological Survey, Public Information Circular No. 43, (2005). 120-123.

34. USEPA, AP 42, Fifth Edition, *Compilation of Air Pollutant Emission Factors, Volume 1: Stationary Point and Area Sources*, AP42511, Chapter 13. January 1995. Accessed November 22, 2016.

35. I was told this by the Shell area environmental representative Jim Sewell. He explained that as long as combustion flame temperatures being logged on a circular plot chart showed above desired minimum value, his company's field personnel were confident of efficient combustion.

36. Criteria pollutants are those specifically listed in the CAA requiring regulatory attention because of their perceived impact upon the environment.

37. Wyoming Department of Environmental Quality Air Quality Division, Letter to Operators: *"Interim Policy on Demonstration of Compliance with WAQSR Chapter 6,*

Section 2(c)(ii) for Sources in Sublette County," July 21, 2008. WYDEQ Operator Permit Archive, Cheyenne, WY.

38. V.R. Kotamarthi, and D.J. Holdridge, *"Process-Scale Modeling of Elevated Wintertime Ozone in Wyoming,"* Argonne National Laboratory Climate Research Section, Environmental Sciences Division for BP America, December 2007. ANL/EVS/R-07/7. Accessed February 16, 2013. http://www.osti.gov/bridge/product.biblio.jsp?query_id=0&page= 0&osti_id=924694&Row=0&formname=basicsearch.jsp

39. EnCana Oil & Gas (USA) Inc. Letter to WYDEQ-AQD: *"Demonstration of Emission Reductions, Emissions Offset, and Emission Credit for Sublette County Emission Sources,"* WYDEQ Operator Permit Archive, Cheyenne, WY, Dec 23, 2008.

40. Wyoming Department of Environmental Quality, Air Quality Division. "DEQ Notice of Public Hearing: Public Hearing on EnCana 32 Well proposal," Jan. 5, 2009, WYDEQ Operator Permit Archive, Cheyenne, WY. 7-8.

41. BP America Production Company, Letter to WYDEQ-AQD:*"Reasons to Trade VOC Offsets for NOx Offsets,"* Attachment A to Permit Applications Submitted to WYDEQ-AQD, 2008. WYDEQ Operator Permit Archive, Cheyenne, WY.

42. Mead Gruver, "Wyo. Ozone season ends, 13 days topped EPA limit," *Associated Press, Bloomberg Businessweek,* April 14, 2011.

43. Ecobadge and Zikua are trademarked products marketed by Vistanomics, Inc. located at 3450 Ocean View Blvd, Glendale, CA. They are dandy inex- pensive tools for amateurs to use in obtaining basic measurements of ozone levels around local communities. The company owner and president is Gary Short who was very helpful and quite interested in the project I set up around Pinedale.

44. Brittni R. Emery, Derek C. Montague, Robert A. Field, and Thomas R. Parish, "Barrier Wind Formation in the Upper Green River Basin of Sublette County, Wyoming, and Its Relationship to Elevated Ozone Distributions in Winter," *Journal of applied Meteorology and Climatology*, Vol.53. (2015): 2427-2442.

45. I wrote an extensive summary of this forum hosted by John Corra and sent to Mary Flanderka. One of the more bizarre happenings was a panelist from the Colorado equivalent of WYDEQ who sang praises of his Wyoming counterpart for its groundbreaking activities in air quality control. It appeared he had been recruited to be a cheerleader for WYDEQ.

46. Field, R. A., Soltis, J., McCarthy, M.C., Murphy, S., Montague, D.C., March 2015, Influence of oil and gas field operations on spatial and temporal distributions of atmospheric non-methane hydrocarbons and their effect on ozone formation in winter. *Journal of Atmospheric and Chemistry Physics*, Copernicus Publications of the European Geosciences Union, 15: 3527-3542.

47. Ibid, 3537

48. Ibid, 3539

49. Wyoming Department of Environmental Quality/Air Quality Division, Permit Application Analyses, WYDEQ Operator Permit Archive (2004-2011), Cheyenne, WY.

50. U.S. Department of The Interior, Bureau of Land Management (PAPA 2006c), "Draft Supplemental Environmental Impact Statement for the Pinedale Anticline Oil and Gas Exploration and Development Project, Air Quality Impact Analysis Technical Support Document," Sublette County Wyoming, Volume 2, Appendix F – Project Emissions Inventory, Tables F.1.2 through F.1.5, "Actual Emissions Inventory," December 2006. F1-7 through F1-32.

51. U.S. Department of The Interior, Bureau of Land Management (PAPA 2008c), "Final Supplemental Environmental Impact Statement for the Pinedale Anticline Oil and Gas Exploration and Development Project. Sublette County, Wyoming, Air Quality Impact Analysis Technical Support Document," Appendix F – Project Emissions Inventory, Tables F.1.2 and F.1.3, 2005 Actual Emissions Inventory, June 2008. F-7 through F-25.

52. U.S. Department of The Interior, Bureau of Land Management (PAPA 2006d), "Draft Supplemental Environmental Impact Statement for the Pinedale Anticline Oil and Gas Exploration and Development Project, Sublette County, Wyoming, Air Quality Impact Analysis Technical Support Document," Volume 2, Appendix G – Cumulative Emissions Inventory, Table G.8, Table of Excluded Sources, December 2006. G-18 through G-56.

53. U.S. Department of The Interior, Bureau of Land Management (PAPA 2008d), "Final Supplemental Environmental Impact Statement for the Pinedale Anticline Oil and Gas Exploration and Development Project, Sublette County, Wyoming, Air Quality Impact Analysis Technical Support Document," Appendix G – Cumulative Emissions Inventory, Table G.8 "Table of Excluded Sources," June 2008. G-17 through G-35.

54. United States Department of the Interior, Bureau of Land Management Wyoming State Office Cheyenne (PAPA 2008e), Wyoming Pinedale Field Office Pinedale, Wyoming, "Final Supplemental Environmental Impact Statement Pinedale Anticline Oil and Gas Exploration and Development Project Sublette County, Wyoming," BLM/WY/PL-08/022+1310, (Volume 1 of 2), Figure 4.3-1, June 2008. 4-13.

55. U.S. Department of The Interior, Bureau of Land Management (PAPA 2006a), Draft Supplemental Environmental Impact Statement for the Pinedale Anticline Oil and Gas Exploration and Development Environmental Impact Statement, Air Quality Impact Analysis Technical Support Document, Sublette County, Wyoming," Volume 1, Appendix A - Air Quality Impact Analysis Protocol, Sect. 1.2 - Relationship to Existing Plans and Documents, TRC Environmental Corp., August 2006. A-4

56. U.S. Department of The Interior, Bureau of Land Management (PAPA 2008a), "Air Quality Impact Analysis Technical Support Document for the Final Supplemental Environmental Impact Statement for the Pinedale Anticline Oil and Gas Exploration and Development Project, Sublette County, Wyoming," Chapter 1.0 – Introduction, Section 1.2 – Relationship to Existing Plans and Documents. June 2008. 6.

57. U.S. Department of The Interior, Bureau of Land Management, "Final Supplemental Environmental Impact Statement for the Pinedale Anticline Oil and Gas Exploration and Development Project, Sublette County Wyoming," Volume 1, Chapter 3 - Affected Environment, June 2008. 3-5,3-74,75.

58. U.S. Department of The Interior, Bureau of Land Management (PAPA 2006a), "Draft Supplemental Environmental Impact Statement for the Pinedale Anticline Oil and Gas Exploration and Development Environmental Impact Statement, Air Quality Impact Analysis Technical Support Document, Sublette County, Wyoming," Volume 1, Appendix A – Air Quality Impact Analysis Protocol, Sect. 1.2 – Relationship to Existing Plans and Documents, TRC Environmental Corp., August 2006. A-4.

59. U.S. Department of The Interior, Bureau of Land Management (PAPA 2006b), "Draft Supplemental Environmental Impact Statement for the Pinedale Anticline Oil and Gas Exploration and Development Project, Sublette County, Wyoming," Chapter 3 – Affected Environment, Section 3.11 – Air Quality, Sub-Section 3.11.2 – Impacts to Air Quality from Existing Wellfield Activities, December 2006. 3-61 through 3-63.

60. U.S. Department of The Interior, Bureau of Land Management (PAPA 2008a), "Air Quality Impact Analysis Technical Support Document for the Final Supplemental Environmental Impact Statement for the Pinedale Anticline Oil and Gas Exploration and Development Project, Sublette County, Wyoming," Chapter 1.0 – Introduction, Section 1.2 – Relationship to Existing Plans and Documents, June 2008. 6.

61. U.S. Department of The Interior, Bureau of Land Management (PAPA 2008e), "Final Supplemental Environmental Impact Statement for the Pinedale Anticline Oil and Gas Exploration and Development Project, Sublette County Wyoming," Volume 1, Chapter 3 – Affected Environment, June 2008. 3-5,3-74,75.

62. Table 5 presents compressor station statistics in the context of early volumes and subsequent increases or decreases as declared in the EIS documents and formal WYDEQ/AQD permitted volumes.

63. U.S. Department of The Interior, Bureau of Land Management (PAPA 2008b), "Air Quality Impact Analysis Technical Support Document for the Final Supplemental Environmental Impact Statement for the Pinedale Anticline Oil and Gas Exploration and Development Project, Sublette County, Wyoming, Chapter 2 – Emissions Inventory, Section 2.1 – Project Emissions, Sub-Section 2.1.3 – Total Field

Emissions, Table 2.1 – Estimated Potential Emissions by Alternative (tpy), Pinedale Anticline Project, June 2008. 11.

64. U.S. Department of The Interior, Bureau of Land Management (PAPA 2006f), "Revised Draft Supplemental Environmental Impact Statement for the Pinedale Anticline Oil and Gas Exploration and Development Project. Sublette County, Wyoming," Chapter 4 – "Summary of Impacts Common to All Alternatives – Natural Gas Development in the PAPA," December 2007. 4-71.

65. U.S. Department of The Interior, Bureau of Land Management (Jonah 2005b), "Draft Environmental Impact Statement, Jonah Infill Drilling Project, Sublette County, Wyoming," Volume 1, Chapter 2 – "Public Participation, Issues and Concern, and Alternatives," Section 2.14 BLM Preferred Alternative, Section 2.14.1 Outcome based Performance Objectives, February 2005. 2-26.

66. U.S. Department of The Interior, Bureau of Land Management (Jonah 2006d), "Final Jonah Infill Development Project Environmental Impact Statement, Part II, Substantive Comments Received During Public Comment Analysis Process of the Jonah Infill Drilling Project," Table II-B "Comments and BLM Reply to Commenter L-82, January 2006. 251.

67. U.S. Department of The Interior, Bureau of Land Management, "Final Environmental Impact Statement, Jonah Infill Drilling Project, Sublette County, Wyoming, Comment Analysis Report, Part II: Substantive Comments Received During Public Comment Analysis Process of the Jonah Infill Drilling Project, Table II-A: "Persons Submitting Comments on the JIDP DEIS," Submittal ID L-82, January 2006. 249-254. Accessed April 7, 2014. jonah.Par.7869.File.dat/27Comments_Part2.pdf

68. U.S. Department of The Interior, Bureau of Land Management, Cheyenne WY State Office, Letter to the Author: "Record of Decision Requirements and Conditions for the Pinedale Anticline Oil and Gas Exploration and Development Project and the Jonah Infill Drilling Project," January 30, 2012.

69. Enger, Erika Z., Attorney, EnCana Oil and Gas (USA) Inc., "Operation Reports and Emissions Reduction Report-Jonah Infill Drilling Project, Sublette County, WY," Letter to Jim Lucas, Project Coordinator, Bureau of Land Management, Jonah Interagency Mitigation and Reclamation Office, Author's personal archive, January 28, 2009. 4,5,6.

70. Jeff Tollefson, "Methane leaks during production may offset climate benefits of natural gas," *Nature News*, Feb 7, 2012.

71. Carmel Kail, e-mail message to WOGCC, March 17-18, 2011.

72. U.S. Environmental Protection Agency, National Risk Management Research Center, Office of Research and Development, Greenhouse Gas Technology Center, Southern Research Institute, ETV Joint Verification Statement: "Emissions Control of Criteria pollutants, Hazardous Pollutants, and Greenhouse Gases," Application:

Natural Gas Dehydration-Quantum Leap Dehydrator; Statement of Evaluation Testing conducted under the Environmental Technology Verification (ETV) Program, 2003. S-2.

73. U.S. Department of The Interior, Bureau of Land Management (PAPA 2007b), "Revised Draft Supplemental Environmental Impact Statement for the Pinedale Anticline Oil and Gas Exploration and Development Project, Sublette County, Wyoming," Chapter 3 - Affected Environment, Section 3.11.2 Impacts to Air Quality from Existing Wellfield Activities, December 2007. 11,3-71.

Epilogue

1. Stephen Schneider, *Science as a Contact Sport*, National Geographic: Washington D.C., 2009. 69.
2. My data archiving assistant Evie Stura-Carr is an employee at Stanford University where Stephan Schneider was a professor. She bought his book for me and had Schneider autograph it. He wrote: *"To Perry, with appreciation for your work to keep the planet healthier. With Regards, Steve Schneider. 15 Jan, 2010."* As I read through the book I saw so many of my own experiences playing out at much loftier levels and I was on the verge of writing him a note stating that revelation and thanking him for making me feel relevant. Sadly, I waited too long because I read he had passed away suddenly while aboard a plane carrying him home from a recent conference in Sweden.
3. Associated Press, "Winter ozone problem returns to western Wyoming gas fields," *Jackson Hole News & Guide*, March 21, 2017.
4. U.S. Department of The Interior, Bureau of Land Management Draft Environmental Impact Statement, Normally Pressurized Lance Natural Gas Development Project, Wyoming – Pinedale Field Office, Executive Summary, July 2017. ES-1.
5. Ibid, Chapter 3 – Affected Environment, Section 3.2.3 – Local Topography and Climate, Figure 3.1 – Distribution of Surface Wind Direction and Wind Speed for the Big Piney-Marbleton Airport for 2000-2010. 3-9.
6. Abraham Lustgarten, "EPA: Chemicals Found in Wyoming Drinking Water Might Be from Natural Gas Drilling," Scientific American/ProPublica, August 26, 2009. Accessed Nov. 5, 2017. https://www.scientificamerican.com/article/chemicals-found-in-drinking-water-from-natural-gas-drilling/
7. U.S. Dept. of Health and Human Services, Agency for Toxic Substances and Disease Registry (ATDSR), "Evaluation of Contaminants in Private Residential Well Water - Pavillion, Wyoming, Fremont County, Health Consultation," August 31, 2010. Accessed Nov. 5, 2017. https://www.atsdr.cdc.gov/hac/PHA/Pavillion/Pavillion_HC_Well_Water_08312010.pdf

8. Abraham Lustgarten, "EPA Abandons Fracking Study In Pavillion, Wyoming Following Similar Closed Investigations," Huffington Post/ ProPublica, July 3, 2013. Accessed Nov. 5, 2017. https://www.huffingtonpost.com/2013/07/03/epa-fracking-study-pavillion-wyoming_n_3542365.html

9. Sharon Kelley, "In Pavillion, Wyoming Water Contamination Case, Questions Continue to Swirl About Oil and Gas Industry's Role," DESMOG, February 6, 2014. Accessed Nov. 5, 2017. https://www.desmogblog.com/2014/02/06/pavillion-wyoming-water-contamination-case-questions-continue-swirl-about-oil-and-gas-industry-s- role

10. Elizabeth Shogren, "Fracking linked to groundwater contamination in Pavillion, Wyoming," High Country News, March 30, 2016. Accessed Nov 5, 2017. contamination-groundwater-pavillion-wyoming

11. Oliver Milman, Scientists find fracking contaminated Wyoming water after EPA halted study," The Guardian, April 7, 2016. Accessed Nov. 5, 2017. https://www.theguardian.com/us-news/2016/apr/07/wyoming-fracking-water-contamination-dangerous-chemicals

12. Mike Koshmrl, "Hoback Rim drilling considered by BLM," Jackson Hole News and Guide, September 20, 2017. Accessed September 25, 2017. http://www.jhnewsandguide.com/news/environmental/hoback-rim-drilling-considered-by-blm/article_d8eb1d8e-185d-5cdd-ac38-eb42fc858eb1.html

13. Heather Richards, "Cheney bill seeks to put oil and gas development in state hands," Casper Star Tribune, September 10, 2017. Accessed Oct. 18, 2017. http://trib.com/business/energy/cheney-bill-seeks-to-put-oil-and-gas-development-in/article_79bb77ea-5306-56bc-96f5-825331eb6bde.html

14. Mike Koshmrl, "Feds boost drilling sale in Hoback deer Path," Jackson Hole News and Guide, June 6, 2018. Accessed June 11, 2018. https://www.jhnewsandguide.com/news/environmental/article_a18369ff-6472-5fe4a3bb6d7185aec59c.html?utm_medium=social&utm_source=email&utm_campaign=user-share

15. Email from Derek Montague to author, August 8, 2017.

www.ingramcontent.com/pod-product-compliance
Lightning Source LLC
Chambersburg PA
CBHW061501180526
45171CB00001B/1